NONEQUILIBRIUM STATISTICAL MECHANICS AND NONLINEAR PHYSICS

NONEQUILIBRIUM STATISTICAL MECHANICS AND NONLINEAR PHYSICS

XV Conference on Nonequilibrium Statistical Mechanics and Nonlinear Physics

Mar del Plata, Argentina *4 – 8 December 2006*

EDITORS

Orazio Descalzi
University of los Andes
Santiago, Chile

Osvaldo A. Rosso
University of Buenos Aires
Buenos Aires, Argentina

Hilda A. Larrondo
National University of Mar del Plata
Mar del Plata, Argentina

All papers have been peer reviewed.

SPONSORING ORGANIZATIONS
National Council of Scientific and Technical Investigations (CONICET), Argentina
National Agency of Scientific and Technological Promotion (ANPCyT), Argentina
Latin American Center of Physics (CLAF)
Air Force Office of Scientific Research (AFOSR), U. S. A.
National Science Foundation Office of International Science and Engineering (OISE), U.S.A.
Naval Office Global Research, U.K.
Commission of Scientific Investigations of the Buenos Aires State (CICPBA), Argentina
Municipality of General Pueyrredón, Municipal Entity of Tourism (EMTUR), Argentina
Faculty of Engineering, University of los Andes, Chile
Faculty of Sciences, Buenos Aires University, Argentina
Faculty of Engineering, National University of Mar del Plata, Argentina
Faculty of Law, National University of Mar del Plata, Argentina
Center of Nonlinear Physics and Complex Systems (CFNL), Chile
Ring of Science and Technology (ACT15) Chile

AMERICAN INSTITUTE OF PHYSICS

Melville, New York, 2007
AIP CONFERENCE PROCEEDINGS ■ 913

Editors

Orazio Descalzi
Faculty of Engineering
University of los Andes
Av. San Carlos de Apoquindo 2200
Las Condes, Santiago
Chile
E-mail: oraziod@uandes.cl

Osvaldo A. Rosso
Chaos & Biology Group
Institute of Calculus, Faculty of Science
University of Buenos Aires
Building II, 2nd Floor
(1428) Buenos Aires
Argentina
E-mail: oarosso@fibertel.com.ar

Hilda A. Larrondo
Department of Physics
Faculty of Engineering
National University of Mar del Plata
Av. Juan B. Justo 4302
(7600) Mar del Plata
Argentina
E-mail: larrondo@fi.mdp.edu.ar

L.C. Catalog Card No. 2007927126

ISBN 978-0-7354-0421-2
ISSN 0094-243X

Printed in the United States of America

CONTENTS

ORAL CONTRIBUTIONS

Preface

The *Fifteenth Conference on Nonequilibrium Statistical Mechanics and Nonlinear Physics (MEDYFINOL '06)* took place, surrounded by a vast and enthusiastic audience, in Mar del Plata, Argentina, from December 4 to 8, 2006. This MEDYFINOL conference series was conceived in 1986 as a means of collecting researchers in various branches of the physics of nonequilibrium and dynamical systems, mostly in Argentina and neighboring countries, aiming at broadening the scope to attract colleagues from all over the world.

Twenty years have elapsed and we can ascertain that MEDYFINOL is firmly established in the international community as one of the most important regular meetings in the field. This 2006 Conference continues with the tradition of covering a large variety of topics, involving general theory and applications to specific and novel problems in many subfields, making room to the exchange of ideas and results of recent research among a diversity of specialists.

As MEDYFINOL "founding fathers" we are enormously proud and gratified for having the privilege of attending this particular 2006 meeting. The present Conference, to our belief, meets the highest standards of the series. Most specially, we feel obliged to the members of the Local Organizing Committee, Hilda A. Larrondo, Osvaldo A. Rosso and Orazio Descalzi, for successfully tackling a quite difficult task that demanded much time and effort and resulted in an exciting Conference, with an excellent survey of invited talks, along with first quality oral and poster presentations.

Prof. Dr. Susana Hernandez
Prof. Dr. Araceli Proto
Prof. Dr. Angel Plastino

Editorial

The XV Conference on Non equilibrium Statistical Mechanics and Nonlinear Physics (MEDYFINOL'06) was held in Mar del Plata, Argentina, from December 4 to December 8, 2006. It was hosted by three Latin-American scientific institutions: Faculty of Exact and Natural Sciences, University of Buenos Aires, Argentina; Faculty of Engineering, University of Mar del Plata, Argentina; and Faculty of Engineering, University of los Andes, Chile. It was very promising to see many people, from young students to well-recognized senior scientists, participating in the meeting (see photo). The conference was attended by more than 160 participants from 16 countries: Argentina, Australia, Brazil, Belgium, Bolivia, Chile, China, Colombia, France, Germany, Italy, Mexico, Spain, United States, United Kingdom and Uruguay.

The MEDYFINOL is a traditional and regular meeting of the statistical and nonlinear physics Latin-American community. This XV MEDYFINOL follows a series of meetings that took place in La Serena, Chile (2004), Colonia del Sacramento, Uruguay (2002), Huerta Grande, Córdoba (1999), Canela, Brazil (1997), Tucumán, Argentina (1996), San Carlos de Bariloche, Argentina (1995), Montevideo, Uruguay (1994), Mar del Plata, Argentina (1993), Mar del Plata, Argentina (1991), Santiago de Chile, Chile (1990), Puerto Iguazú, Argentina (1989), Mar del Plata, Argentina (1988), Lujan, Argentina (1987), and Lujan, Argentina (1986).

Statistical Mechanics and Nonlinear Physics are research areas that have developed a multidisciplinary character with many applications in biology, physiology, ecology, economics, sociology, etc. The scientific problems discussed in the meeting included biological applications of statistical physics, econophysics and sociologically related problems, chaos and nonlinear dynamics, pattern formation and spatio-temporal complexity, fluids and granular media, nonextensive thermostatistics, stochastic processes, lasers and nonlinear optics, synchronization, wavelets and nonlinear time series analysis.

The purpose of the MEDYFINOL series of meetings is to keep the Latin-America scientific community updated with new developments and tendencies in the statistics and nonlinear physics fields, to incentive collaborative international science programs, and to identify and discuss the most relevant issues in the area. The success of the XV MEDYFINOL meeting can be attributed to the presence of eminent speakers, the high quality contributed papers, as well as significant student participation. As organizers, we hope that the XV MEDYFINOL provided an opportunity to develop scientific collaborations among Latin-American scientists, and contributed to the nucleation of new inter-institutional collaborating groups.

The mini-workshop Brain Physics & Mind Dynamics was held during the MEDYFINOL'06, in Mar del Plata, Argentina on December 6th, 2006, co-organized by Dr. Dante Chialvo and Dr. Osvaldo A. Rosso. This meeting was devoted to examine recent advances on the theoretical understanding of the brain spatio-temporal patterns as the physical realization of the mind dynamics. This workshop worked to advance discovery in this area by bringing together researchers new to the field, as well as senior researchers who worked together to discuss the present status of the field, challenges and possible collaborations ahead, and to design a research agenda that will advance the field.

Preceding the MEDYFINOL'06 the satellite conference Reaction-Diffusion Systems: Theory and Experiments II, took place in El Calafate, Argentina, from November 29 to December 2, 2006. It was hosted by the Facultad de Ingeniería, Universidad de Los Andes, Chile, and co-organized by Dr. Orazio Descalzi, Dr. Helmut R. Brand and Dr. Alex Blumen. The purpose of the satellite conference was to bring together specialists in reaction-diffusion systems and related fields and allow the free flowing exchange of ideas and intense discussions within a moderate size group.

It is well known that it is very difficult to conduct scientific research in developing countries. In our case, the critical economic situation in Argentina and Chile during the gestation and realization of the meeting made the realization of the conference extremely challenging. We wish to thank the following international agencies, as well as local agencies, for their economic help and support without them this conference would not have been possible: Consejo Nacional de Investigaciones Científicas y Técnicas (CONICET), Argentina; Agencia Nacional de Promoción Científica y Tecnológica (ANPCyT), Argentina; Centro Latinoamericano de Física (CLAF); Air Force

Office of Scientific Research (AFOSR), United States, Air Force Research Laboratory, International Field Office, USA; National Science Foundation Office of International Science and Engineering (OISE), USA; Office Naval Research Global, UK; Comisión de Investigaciones Científicas de la Provincia de Buenos Aires (CICPBA), Argentina; Municipalidad de General Pueyrredón, Ente Municipal de Turismo (EMTUR), Argentina; Facultad de Ingeniería, Universidad de Los Andes, Chile; Facultad de Ciencias Exactas y Naturales, Universidad de Buenos Aires, Argentina; Facultad de Ingeniería, Universidad Nacional de Mar del Plata, Argentina; Facultad de Derecho, Universidad Nacional de Mar del Plata, Argentina; Centro de Física No Lineal y Sistemas Complejos (CFNL), Chile; Anillo de Ciencia y Tecnología ACT15 "Dinámica, singularidades y geometría de la materia fuera del equilibrio", Chile.

In this 20th anniversary of MEDYFINOL conference we want specially to thank Prof. Dr. Susana Hernández, Prof. Dr. Araceli Noemí Proto and Prof. Dr. Angel Plastino. They were the first Organizing Committee, who started MEDYFINOL in Luján, Argentina in 1986. It was an honor to have them as our Honorary Organizing Committee and MEDYFINOL conference success shows that their pioneering work was effective to increase communication and interaction between all researchers in Statistical Mechanics and Nonlinear Physics.

Dr. Orazio Descalzi
Dr. Hilda A. Larrondo
Dr. Osvaldo A. Rosso

MEDYFINOL' 06
20th Anniversary
Mar del Plata, Argentina
December 4th to 8th, 2006

ORAL CONTRIBUTIONS

Rescuing MaxEnt

A. Plastino*, A. R. Plastino† and B. H. Soffer**

*National University La Plata & IFLP-CONICET, C.C. 727, 1900 La Plata, Argentina
†Physics Department, University of Pretoria, Pretoria 0002, South Africa
**Department of Electrical Engineering, University of California at Los Angeles

Abstract. Relevant probability distributions (PDs) in sciences can be obtained by recourse to the constrained maximization of appropriately devised information-entropic functionals. The associated Maximum Entropy formalism (MaxEnt) and its application to the many fields has been the subject of intense activity. However, there lurks the ominous possibility that MaxEnt might be saddled by a basic ambiguity: *that any PD* might be derivable from the maximization of an arbitrary entropic measure if only an adequate constraint is provided. However, the aforementioned, troublesome ambiguity vanishes from MaxEnt if *mono-parametric families of PDs* are considered.

Keywords: MaxEnt, Entropy, variational principle, parametric families
PACS: 05.40.-a, 05.20.Gg

INTRODUCTION

The principle of maximum entropy (MaxEnt) is a method for analyzing the available information so as to determine a unique epistemic probability distribution (PD) [1, 2]. MaxEnt uses a reasonably smooth entropic measure S to rank probability distributions and states that the least biased distribution that encodes certain given information (the "prior information") is that which maximizes S while remaining consistent with the given a priori information. MaxEnt was first expounded by Jaynes in 1957 by introducing what is now known as *Maximum Entropy Thermodynamics*, as an interpretation of the Gibbs algorithm of statistical mechanics [3]. He suggested that thermodynamics, and in particular thermodynamic entropy, should be seen just as the particular application of a general tool of inference and information theory. MaxEnt is like other Bayesian methods in that it makes explicit use of prior information. This is an alternative to the methods of inference of classical statistics [1, 2]. The application of MaxEnt to the description of a panoply of systems and processes in science thus became a staple source of interest during the last 50 years [1, 4, 5, 6, 7, 8, 9, 10, 11, 12, 13]. Some hints on the origins of the approach can be found in the work of Gibbs [3], although it was Jaynes who elevated the principle of maximum entropy to the status of a foundational starting point for the development of statistical mechanics, and the first to recognize its relevance as a general statistical inference principle [2] that provides the most unbiased description of a system compatible with the available data by incorporation of the actual prior information at hand *and nothing else*. Novel MaxEnt versions (and their associated thermo-statistical formalisms) based upon generalized or alternative information-entropic functionals (to that of Shannon's used by Jaynes) have attracted the intense attention of hundreds of researchers during the last 20 years [8, 9, 10, 11, 12, 13, 14, 15, 16, 17, 18, 19, 20]. In particular, power-law statistical ensembles, and their associated thermo-statistics are nowadays the focus of frantic activity because of their utility for describing many systems and processes in science [10, 11, 12, 13] that are beyond the reach of Shannon's measure. Parameterizing the power-law distributions in terms of q-distributions (that is, distributions maximizing the non-extensive information measure S_q introduced by Tsallis [10], which depends on a real parameter q) makes it possible to recover the standard exponential canonical distribution in the limit $q \to 1$. *The crux of the problem we wish here to discuss* arises in the wake of Ref. [19], where it is pointed out that MaxEnt suffers from a basic weakness since it is marred by an intrinsic ambiguity in the forms of i) the entropic measure and ii) the constraints used in the entropic-information variational principle [19]. A seemingly strong objection thus emerges that boils down to the following statement: almost any probability distribution (PD) p can be regarded as the result of the constrained maximization of a given entropic functional $S[p]$, if the relevant constraint is adroitly chosen. (The fact that a single instance of a probability distribution p is not enough to fix the form of the entropic functional had indeed been previously noticed by various researchers [17, 18].)

Sudden insights leading to brief, devastating arguments against fashionable scientific doctrines naturally arise great interest. The History of Science presents various such episodes. One can mention, for instance, Kelvin's arguments against Darwinian evolution, or Russell's celebrated paradox in set theory. These kind of "brief refutations" (even in

CP913, *Nonequilibrium Statistical Mechanics and Nonlinear Physics, XV Conference*
edited by O. Descalzi, O. A. Rosso, and H. A. Larrondo

those cases where they turned out to be incorrect) always raised interesting points and stimulated further research. We believe that the aforementioned anti-MaxEnt argument (although in a rather modest way) is no exception to this rule. Indeed, a quite interesting and provocative issue gets thus advanced that undoubtedly merits close scrutiny. *Our objective here* is to discuss in detail such alleged ambiguities in the MaxEnt-implementation by focusing attention on MaxEnt representations of mono-parametric PD-families. Within such a family the single parameter characterizing a member of it can be identified with the Lagrange multiplier associated with the single relevant constraint. Although this is certainly a restricted scenario, involving just one constraint besides normalization, it comprises all the *essential facets* of the novel applications of generalized MaxEnt formalisms that have been recently reported in the literature [21]. Our materials are organized as follows. Basic MaxEnt properties are briefly recapitulated in Section 2. The alleged MaxEnt malady and its "cure" are given in Sections 3 and 4, while some conclusions are drawn in Section 5.

MAXENT METHOD: A BRIEF REVIEW

Assume we wish to infer a normalized probability density $f(\mathbf{y})$, where $\mathbf{y}$ denotes a point in the relevant space where the probability density is defined, and $d\mathbf{y}$ the space's volume element. Let us consider an entropic measure of the form

$$S[f] = \int C(f(\mathbf{y}))\, d\mathbf{y}; \text{ and a constraint } \Psi = \langle \phi \rangle = \int \phi(\mathbf{y}) f(\mathbf{y})\, d\mathbf{y}, \tag{1}$$

playing the role of our "prior information". The optimization of the entropic measure (1) under the constraints imposed by normalization and the mean value above leads to the variational problem,

$$\delta\left[S - \beta\Psi - \alpha \int f d\mathbf{y}\right] = 0; \text{ with solution } C'(f) - \beta\phi - \alpha = 0, \tag{2}$$

where α and β are appropriate Lagrange multipliers. We assume that C' above is an invertible function, i.e., $[C']^{-1}$ exists. One makes now judicious use of the inverse (let us call it $G(x)$) of $C'(x)$. G satisfies

$$G(C'(x)) = C'(G(x)) = x, \tag{3}$$

and, in terms of this known form G, the probability density maximizing the entropy under the alluded constraints can be cast as

$$f(\mathbf{y};\beta) = G[\alpha + \beta\phi(\mathbf{y})]. \tag{4}$$

Due to the normalization requirement on f, the two Lagrange multipliers are not independent, and we can regard the Lagrange multiplier α as a function of the Lagrange multiplier β,

$$\int f d\mathbf{y} = 1 \longrightarrow \alpha = \alpha(\beta). \tag{5}$$

It is useful to define the Jaynes' parameter (Legendre transform of S) [8, 9]

$$\lambda_J = S - \beta\Psi. \tag{6}$$

The entropy S, the mean value Ψ, the Jaynes' parameter λ_J and the Lagrange multiplier β are connected by the following important relations,

$$\frac{dS}{d\Psi} = \beta, \text{ and } \frac{d\lambda_J}{d\beta} = -\Psi. \tag{7}$$

The mono-parametric family of maximum entropy distributions (4) can be regarded as parameterized either by the Lagrange multiplier β or by the mean value Ψ. The last parameterization is used in the first of Eqs. (7), where the quantities involved are regarded as functions of Ψ. On the other hand, the first parameterization is used in the second equation (7), where the quantities involved are regarded as functions of β. It is instructive to consider the particular example provided by the (classical) Gibbs' canonical ensemble in statistical mechanics, where the entropic functional is given by $C(x) = -x\ln x$. Consequently, we have

$$S = -\int f \ln f \, d\Omega, \tag{8}$$

where $d\Omega$ is the phase space's volume element. The relevant mean value is the mean energy $E = \langle H \rangle$ (H being the Hamiltonian of the system under study) and the Lagrange multiplier β is to be identified with the inverse temperature $1/kT$. For the canonical distribution function we have $G(x) = \exp[-1-x]$, and thus $f_G = \exp[-1-\alpha-\beta H] = \frac{1}{Z}\exp[\beta H]$, were Z is the canonical partition function given by $Z = \exp[1+\alpha] = \int \exp[-\beta H] d\Omega$. We wish however to address more general entropic functionals, of the form (1). Remark that even more general entropies of the form

$$S^*[f] = D[S[f]] = D\left[\int C(f(\mathbf{y}))\, d\mathbf{y}\right], \tag{9}$$

are not needed, since it is obvious that, if $D(x)$ is a monotonous function, the maximization under given constraints of $S[f]$ yields the same maximum entropy distributions as the maximization of $S^*[f]$ does. Consequently, as far as the maximum entropy principle is concerned, the functionals $S[f]$ and $S^*[f]$ describe the same situation. Thus, it is our concern here that of ascertaining whether two *different* entropic functionals of the form (1) can equally describe the same scenario.

MAXENT REPRESENTATION OF PARAMETRIC PD-FAMILIES

In the most interesting applications of Jaynes' maximum entropy approach (at least from a fundamental point of view) one is interested in a maximum entropy representation of a *mono-parametric family* of probability distributions, rather than in a maximum entropy representation of *just one, single PD*. For instance, within Jaynes' approach to statistical mechanics one considers (for a given Hamiltonian system) a MaxEnt representation of the complete family of Gibbs' canonical distributions, which can be regarded as a mono-parametric family characterized by one parameter: the inverse temperature β. Certainly, people would not be satisfied with a MaxEnt description of just *one member* of the Gibbs' canonical family, corresponding to a certain fixed temperature. Within the context of the power-law q-thermostatistics there are several examples of systems or processes described by a probability distribution depending on a parameter [10, 12, 13]. For instance, the (non-Gaussian) velocity distribution of *Hydra viridissima* cells in cellular aggregates fits well a q-distribution function involving a parameter (playing a role formally analogous to the temperature) describing the effective mobility of the *Hydra* cells [12]. Now, let us consider a mono-parametric family of PDs

$$f = f(\mathbf{y}; \lambda), \tag{10}$$

parameterized by a parameter λ. If the family of PDs (10) is to admit a MaxEnt representation, it has to be possible to perform an appropriate re-parameterization

$$\beta = \beta(\lambda), \tag{11}$$

that brings the distribution (10) into the form [Cf. equation (4)]

$$f(\mathbf{y};\beta) = G[\alpha(\beta) + \beta\phi(\mathbf{y})] \Rightarrow G^{-1}(f) = \alpha(\beta) + \beta\phi(\mathbf{y}). \tag{12}$$

Now, from the representation (12) of the parametric family (10) of distributions, it is possible to re-construct the associated entropic measure. From (3), i.e., $G(C'(x)) = C'(G(x)) = x$, we determine the form of the function $C(x)$,

$$C'(x) = G^{(-1)}(x) \longrightarrow C(x) = \int^x G^{(-1)}(u)du, \tag{13}$$

which, in turn, determines the entropic measure $S[f] = \int C(f(\mathbf{y}))\,d\mathbf{y}$. Our main question is whether a parametric family of distributions (10) can admit *more that one* (non-equivalent) MaxEnt representation of the type (4). From (12) we get (the prime denotes phase-space derivatives)

$$\beta^{-1}\left[G^{(-1)}(f) - \alpha\right] = \phi(\mathbf{y}) \text{ and } \left[G^{(-1)}(f)\right]' \nabla f = \beta \nabla\phi(\mathbf{y}) \text{ or,} \tag{14}$$

$$h(f)\nabla f = \beta\nabla\phi \text{ (defining } h(x) = d[G^{(-1)}(x)]/dx). \tag{15}$$

Now, let us assume that the parametric family (10) admits two maximum entropy representations (which we are going to designate by the subindexes 1 and 2). We are going to have (remember that $\beta_{1,2}$ are functions only of the parameter λ while $\phi_{1,2}$ depend solely on the phase space coordinates $\mathbf{y}$)

$$h_1(f)\nabla f = \beta_1 \nabla \phi_1; \quad h_2(f)\nabla f = \beta_2 \nabla \phi_2. \tag{16}$$

Consequently,

$$\frac{\beta_1 h_2}{\beta_2 h_1}\nabla \phi_1 = \nabla \phi_2; \text{ implying } \frac{\beta_1 h_2}{\beta_2 h_1} = d_1(\mathbf{y}), \tag{17}$$

since the right hand side of the first Eq. in (17) does not depend on λ. Thus, introducing the two definitions

$$g = h_2/h_1 \text{ and } d_2 = \beta_2/\beta_1 \text{ one finds } g(f) = d_1(\mathbf{y})\,d_2(\lambda). \tag{18}$$

Hence, applying the function g to $f(\mathbf{y};\lambda)$ we obtain the *factorized* function above that is equal to the product of two factors, one depending solely on $\mathbf{y}$, and the other depending solely on the parameter λ. Now, in general, for a given family $f(\mathbf{y};\lambda)$ the only function $g(f)$ that leads to the alluded factorization is the constant function,

$$g = (h_2/h_1) = c_0. \tag{19}$$

In general thus, up to appropriate multiplicative and additive constants, the forms of the entropic functional and the constraint leading to a maximum entropy representation of a mono-parametric family of probability distributions are uniquely determined.

SINGLE VS. MONO-PARAMETRIC FAMILIES OF DISTRIBUTIONS

In the presence of i) a single, individual probability distribution $p(\mathbf{y})$ and ii) an invertible function $G(x)$ (associated with a MaxEnt treatment through equations (1-3)), it is easy to construct the function

$$\tilde{\phi}(\mathbf{y}) = (\tilde{\beta})^{-1}[G^{(-1)}p(\mathbf{y}) - \tilde{\alpha}], \tag{20}$$

such that $p(\mathbf{y})$ can be formally regarded as the PD maximizing the entropic measure under the constraints imposed by normalization and the mean value of $\tilde{\phi}(\mathbf{y})$

$$p(\mathbf{y}) = G[\tilde{\alpha} + \tilde{\beta}\tilde{\phi}(\mathbf{y})]. \tag{21}$$

In (20) the constant $\tilde{\beta}$ plays the role of the Lagrange multiplier associated with the $\langle\tilde{\phi}(\mathbf{y})\rangle$ constraint. Equations (20-21) constitute the gist of the "anti-MaxEnt" argument advanced in [19]. In general, there is a considerable freedom for the selection of the values of the constants $\tilde{\alpha}$ and $\tilde{\beta}$. The construction based upon equations (20-21) will indeed work for a single, individual probability distribution $p(\mathbf{y})$. Thus, when we have one such PD $p(\mathbf{y})$, and a given entropic functional, it is possible, under quite general conditions, to devise an appropriate constraint, such that the PD $p(\mathbf{y})$ is recovered from the constrained optimization of the alluded entropic functional. However, the construction based on Eqs. (20-21) will not lead, in general, to a legitimate MaxEnt representation of a *mono-parametric* family of probability densities $f(\mathbf{y};\lambda)$. If this were possible Eq. (20) would write

$$\tilde{\phi}(\mathbf{y}) = [\tilde{\beta}]^{-1}[G^{(-1)}f(\mathbf{y};\lambda) - \tilde{\alpha}], \tag{22}$$

and, in general, the constraint function ϕ would, in turn, depend itself on the parameter λ. However, in a useful MaxEnt description the mono-parametric family of distributions has to be derivable from the constrained maximization of a single entropic measure, under the constraint imposed by the mean value of a single, previously specified quantity. Indeed, in a MaxEnt representation the parametric dependence on λ is always to be mapped upon a parametric dependence on β, the Lagrange multiplier associated with the relevant mean value. As made explicit in Section III, this can only be achieved for a particular (up to multiplicative and additive constants) choice of the entropic functional and of the relevant constraint. The lesson of the preceding Section is that, *if* the distribution depends *both* on $\mathbf{y}$ *and* on a parameter λ, no ambiguity à la [21] can arise. It has to be once again stressed that within a proper MaxEnt scheme the quantity whose mean value defines the relevant constraint must not itself depend upon the parameter

λ (or, equivalently, upon the Lagrange multiplier β). The value of this Lagrange multiplier must emerge from the implementation of the MaxEnt procedure and not be added "by hand", from the very beginning, into the structure of the constraint. If one chooses to do that, one may as well take the PD under consideration for granted, and not bother with a MaxEnt representation.

An alternative way to look at the issue reads as follows: let A denote a quantity whose mean value we regard as prior information for the MaxEnt procedure. Given the (*a priori unknown*) probability distribution f describing a system, suppose that we ask an experimental physicist to determine the mean value $\langle A \rangle$. Taking that experimental result as constraint, we can then construct (infer) a maximum entropy distribution characterized by a Lagrange multiplier β. However, if the quantity A depends itself on β, we cannot tell the experimentalist precisely what to measure if we do not know beforehand the value of β or, what in this context is essentially the same, without knowing beforehand the very distribution f that we are assumed to infer!

There is also another important reason for discarding the case of constraints explicitly dependent on the Lagrange multipliers. In general, if one allows for such a dependence, the Jaynes' relations (6-7) connecting the Lagrange multipliers, the entropy, and the constraints, are not valid any longer. But these Jaynes relations constitute one of the most important and useful features of the MaxEnt formalism! In particular, in the case of thermal physics, they provide the connection between the statistical mechanical formalism and macroscopic thermodynamics [2].

CONCLUSION

We have discussed some facets of the (generalized) MaxEnt approach in connection with possible ambiguities in the forms of i) the entropic-information functionals and ii) the relevant constraints. The main conclusion to be drawn from this discussion is that in general, and up to multiplicative and additive constants, the forms of the entropic functional and the constraint yielding a MaxEnt representation of a given mono-parametric family of probability distributions are *uniquely determined by the particular form of the alluded family of probability distributions* . The above conclusions hold for mono-parametric families of probability distributions. On the other hand, when dealing with just a single, individual instance of a probability distribution, it is possible to recover, via the MaxEnt approach, the very same distribution using different forms for the entropic (information) measure and for the relevant constraint. However, this situation is not as awkward as it might appear at first glance. The information measure and the constraint are "appropriately" devised or "chosen" by the practitioner to suit the physical, or even the non-physical facets of the scenario at hand. The factors that determine which choice of information measure and of relevant constraints are the most suitable include i) extensive experience with such a physical situation and ii) a deep understanding of the associated physics, with an eye on computational complexities. For non-physical scenarios, the same experience, via physical analogies, will determine the optimal choices.

Summing up, we have focused our attention on maximum entropy representations of families of probability distributions depending on one parameter. This scenario leads to MaxEnt probability distributions involving one constraint (besides normalization). This particular case incorporates many of the most interesting applications of generalized maximum entropy ideas reported in recent literature (see, for instance, [10, 11, 12, 13]). Even in Hamiltonian statistical mechanics, this restricted setting, that leads to the canonical ensemble, constitutes the basis for an important body of physical and mathematical knowledge. It must be kept in mind that the present discussion does not contradict one important conclusion that can be drawn from the work reported in references [17, 18, 19]. From an isolated instance of an experimentally observed probability distribution it is not possible to determine, in a unique way, the forms of the entropy and the canonical constraint that lead, via Jaynes' prescription, to the alluded distribution. For example, the observation of a particular power-law in a given context does not constitute enough evidence to claim that the q-nonextensive thermostatistics is at work. However, when dealing with a mono-parametric family of distributions, the forms of the entropic measure and of the relevant constraint are essentially unique. In particular, in the case of the non extensive q-statistical formalism, the value of the non extensivity parameter q is determined by the form of the family of probability distributions under consideration.

REFERENCES

1. E. T. Jaynes, *Probability Theory: The Logic of Science*, Cambridge University Press, Cambridge, 2005.
2. E. T. Jaynes *Papers on probability, statistics and statistical physics* , edited by R. D. Rosenkrantz, Dordrecht, Reidel, 1987.
3. J. Willard Gibbs, *Elementary Principles in Statistical Mechanics*, New Haven, Yale University Press, 1902.

4. B. R. Frieden, *Science from Fisher Information*, Cambridge University Press, Cambridge, 2004.
5. B. R. Frieden and B. H. Soffer, *Phys. Rev. E* **52**, (1995) 2274-2283 (1995).
6. C. Beck and F. Schlogl, *Thermodynamics of Chaotic Systems*, Cambridge University Press, Cambridge, 1993.
7. S. Sieniutycz and H. Farkas, *Variational and Extremum Principles in Macroscopic Systems*, Elsevier, Amsterdam, 2005.
8. A. Plastino and A. R. Plastino, *Phys. Lett. A* **177**, 177-183 (1993).
9. A. Plastino and A. R. Plastino, *Phys. Lett. A* **226**, 257-263 (1997).
10. M. Gell-Mann and C. Tsallis, Eds. *Nonextensive Entropy: Interdisciplinary applications*, Oxford University Press, Oxford, 2004.
11. T. D. Frank, *Nonlinear Fokker-Planck Equations*, Springer-Verlag, Berlin, 2005.
12. A. Upadhyaya, J. P. Rieu, J. A. Glazier, Y. Sawada, *Physica A* **293** 549-556 (2001).
13. L. Borland, *Phys. Rev. Lett.* **89** 098701 (2002).
14. T. Yamano, *Phys. Lett. A* **308**, 364-371 (2003).
15. K. S. Fa and E. K. Lenzi, *Chaos, Solitons, and Fractals* **20**, 227-234 (2004).
16. B. M. Boghosian, *Phys. Rev. E* **53**, 4754-4756 (1996).
17. L. Borlnad, *Phys. Lett. A* **245**, 67-73 (1998).
18. G. Kaniadakis and G. Lapenta, *Phys. Rev. E* **62**, 3246-3255 (2000).
19. D. H. Zanette and M. A. Montemurro, *Phys. Lett. A* **324**, 383-390 (2004).
20. E. Curado and A. Plastino, *Phys. Rev. E* **72**, 047103 (2005).
21. A. R. Plastino, A. Plastino, and B. H. Soffer, *Phys. Lett. A* **363**, 48-52 (2007).

Convection in a Viscoelastic Magnetic Fluid

D. Laroze and J. Martínez-Mardones

Instituto de Física, Universidad Católica de Valparaíso,Casilla 4059,Valparaíso, Chile

Abstract. In this work we report theoretical and numerical results on convection for a binary magnetic mixture in a viscoelastic carrier liquid. We focus in the stationary and oscillatory convection for idealized boundary conditions and we obtain explicit expressions of convective thresholds in terms of the control parameters of the system. The effect of the magnetophoresis and Kelvin force are emphasized. Finally, we analyze the effect of the rheology on instability thresholds for a diluted suspensions.

Keywords: Thermal convection, magnetic fluid, viscoelastic fluid
PACS: 47.20.2k, 44.27.1g, 75.50.Mm

INTRODUCTION

Ferrofliuds are superparamagnetic fluids formed by a stable colloidal suspension of ferromagnetic nanoparticles dispersed in a carrier liquid, and there are two main features that distinguish ferrofluids from ordinary fluids, the polarization force and the body couple. Beside, one of the most important application of ferrolfuids is in biomedicine where the carrier liquid is the blood [1]; and it has a rheological proprieties. Hence, the detail study of viscoelastic magnetic fluid is quite important and in order.

The first continuum description of the magnetic fluids was due to Neuringer et al [2]. Later, Finlayson [3] studied the convective instability of a magnetic fluid for a fluid layer heated from below in the presence of a uniform vertical magnetic field. He discussed both shear free and rigid horizontal boundaries using the linear stability method. In addition, Ryskin et al. [4] using the nonequilibrium thermodynamics have derived a complete set of equations to describe ferrofluids in an external magnetic field. They did so in terms of a binary mixture where the magnetophoretic effect, as well as magnetic stresses, have been taken into account in the static and dynamic parts of the ferrofluid equations.

On the other hand, the viscoelastic properties of fluids appear in the constitutive equation, which relates the stress and strain rate tensors. Finding this relation, which should generalize the linear dependence characteristic of Newtonian fluids, is the main purpose of rheology. The simplest constitutive equation capable of describing realistically the viscoelastic properties is given by the so-called Oldroyd model [5]. In this model, the stress tensor is decomposed into both a polymeric contribution and a solvent contribution. The convection such fluids has been performed by various authors, either with free-free or rigid-rigid boundary conditions [6, 7, 8, 9]. It has been found that, besides the usual stationary convection, oscillatory states can also be obtained at onset. Which type of convection stationary or oscillatory appears first is determined by the values of the rheological parameters.

The purpose of this paper is to communicate our analysis of the viscoelastic properties influence in the convective thresholds of a magnetic fluid for idealized boundary conditions. The linear stability analysis of the conduction state is performed in the framework of Navier-Stokes equations and we obtain explicit expressions of convective thresholds in terms of the control parameters of the system. The paper is organized as follows: In Sec. II, the basic hydrodynamic equations for viscoelastic magnetic fluid convection are presented.In Sec. III the linear stability analysis of the conduction state is performed. Finally, conclusions are presented in Sec. IV.

THEORETICAL MODEL

We consider a layer of incompressible binary magnetic fluid, of thickness d, parallel to the xy-plane, with very large horizontal extension, in a gravitational field $\mathbf{g}$ and submitted to a vertical temperature gradient. The magnetic fluid properties can be model as electrically nonconducting superparamagnets, and it is assumed to be placed in a magnetic field $\mathbf{H}$ parallel to $\hat{\mathbf{z}}$, which would be homogeneous if the magnetic fluid were absent. Let us choose the z-axis such that

CP913, *Nonequilibrium Statistical Mechanics and Nonlinear Physics, XV Conference*
edited by O. Descalzi, O. A. Rosso, and H. A. Larrondo

$\mathbf{g} = -g\hat{\mathbf{z}}$ and that the layer has its interfaces at $z = -d/2$ and $z = d/2$. A static temperature difference across the layer is imposed, $T(z = -d/2) = T_0 + \triangle T$ and $T(z = d/2) = T_0$. Under the Boussinesq approximation, the dimensionless balance perturbation equations of the conduction states and the Maxwell equations read as

$$\nabla \cdot \mathbf{v} = \mathbf{0} \tag{1}$$

$$(\partial_t + \mathbf{v} \cdot \nabla)(\theta - M_4 \partial_z \phi) = (1 - M_4)w + \nabla^2 \theta + F\nabla^2(c - M_2 \partial_z \phi) \tag{2}$$

$$(\partial_t + \mathbf{v} \cdot \nabla)c = L\nabla^2(c + \theta - M_2 \partial_z \phi) \tag{3}$$

$$P^{-1}(\partial_t + \mathbf{v} \cdot \nabla)\mathbf{v} = -\nabla p_{eff} + \nabla \cdot \tau + R_a f(\theta, c, \phi)\hat{\mathbf{z}} + R_a M_1(\theta - \psi_M c)\nabla(\partial_z \phi) \tag{4}$$

$$(\partial_{zz} + M_3 \nabla_\perp^2)\phi - \partial_z(\theta - \psi_M c) = 0 \tag{5}$$

$$(1 + \Gamma D_t)\tau = (1 + \Lambda \Gamma D_t)\mathbf{D} \tag{6}$$

$$\nabla^2 \phi_{ext} = 0 \tag{7}$$

where $\mathbf{v} = (u, v, w)^T$ is the velocity field, θ is the temperature, c is the concentration, p_{eff} is the effective pressure, τ is the extra stress tensor, ϕ and ϕ_{ext} are the internal and the external scalar magnetic potential, respectively and $f(\theta, c, \phi) = (1 + M_1)\theta - (\psi + M_1 \psi_M)c - (M_1 - M_5)\partial_z \phi$. Also the following groups of dimensionless numbers have been introduced: a) (pure fluids) the Rayleigh number $R_a = \alpha_T g \triangle T d^3 / \kappa \nu$ accounting for buoyancy effects and the Prandtl number $P = \nu/\kappa$, relating viscous and thermal effects; b) (binary mixtures) the Lewis number $L = D_c/\kappa$, relating diffusion with thermal diffusivity, the separation ratio $\psi = \alpha_c D_T / (\alpha_T D_c)$ and the Dufuor number $F = D_T^2/(D\kappa)$; c) (magnetic fluid) the strength of magnetic force relative to buoyancy $M_1 = \beta \chi_T^2 H_0^2 / (\rho_0 g \alpha_T (1 + \chi)))$, the the magnetophoretic number which gives rise to a field dependence of heat and concentration currents $M_2 = D\chi_c \chi_T H_0^2 / (\rho_0 g \alpha_T (1 + \chi))$, the nonlinearity of magnetization $M_3 = 1 - (\chi_H H_0^2)/(1 + \chi)$ is a measure of the deviation of the magnetization curve from the linear behavior $M_0 = \chi_H H_0$, the relative strength of temperature dependence of the magnetic susceptibility $M_4 = \chi_T^2 H_0^2 T_0 / c_0(1 + \chi)$, the he ratio of magnetic to thermal buoyancy $M_5 = \alpha_H \chi_T H_0^2 / (\alpha_T (1 + \chi))$ and the magnetic separation ratio $\psi_M = -\chi_c D_T / (\chi_T D_c)$; and d) (viscoelastic) the Deborah number $\Gamma = \lambda_1 \kappa / d^2$ and the ratio between retardation and stress relaxation times $\Lambda = \lambda_2 / \lambda_1$.

In addition, the symbol D_t in the equation (6) denotes an invariant (frame-indifferent) time derivative, defined as $D_t \tau = (\partial_t + \mathbf{v} \cdot \nabla)\tau + \tau \cdot \mathbf{W} - \mathbf{W} \cdot \tau + a(\tau \cdot \mathbf{D} + \mathbf{D} \cdot \tau)$, where $\mathbf{W}$ and $\mathbf{D}$ are the skew-symmetric and symmetric parts of the velocity field gradient, respectively; also a is a phenomenological parameter in the range -1 to +1.

Let us comment about of the numerical values of the parameters, the parameter R_a may be changed in several orders of magnitude while Γ, which relates a typical relaxation time to the thermal diffusion time, usually changes by one order of magnitude. For $\Lambda = 1$ one recovers a Newtonian fluid while for $\Lambda = 0$ a Maxwellian fluid is obtained. The magnetic numbers have the following order of magnitude $M_1 = 10^{-4} - 10^1$, $M_3 \approx 1.1$, $M_4 \approx M_5 \approx 10^{-6}$ [4]. The values of M_4, M_5 and F are very small, so in our calculation we will not take them into account.

LINEAR STABILITY ANALYSIS

For calculating the linear stability analysis, we only need the linear parts of equations (1)-(6), the effective pressure and two components of the velocity field could easily be eliminated by applying the rotor and double rotor opereator in the Navier-Stokes equation and then considering only the z-component of the resulting equations. The space-temporal dependencies are separated by the usual normal mode expansion

$$(c, \theta, \phi, w)(\mathbf{r}, t) = (\Psi, \Theta, \Phi, W)(z) \exp[i\wp] \tag{8}$$

with $\wp = \mathbf{k} \cdot \mathbf{r}_\perp + \sigma t$, being $\mathbf{k}$ the horizontal wavenumber vector and σ denote the growth factor of a perturbation. Therefore, this leads to the following coupled ordinary differential equations:

$$\left(D^2 - k^2 - \frac{\sigma}{L}\right)\Psi + (D^2 - k^2)(\Theta - M_2 D\Phi) = 0 \tag{9}$$

$$(D^2 - k^2 - \sigma)\Theta + W = 0 \tag{10}$$

$$D^2\Phi - M_3 k^2 \Phi - D\Theta + \psi_M D\Psi = 0 \tag{11}$$

$$\aleph\left(D^4 - \left(2k^2 - \frac{\Im\sigma}{P}\right)D^2 + k^2\left(k^2 + \frac{\Im\sigma}{P}\right)\right)W = R_a k^2 \Im\, G(\Theta, \Psi, \Phi) \tag{12}$$

where $D^n f = \partial_z^n f$, $\aleph = 1 + \sigma\Lambda\Gamma$, $\Im = 1 + \sigma\Gamma$ and $G(\Theta,\Psi,\Phi) = (1+M_1)\Theta - (\psi + M_1\psi_M)\Psi - M_1 D\Phi$. In order to solve the set of differential equations (9)–(12) and then to find analytical expressions the following boundary conditions $W = D^2W = \Theta = D\Psi = \Phi = 0$ are imposed in $z = \pm 1$. So, with them the eigenvalue problem gives the Rayleigh number.

Stationary Convection

The stationary bifurcation is obtained for $\sigma = 0$ and for the fundamental mode, gives the marginal curve for stationary convection:

$$R_{as} = \frac{(k^2 M_3 + \eta)(\pi^2 + k^2)^3}{k^2\left[\pi(\psi + 2M_1\psi_M)(\pi - M_2) + k^2 M_3 \lambda + \eta\right]} \tag{13}$$

where $\eta = \pi(\pi - M_2\psi_M)$ and $\lambda = 1 + \psi + M_1(1 + \psi_M)$. Note that, the equation (13) reproduce all the know result, for example if we take $\psi = 0$ we obtain the expression of R_a for a simple ferrofluid [3]. The minimum of the marginal curve ($\partial_k R_a = 0$) gives the critical wavenumber k_c and the critical Rayleigh number, R_{ac}, can be obtained by replacing it in equation (13). We remarks that, when the M_1 and ψ_M increases the critical Rayleigh number decreases; therefore, the magnetic and binary effect are destabilize effect.

Oscillatory Convection

In order to analyze the rheological properties our first approach is only studied the case of pure fluid, so at follow the concentration effects are spurned. In such case, from corresponding equations (10)–(12) we obtain a dispersion relationship given by:

$$a_0 + a_1\sigma + a_2\sigma^2 + a_3\sigma^3 + a_4\sigma^4 = 0 \tag{14}$$

where a_j are complex functions of the control parameters. For simplicity, we assume a Maxwell fluid $\Lambda = 0$, so convection thresholds are obtained after replacing $\sigma = i\varpi$ in the equation (14). Its complex part must vanish and the real part gives the marginal stability curve $R_a(k, \varpi)$. After some manipulations we find

$$\varpi^2 = \frac{P\left(\Gamma(k^2 + \pi^2) - 1\right) - 1}{\Gamma^2} \tag{15}$$

$$R_{ao} = \frac{(k^2 M_3 + \pi^2)(\pi^2 + k^2)(1 + P + \Gamma(\pi^2 + k^2))}{k^2 P\Gamma\left(k^2 M_3(1 + M_1) + \pi^2\right)} \tag{16}$$

We remark that, the frequency has not magnetic effects dependence. The critical condition $\partial_k R_a = 0$ gives the critical wavenumber k_c and replacing it in equations (15) and (16) are obtained the critical frequency, ϖ_c, and the critical Rayleigh number, R_{aoc}, respectively.

The main results are displayed in Fig.1 where we plot the ratio between the critical oscillatory Rayleigh number R_{aoc} and the critical stationary Rayleigh number R_{asc} as function of the Deborah number Γ for $P = 10^2$, $M_3 = 1.1$ and $M_1 = (1, 10^2)$. We note that, when Γ increases the critical Rayleigh number decreases and when M_1 decreases it increases as expected. The inset show the logarithm function of the ratio between the critical oscillatory Rayleigh

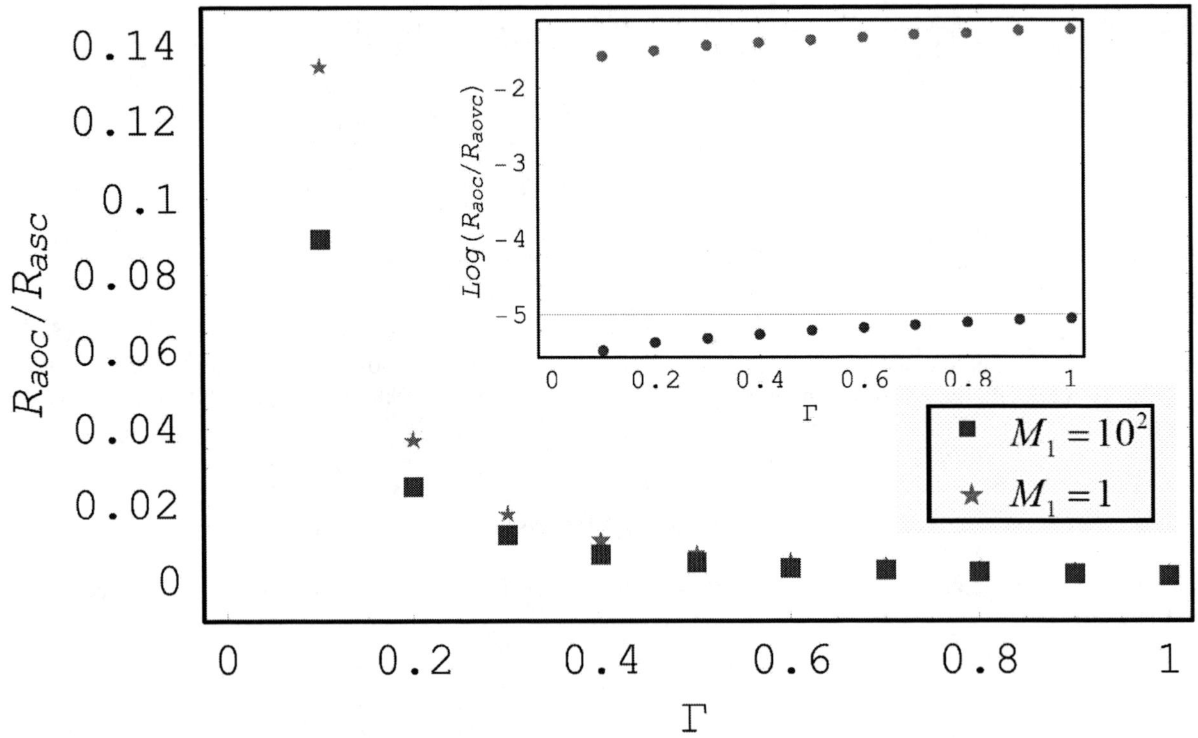

FIGURE 1. Ratio of critical Rayleigh numbers R_{aoc}/R_{asc} as function Γ at $M_3 = 1.1$ and $P = 10^2$ for different values of M_1. The inset show the ratio of critical Rayleigh numbers R_{aoc}/R_{aovc} for same values set.

number for magnetic case R_{aoc} and the critical oscillatory Rayleigh number for non-magnetic case R_{aovc} as function of Γ for the same fixed parameters; and we observe that the magnetic case is drastically more unstable than the non-magnetic case. Finally, we remark that, the main difference between the Newtonian fluid and the viscoelastic fluid is that only in the last one can be appear oscillation. We intend to deal with more realistic rigid-rigid boundary conditions in a future work.

CONCLUSIONS

In the present work, we analyzed the instability thresholds for binary magnetic fluid liquid mixture with viscoelastic properties. We determine the stability thresholds for stationary and oscillatory convection for idealized boundary conditions. In the case of Maxell-Kelvin phenomenon, for oscillatory convection we find the analytical expression for Rayleigh number and the frequency. The results show that the critical value of Rayleigh number decreases when the rheological parameter increases.

ACKNOWLEDGMENTS

This research received financial support from MECESUP FSM-9901, MECESUP USA-0108 and the Millennium Science Initiative from Chile. Grant Ring "Nano-bio computer Lab" of Bicentennial Program of Sciences and Technology-Chile is gratefully acknowledged.

REFERENCES

1. E. H. Kim, Y. Ahn and H. S. Lee, *J. Alloys and Compounds*, in press (2006).
2. J. L. Neuringer and R. E. Rosensweig, *Phys. Fluids* **7**, 1927–1937 (1964).
3. B. A. Finlayson, *J. Fluid Mech* **40**, 753–767 (1970).
4. A. Ryskin and H. Pleiner, *Phys. Rev. E* **69**, 046301 (2004).
5. D. D. Joseph, *Fluid Dynamics of Viscoelastic Liquids*, Springer, New York, (1990).
6. C. M. Vest and A. Arpaci, *J. Fluid Mech.* **36**, 613–623 (1969).
7. M. Sokolov and R. I. Tanner, *Phys. Fluids* **15**, 534-539 (1972).
8. R. W. Kolkka and G. R. Ierly, *J. Non-Newtonian Fluid Mech.* **25**, 209-237 (1987).
9. J. Martinez-Mardones and C. Perez-Garcia, *J. Phys. Condens. Matter* **2**, 1281–1290 (1990).

Secondary Waves in Ribbing Instability

Marta Rosen and Mariano Vazquez

Laboratorio de Medios Porosos, Facultad de Ingenieria, Universidad de Buenos Aires. Paseo Colón 850, (1063) Buenos Aires, Argentina

Abstract. Many natural and technological processes involve phenomena dominated by interfacial mechanics, i.e., occurring within regions of intersection between several fluid and/or solid phases. In addition to capillary and gravitational effects, interfacial phenomena typically involve the interplay of complex processes such as dynamic contact lines, surface active materials, adhesion, temperature and/or compositional gradients, evaporation, etc. The aim of this work is to analyze the answer of the coating system with one cylinder, when viscoelastic polymers are used. We report new experimental results concerning different dynamical regimes, including traveling waves, obtained in that simple configuration and with free boundary conditions.
In the experiments with a non-Newtonian fluid (viscoelastic) with high molecular weight polymer (*PIB_H*), propagative modes have been observed for $Ca > Ca*$. In this case, as the distance to the threshold increases, the digitations standard, initially stationary, presents propagative modes only in some of the regions over the cylinder, and finally the traveling wave is established throughout the whole cylinder without showing any preferential direction. Propagative states have already been observed in the case of two rigid cylinders rotating with opposed surface speeds. However, in that case, the presence of defects does not produce a well-defined traveling wave and the system shows a rapid transition to chaos. Analyzing the space-temporal diagrams, the wave phase velocity could be measured with a methodology that allows calculating systematically this value, and we find a linear correlation between the capillary number Ca and the phase velocity Vf. Besides, a detailed description of bifurcations has been made until reaching a chaotic state where it is possible observes the coexistence of traveling wave zones.

Keywords: Hydrodynamics instability, traveling waves.
PACS: 4720.k, 8310.ji

INTRODUCTION

The aim of this work is to study the dynamical behavior of the directional viscous fingering (ribbing instability) in the nonlinear regime. Many natural and technological processes involve phenomena dominated by interfacial mechanics that is, occurring within regions of intersection between several fluid and/or solid phases. In addition to capillary and gravitational effects, interfacial phenomena typically involve the interplay of complex processes such as dynamic contact lines, surface active materials, adhesion, temperature and/or compositional gradients, evaporation, etc.

The exact mechanics governing these processes, such as the hydrodynamic that controls wetting, or the macroscopic action of surfactants in the evolution of films or coatings, is still not fully understood. Basically, there is a strong interest to know the non linear aspect of the evolution of the interface previous (and during) to coating. Moreover, is necessary to predict the flow behavior during application and the drying process.

Basic question involve the wetting and spreading of thin liquid layers as they flow on geometrically complex surfaces. At the same time, some physical properties of a liquid, such as viscosity and surface tension, are continually changing, as it dries. The problem can be represented by a set of coupled non linear partial differential equations in space and time that can be solved numerically.

Final coating defects, such as sagging, blisters and ribs, can be visualized, as they develop in the simulation output, and their origin can be related to rheology, processing and geometric parameters. Rheology plays a major role in the development and onset of such hydrodynamics instability. In the case of viscoelastic fluids, it has been noticed [1] that the presence of elastic effects has a dramatic influence on the instabilities and change new purely elastic interfacial behaviours due to the influence of normal stress [2].

From a critical value of the Capillary Number $C_a* = \frac{\mu v}{\sigma}$ where μ is the dynamical viscosity σ the surface tension and v the mean roll speed, instability develops on the air-fluid interface (ribbing instability). These configurations are similar to the ones observed in experiments of directional-solidification in liquid crystals and eutectic alloys.

Our interest is to understand the particular behaviour that appears when the system is far away from the onset of this ribbing instability. In the Fig.1 we can observe the wave pattern corresponding to a system of two cylinders where,

CP913, *Nonequilibrium Statistical Mechanics and Nonlinear Physics, XV Conference*
edited by O. Descalzi, O. A. Rosso, and H. A. Larrondo

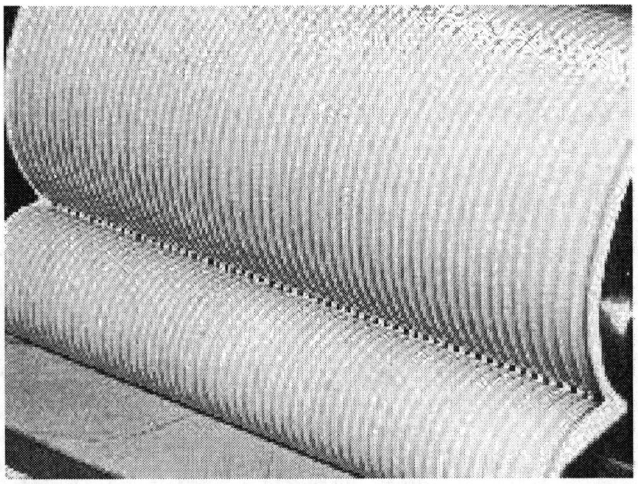

FIGURE 1. Ribbing instability formation in a two-roller system.

FIGURE 2. Experimental system diagram.

when the critical condition (C_a*) is reached the formed structure is transmitted over a second cylinder.

EXPERIMENTAL SETUP

In the Fig.2 (plane-cylinder system) there is a schema of the experimental system where the instability was studied. This journal bearing geometry is adequate to study the instability threshold and the posterior evolution. The system consist on a steel roller ($R = 37.5$ mm, $L = 380$ mm) and a 10-mm thick glass plate mounted on top. The temperature was controlled via a thermostat zed bath. The gap h was varied between 0.2 mm to 0.45 mm. So, the geometrical parameter, aspects relation $\Gamma = h/R$ varied from $1.2x10^2$ to $9x10^{-3}$. When $C_a = C_a*$, there appears what we call primary instability. We have an example of the evolution with Ca of this primary structure for a Newtonian fluid in Fig. 3. As we can observe, change the wave length, yet they continue being stationary; ε is the distance to threshold $\varepsilon = \frac{C_a - C_a*}{C_a*}$ and is controlled by modifying thr cylindr rotation velocity.

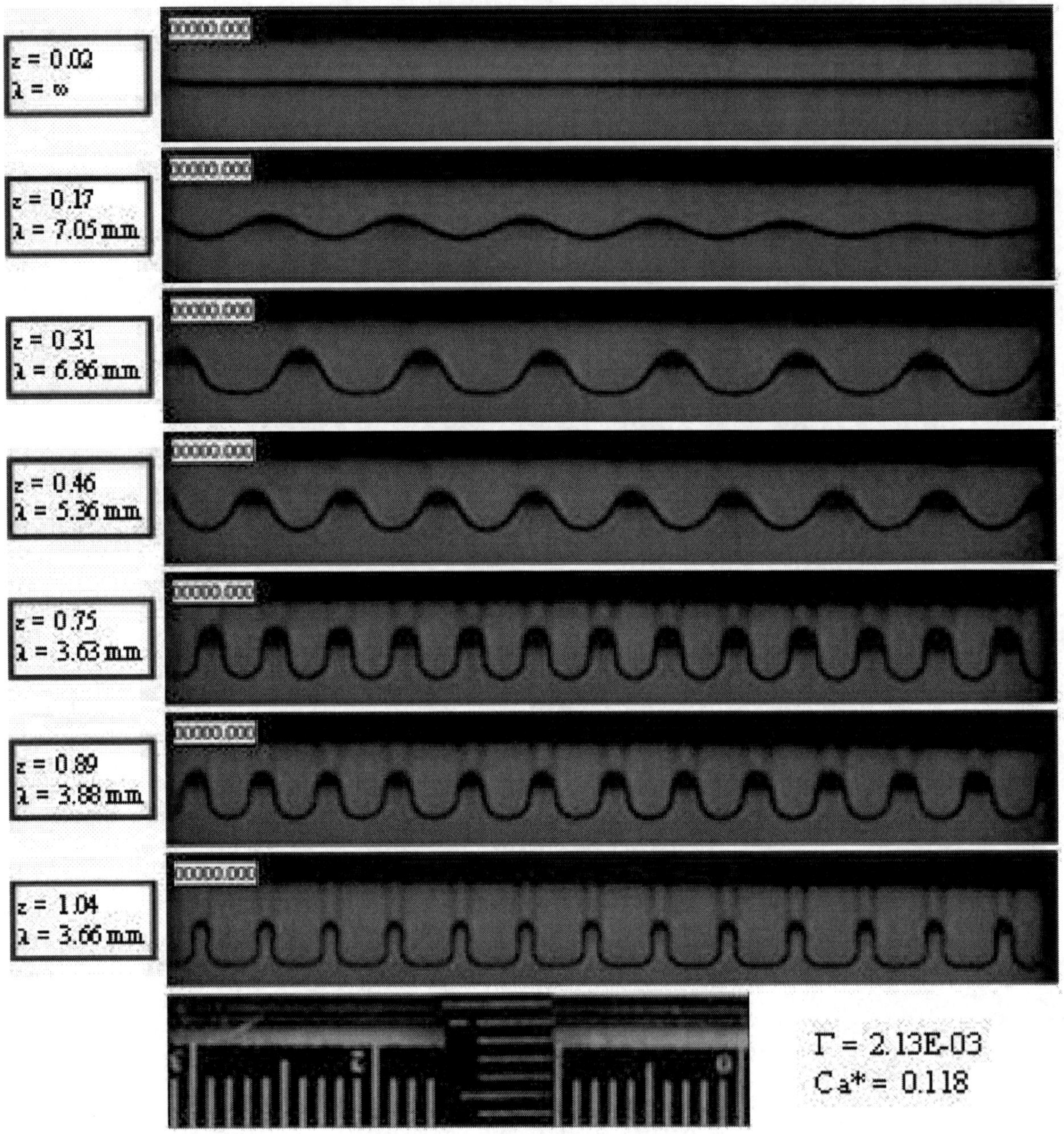

FIGURE 3. Ribbing instability evolution obtained with the same experimental set-up for a Newtonian fluid (vaseline) is a measures of the threshold distance.

TRAVELING WAVES

The experiment was carrying out with non Newtonian fluid, a viscoelastic polymer with high molecular weight, constant viscosity in our range of work, and strong elastic properties. The PIB polymer (6500 wppm), has a molecular weight of $4.70x10^6$ g/mol, the apparent viscosity is 68 mPa.s, the surface tension 29.6 mN/m.

As it can be seen in Fig.4, we find the unexpected appearance of traveling waves. Such appearance, when the system moves away from the threshold, has been observed with Newtonian fluids in experiments [3] made with an eccentrically cylinder system, in which the external cylinder velocity v_e is fixed at a small value, and the internal cylinder velocity v_i in counter-rotation, is varied. Thus, after the first sub-critical bifurcation towards a stationary cell

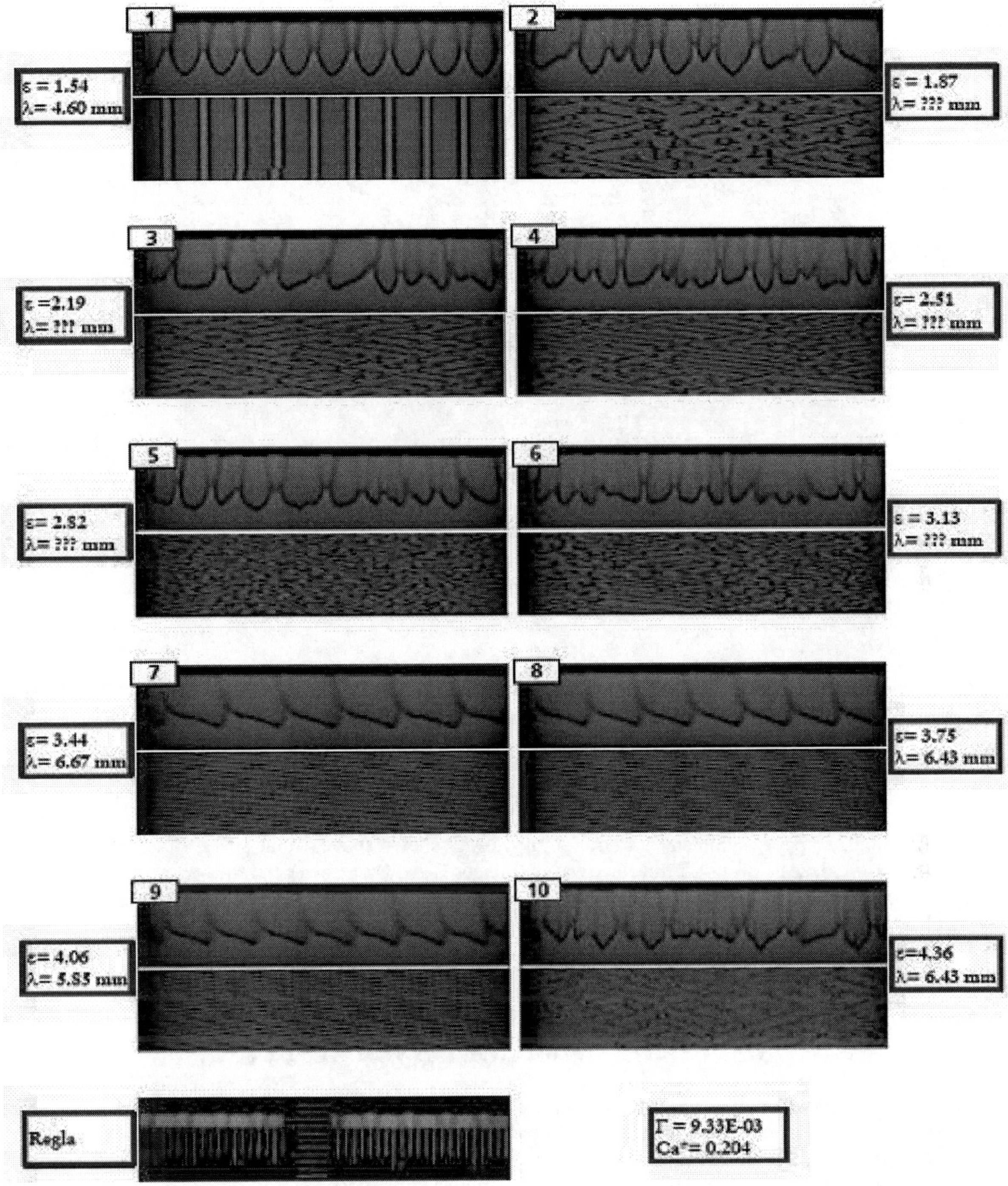

FIGURE 4. Images of instability evolution. Traveling wave and their corresponding spatio-temporal graphics.

pattern (primary instability) is overcome, there appears a secondary bifurcation corresponding to a state in which a uniform traveling cell movement exists.

Fig.4, show the evolution of the system when C_a increases above C_a* until the chaotic regions appear to later give rise to the establishment of the traveling waves. As we can observe, initially the digitations pattern is stable and

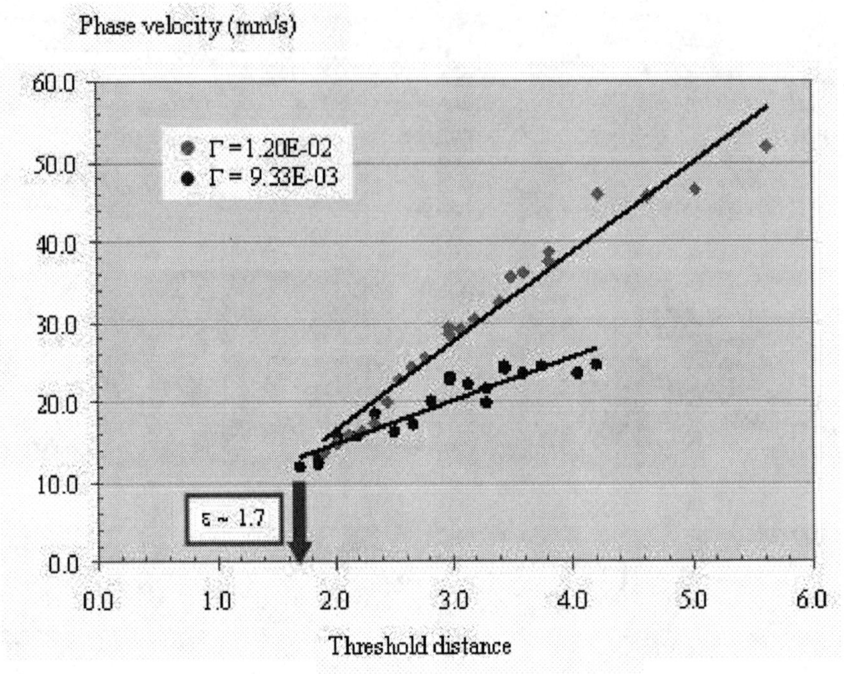

FIGURE 5. Images of instability evolution. Traveling wave and their corresponding spatio-temporal graphics.

propagative modes appear in some regions of the cylinder. After that, the traveling waves settle all along the cylinder without a preferential direction.

PHASE VELOCITY

It is possible visualize a parity symmetry breaking in the images shown above. In the space time graphs the oblique lines indicate the presence of observed traveling waves (images 7 and 8 of Fig.4) and they allow us to calculate the phase velocity. We verified that the pattern wave number decreases linearly with increasing asymmetry. For $\lambda > 1.60$, it is difficult to measure a defined λ.

In Fig.5 a linear correlation between phase velocity v_f and ε may be observed (the lower limit is $\varepsilon = 1.7$). Furthermore, the bifurcation threshold towards the propagative regime for two geometric conditions is visualized.

DISCUSSION OF THE OBTAINED RESULTS

This work presented the development of a study made on the nonlinear regimes of the ribbing instability. In the same system, in the case of Newtonian fluids, the digitations pattern is stable for every $C_a > C_a*$. And, as already remarked, two control parameters are necessary to observe traveling waves with this kind of fluid.

For visco-elastic polymeric solutions, the pattern of primary instability becomes non stationary for values over C_a*, and the subsequent evolution, shows a different behavior before to arrive to fully chaotic state. An approach to analyze the mechanisms that lead to these propagative structures is the method followed by Fauve et al. [4] that uses Ginzburg-Landau's equations formalism.

Some authors propose considered the asymmetry parameter to correlate phase velocity with pattern asymmetry (proportional to amplitude) where the variation in wave numbers, can arise due to the coupling between amplitude and phase. The behavior of these traveling waves can be analyzed using a breaking of parity model where a k and 2k modes interact [5]. Propagative behaviors have been observed in other one-dimensional hydrodynamic systems [6]. For the first time, in this work, a bifurcation for secondary instabilities is observed in the plane-cylinder system with free boundary conditions where the cylinder rotation is the sole control parameter.

REFERENCES

1. A. M. Grillet, A. G. Lee and E. S. G. Shaqfeh, *J.Fluid Mech.* **399**, 49–83 (1999)
2. F. Varela López, L. Pauchard, M. Rosen and M. Rabaud, *J. of Non Newtonian Fluid Mech.* **103**, 123–139 (2002)
3. H. Z. Cummins, L. Fortune and M. Rabaud, *Phys. Rev. E* **47**, 1727–1738 (1993)
4. S. Fauve, S. Douady and O. Thual, *J. Phys. II* **1**, 311–322 (1991)
5. L. Pan and J. R. de Bruyn, *Phys. Rev. Lett.* **70**, 1791–1794 (1993)
6. L. Limat, P. Jenffer, B. Dagens, M. Fermigier and J. E. Wesfreid, *Physica D* **61**, 166–182 (1992)

Paradoxical Way for Losers in a Dating Game

C. M. Arizmendi

Departamento de Física, Facultad de Ingeniería, Universidad Nacional de Mar del Plata, Av. J.B. Justo 4302, 7600 Mar del Plata, Argentina

Abstract. We study the dating market decision problem in which men and women repeatedly go out on dates and learn about each other. We consider a model for the dating market that takes into account progressive mutual learning. This model consists of a repeated game in which agents gain an uncertain payoff from being matched with a particular person on the other side of the market in each time period. Players have a list of preferred partners on the other set. The players that reach higher rank levels on the other set preferences list have also higher probability to be accepted for dating. A question can be raised, as considered in this study: Can the less appreciated players do better? Two different kinds of dating game are combined "à la Parrondo" to foster the less attractive players. Optimism seems to be highly recommendable, especially for losers.

Keywords: Matching Model, Parrondo's Paradox, Dating Market, Bandit Problem
PACS: 02.50.Le, 89.65.-s, 89.65.Gh

INTRODUCTION

Matching problems where the elements of two sets have to be matched by pairs have broad implications in economic and social contexts. As possible applications one could think of job seekers and employers, lodgers and landlords, or simply men and women who want to date. Standard models of matching in economics [1] almost always assume that each agent knows his or her preferences over the individuals on the other side of the market. This assumption, generally associated with neo-classical economics, is too restrictive for many markets, and some interesting work on matching problems with partial information has recently been published [2, 3]. Specifically, perfect information supposition is very far from being a good approximation for the dating market, in which men and women repeatedly go out on dates and learn about each other. Recently, a model for the dating market that takes into account progressive mutual learning was introduced by Das and Kamenica [4]. This model consists of a repeated game in which agents gain an uncertain payoff from being matched with a particular person on the other side of the market in each time period. The problem is related to bandit problems [5], in which an agent must choose which arm of an n-armed bandit to pull in order to maximize long-term expected reward, taking into account the tradeoff between exploring, that is learning more about the reward distribution for each arm, and exploiting, pulling the arm with the maximum expected reward. However, in Das and Kamenica model the arms themselves have agency - they can decide whether to be pulled or not, or whom to be pulled by, and they themselves receive rewards based on who the puller is. This motivates Das and Kamenica formulation of the problem as a "two-sided" bandit problem.

The matching problems describe systems where two sets of persons have to be matched pairwise. Players have a list of preferred partners on the other set. The players that reach higher rank levels on the other set preferences list have also higher probability to be accepted for dating. An interesting question is: Can the less appreciated players do better? In other words: Can the usual dating game losers achieve a better performance? A possible way to accomplish this goal may be the phenomenon known in the literature as Parrondo's paradox [6], devised in 1996 by the Spanish physicist Juan M.R. Parrondo, where the alternation of two fair (or losing) games can result in a winning game. Although initially introduced as individual games, multiplayer versions of the paradox, played by a set of N players, have also been studied [7, 8]. In these collective games, a set of N players are arranged in a ring and each round a player is chosen randomly to play either game A or B. The game A is a simple coin tossing game, where the player wins or loses one unit of capital with probabilities p_A and $1-p_A$, respectively in [7], and a redistribution process where a player is chosen randomly to give away one coin of his capital to another player in [8]. Game A is combined with game B, for which the winning probability depends on the state (winner/loser) of the nearest neighbors of the selected player. A player is said to be a *winner (loser)* if he has *won (lost)* his last game. Recently [9] a new version of collective games was introduced, where besides obtaining the desired result of a winning game out of two fair games, another feature appears: the value of the A/B mixing probability γ determines whether you end up with a winning or a losing game

CP913, *Nonequilibrium Statistical Mechanics and Nonlinear Physics, XV Conference*
edited by O. Descalzi, O. A. Rosso, and H. A. Larrondo

$A+B$. In each round, a player is selected randomly for playing. Then, with probabilities γ and $1-\gamma$, respectively, game A or B is played. Game A is the original coin tossing game. The winning probabilities in game B depend on the collective state of all players. More precisely, the winning probability can have three possible values, determined by the actual number of winners w within the total number of players N, in the following way

$$p_B = winning\,probability\,in\,game B = \begin{cases} p_B^1 & if \quad w > [\frac{2N}{3}], \\ p_B^2 & if \quad [\frac{N}{3}] < w \leq [\frac{2N}{3}], \\ p_B^3 & if \quad w \leq [\frac{N}{3}], \end{cases} \tag{1}$$

where the brackets $[x]$ denote the nearest integer to the number x. The set of values p_B^1, p_B^2, p_B^3 are determined in order to give a fair game and depend on the total number of players N [9].

In this paper we consider a repeated mixing of two different dating games based on both Das and Kamenica learning dating model and on the last version [9] of collective Parrondo games to analyze the possibility that the less attractive players do better. Both dating games are assumed to be fair. For the sake of clarity let us imagine that the different games that we call A and B are played in different places with different rules. In game A, the probability that the man proposal be accepted is modelled by coin tossing, that is $p_A = 1/2$. This does not necessarily mean that every woman flips actually a coin, but it can be thought that all the variables not considered to construct the preference list, such as woman's mood, man's way, or the group size dependance of females selectivity [10] are contributing to the probability of acceptance p_A. On the other hand, the probability of acceptance for game B, p_B, depends on the number of previous winners within all players. This collective influence may seem not so clear at first sight, but it may be thought that in p_B collective moods contribute, such as if all woman's friends are dating, that particular woman is better disposed to accept man's proposal to go out. This is the well known *herd behavior* that is present in stock market bubbles as well as in everyday decision making.

THE MODEL

There are N men and N women, who interact for T time periods. v_j^m is the value of woman j to every man, and v_i^w is the value of man i to every woman. These values are constant through time. In each period, a man i is chosen randomly from the N possible men. The expected $i's$ payoff of dating woman j:

$$payoff_{i,j}^m[t] = Q_{i,j}^m[t] * p_{i,j}^m[t], \tag{2}$$

where $Q_{i,j}^m[t]$ is the man $i's$ estimate of the value of going out with woman j at time t and $p_{i,j}^m[t]$ is the man $i's$ estimate at time t of the probability that woman j will go out with him if he asks her out. In this way man's decision is based on any prior beliefs and the number of rewards he has received. Both the expected value on a date with that particular woman and the probability that she will accept his offer are taken into account by (2). The expected woman $j's$ payoff of dating man i is:

$$payoff_{i,j}^w[t] = Q_{i,j}^w[t], \tag{3}$$

where $Q_{i,j}^w[t]$ is the woman $j's$ estimate of the value of going out with man i at time t. No probability is considered because man i considered as a date to be must propose to the woman j. Since the underlying v_j^m and v_i^w are constant we define $Q_{i,j}^m[t]$ as man i's sample mean at time t of the payoff of going out with woman j:

$$Q_{i,j}^m[t] = \sum (v_j^m + \varepsilon), \tag{4}$$

where the sum is made on the effective dates between i and j and ε is noise drawn from a normal distribution. In the same way, $Q_{i,j}^w[t]$ is woman j's sample mean at time t of the payoff of going out with man i:

$$Q_{i,j}^w[t] = \sum (v_i^w + \varepsilon). \tag{5}$$

In order to deal with the nonstationarity of $p_{i,j}^m[t]$'s, on the other hand, we use a fixed learning rate for updating the probabilities which allows agents to forget the past more quickly:

$$p_{i,j}^m[t] = (1-\eta) p_{i,j}^m[t-1] + \eta \frac{offers_{i,j}\,accepted[t-1]}{offers_{i,j}\,made[t-1]}, \tag{6}$$

where η is a constant parameter.

The Man's Decision Problem

The top ranked woman from the list of preferred partners of i is selected to ask out for a date. The rank levels of the preference list are distributed according to the expected $i's$ payoff of dating woman j (2). The man i acts in a greedy way asking out woman j at the top of his preference list.

The Woman's Decision Problem

The rank levels of the women preference lists are distributed according to the expected woman $j's$ payoff of dating man i (3). The woman's decision problem depends on the game:

Game A

In both games women have to consider the exploration-exploitation tradeoff. Exploitation means maximizing expected reward (greedy choice). Exploration happens when the player selects an action with lower expected payoff in the present in order to learn and increase future rewards. One of the simplest techniques used for bandit problems is the so-called ε-greedy algorithm. This algorithm selects the arm with highest expected value with probability $1-\varepsilon$ and otherwise selects a random arm. We will use slightly changed versions of the ε-greedy algorithm in both games. In game A the exploration-exploitation tradeoff depends on coin tossing, that is the woman accepts the man's i offer to date with probability $p_A = 1/2$ (exploration) or she acts greedily and goes out with her best $payoff_w$ choice with probability $1 - p_A = 1/2$.

Game B

In game B the choice of exploration or greedy behavior depends on the collective state of all men players. A man player is said to be a winner or a loser when he got his date or not, respectively, in his last game. More precisely, the winning or exploration probability can have three possible values, determined by the actual number of winners w within the total number of players N, in the following way

$$p_B = exploration\,probability\,in\,game B = \begin{cases} p_B^1 & if \quad w > [\frac{2N}{3}], \\ p_B^2 & if \quad [\frac{N}{3}] < w \leq [\frac{2N}{3}], \\ p_B^3 & if \quad w \leq [\frac{N}{3}], \end{cases} \tag{7}$$

where the brackets $[x]$ denote the nearest integer to the number x. The woman accepts the man's i offer to date with probability p_B (exploration) or she goes out with her best $payoff_w$ choice with probability $1 - p_B$. The set of values p_B^1, p_B^2, p_B^3 are determined in order to give a fair game and depend on the total number of players N [9].

RESULTS

Our simulations involve a market of $N = 4$ men and $N = 4$ women. The learning rate of probability $p_{i,j}^m[t]$ is $\eta = 0.05$. The set of probabilities for game B, p_B^1, p_B^2, p_B^3 determined to give a fair game [9] for $N = 4$ are $p_B^1 = 0.79$, $p_B^2 = 0.65$ and $p_B^3 = 0.15$. The noise signal is drawn from a normal distribution of standard deviation 0.5. v^m's and v^w's are:

$$v_k^m = v_k^w = N - k + 6, \tag{8}$$

where $1 \leq k \leq 4$. Reported results are obtained with 1000 simulations averages on 10^4 time steps.

Parrondo's Paradox

All players benefit from Parrondo mixing of both games A and B as can be observed in Fig. 1. In this Figure the total matches, that is the number of accepted dating offers from all players, is shown as a function of time for game B

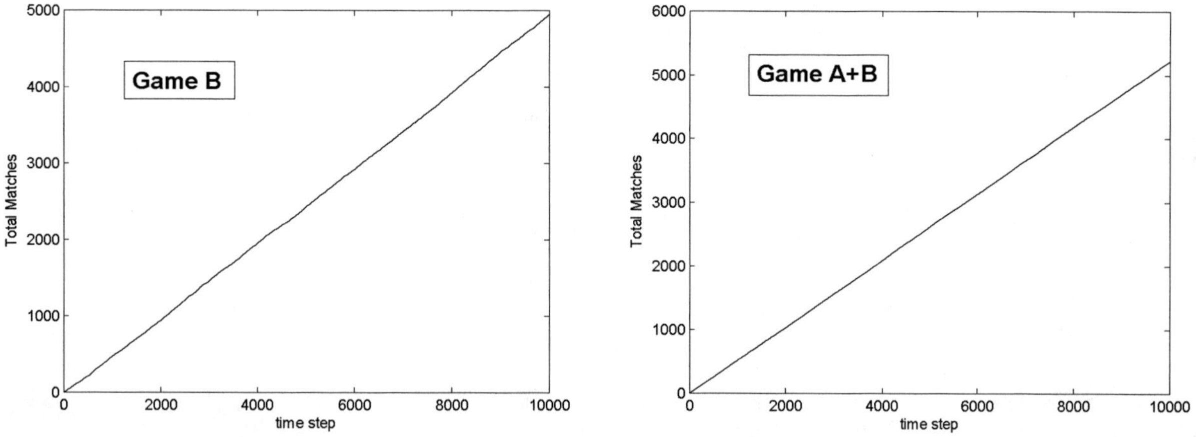

FIGURE 1. Time evolution of total number of offers accepted between dating game B played alone and games AB switched with mixing probability $\gamma = 1/2$. Total matches vs. time (a) Game B; (b) Games AB

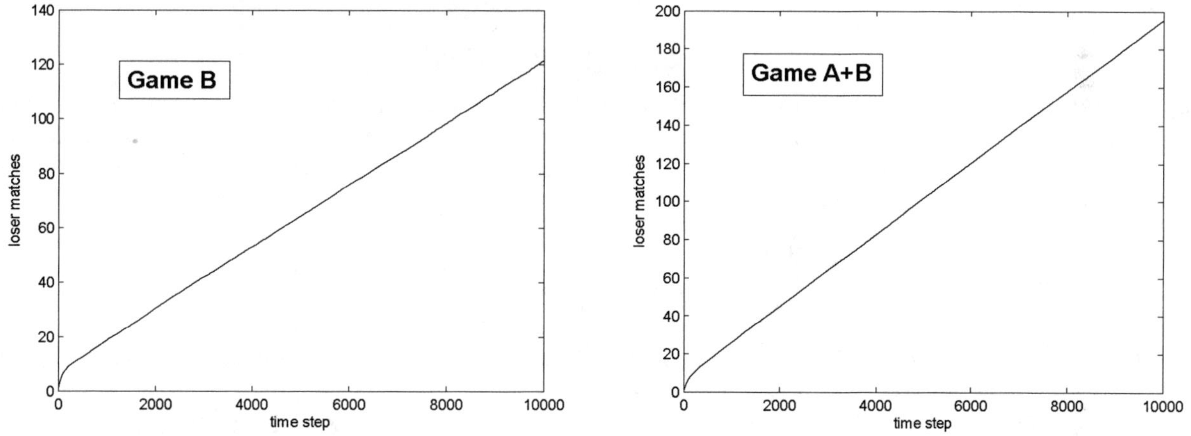

FIGURE 2. Time evolution of number of accepted offers from the last of preferences list (loser) between dating game B played alone and games AB switched with mixing probability $\gamma = 1/2$. Loser matches vs. time (a) Game B; (b) Games AB

played alone and games A and B switched with mixing probability $\gamma = 1/2$. In the first case, the fairness of game B is verified when half of all attempts produce a match. When $A+B$ are played, the paradox produces that more than half of all attempts are successful, as can be seen in Fig. 1b.
On the other hand, less favored players, i.e. the lowest v^w ones, have an evolution that is shown in Fig. 2, for game B played alone and games A and B switched with mixing probability $\gamma = 1/2$. In this Figure the advantage of playing $A+B$ over B for losers can be appreciated. The comparison with game A is esentially the same for all the results studied.

Optimistic Results

Das and Kamenica propose as an alternative method to obtain asymptotic stability in their model for dating couples to suppose that players are initially optimistic and their level of optimism declines over time. As they say "This is another form of patience - a willingness to wait for the best - and it should lead to more stable outcomes" [4]. A systematic overestimate of the probability that the dating offer will be accepted is used to represent optimism.
Let us analyze optimistic players performance at our dating games model. At time t optimistic players use the

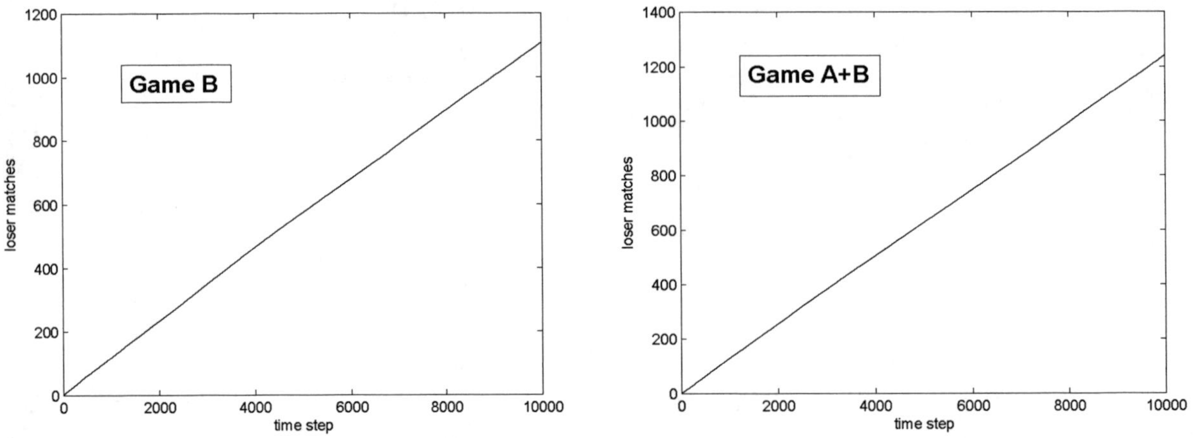

FIGURE 3. Optimistic Players: Time evolution of number of accepted offers from the last of preferences list (loser) between dating game B played alone and games AB switched with mixing probability $\gamma = 1/2$. Loser matches vs. time (a) Game B; (b) Games AB

probability estimate:

$$p'^{m}_{i,j}[t] = \alpha(t) + (1-\alpha(t))p^{m}_{i,j}[t], \tag{9}$$

where $p^{m}_{i,j}[t]$ is updated as before by (6) and $\alpha(t) = (T-t)/T$, with T the total number of time steps in simulations. Figure 3 shows the evolution of loser accepted offers corresponding to optimistic players. The order of loser acceptance increase by a factor 10 comparing optimistic (Fig. 3) and non-optimistic (Fig. 2) losers. On the other hand, the advantage of playing $A+B$ over B for losers is conserved.

N and γ dependence

The results are highly dependent on both, the number N of players and the mixing probability γ of the games A and B. As thoroughly explained on [9], the losing or winning character of the mixed $A+B$ games depends on N and γ. We will present in a more extensive way the dependance on the number N of players and the mixing probability γ of the games A and B elsewhere.

CONCLUSIONS

We find a way for less qualified dating game players to improve their performance by means of a repeated mixing of two different dating games A and B based on a recent dating market model [4] and on a recent collective game version of Parrondo's paradox [9]. In game A, the probability associated to exploration-exploitation tradeoff, that the man proposal be accepted is modelled by coin tossing, that is $p_A = 1/2$. On the other hand, the probability of acceptance for game B depends on the collective state of the men set obtained through the number of previous winners within all players. We show that losers benefit from Parrondo mixing of both games A and B. In the optimistic version of our model, when it is assumed that players are initially optimistic and their level of optimism declines over time, loser acceptance increase by a factor 10 and the *paradoxical* advantage of playing $A+B$ over B or A for losers is conserved. The results are highly dependent on both, the number N of players and the mixing probability γ of the games A and B.

ACKNOWLEDGMENTS

We would like to thank Hugo Lopez Montenegro for many interesting and inspiring discussions at the beginning of this work. This research was partially supported by Universidad Nacional de Mar del Plata and ANPCyT (PICTO 11-495).

REFERENCES

1. A. E. Roth and M. Sotomayor , *Two-Sided Matching: A Study in Game Theoretic Modeling and Analysis*, Econometric Society Monograph Series. Cambridge University Press, Cambridge, UK, 1990.
2. Y.-C. Zhang, Happier World with More Information, *Physica A* **299**, 104-120 (2001).
3. P. Laureti and Y.-C. Zhang, Matching Games with Partial Information, *Physica A* **324**, 49-65 (2003).
4. Sanmay Das and Emir Kamenica, "Two-Sided Bandits and the Dating Market," in *IJCAI-2005*, edited by Leslie Pack Kaelbling, Alessandro Saffiotti, Proceedings of the Nineteenth International Joint Conference on Artificial Intelligence, Edinburgh, Scotland, UK, July 30-August 5, 2005, pp. 947-952.
5. D. A. Berry and B. Fristedt, *Bandit Problems: Sequential Allocation of Experiments*, Monographs on Statistics and Applied Probability. Chapman and Hall, London, UK, 1985.
6. G.P. Harmer, D. Abbott, Losing strategies can win by Parrondo's paradox, *Nature* **402**, 864-864 (1999).
7. R. Toral, Cooperative Parrondo's games, *Fluctuations Noise Lett.* **1** L7-L12 (2001).
8. R. Toral, Capital redistribution brings wealth by Parrondo's paradox, *Fluctuations Noise Lett.* **2** L305-L311 (2002).
9. P. Amengual, P. Meurs, B. Cleuren, and R. Toral, Reversals of chance in paradoxical games, *Physica A* **371**, 641-648 (2006).
10. R. Fisman, S. Iyengar, E. Kamenica, I. Simonson, Gender Differences in Mate Selection: Evidence from a Speed Dating Experiment, *Quarterly Journal of Economics,* **121**, 673-698 (2006).

Cooperation in Diffusive Spatial Games

Mendeli H. Vainstein*, Ana T. C. Silva† and Jeferson J. Arenzon**

*Instituto de Física and International Center of Condensed Matter Physics, Universidade de Brasília, CP 04513, 70919-97, Brasília, DF, Brazil

†Departamento de Física, Universidade Estadual de Feira de Santana, Campus Universitário, km 3, BR 116, 44031-460 Feira de Santana BA, Brazil

**Instituto de Física, Universidade Federal do Rio Grande do Sul, CP 15051, 91501-970 Porto Alegre RS, Brazil

Abstract. Random diffusion is shown to be an important mechanism on fostering cooperative behavior among simple agents (memoryless, unconditional cooperators or defectors) living on a spatially structured environment. In particular, under the Prisoner's Dilemma framework, when allowing the agents to move with the simple "always-move" rule, we find that cooperative behavior is not only possible but may even be enhanced. In addition, for a broad range of densities, mobile cooperators can more easily invade a population of mobile defectors, when compared with the fully viscous, immobile case. Thus, such simple mobility pattern may have played a fundamental role both in the onset and development of cooperative behavior, paving the way to more complex, individual and group, motility rules.

Keywords: Game theory, Prisoner's Dilemma, Cooperation, Diffusion
PACS: 87.23.Ge

Among the top 25 scientific questions listed on a recent issue of the Science magazine (July 1, 2005), representative of many major gaps in our basic knowledge, there was "How did cooperative behavior evolve?". For a long time, this question has been eluding scientists coming from very different fields, ranging from psychological and social sciences to statistical physics, even though remarkable progress and insights have been accumulated in the last decades [1, 2, 3, 4, 5, 6]. Several mechanisms have been identified, like direct, indirect and spatial reciprocity, kinship and group selection, but a complete answer regarding the general conditions under which cooperative behavior may evolve and be sustained is still lacking [7]. Cooperation, although involving a cost to the performer, is ubiquitous in biological populations, even in groups of extremely simple individuals [8, 9, 10, 11, 12, 13, 14, 15, 16, 17, 18] where a mechanism not relying on complex memory functions or cognitive processes must be invoked. One such mechanism is spatial or network reciprocity: when dispersal is very limited (high population viscosity), differently from the standard, random mixing population, the continuing interaction within the local neighborhood increases the probability of future encounters and the maintenance of cooperation. To model such a situation, Nowak and May [19, 20] extended the first considerations on the effects of territoriality in the spread of strategies in the Prisoner's Dilemma game done by Axelrod [2], and showed that with geographical fixation, even simple nice rules like unconditional cooperation are able to survive. In these structured populations, cooperative strategies can build clusters in which the benefits of mutual cooperation can outweigh losses against defectors, maintaining the population of cooperators stable. These spatial games, in which the interactions are localized and non random, have been studied and extended in many ways (see, for example, Refs. [21, 22, 23, 24, 25, 26, 27, 28, 29, 30, 31, 32, 33, 34, 35, 36, 37, 38, 39, 40, 41, 42, 43, 44, 45]). In such spatially structured populations, one may wonder about the effects of mobility: is it possible to evolve and sustain cooperation in a population of mobile agents? In particular, is it possible to have a finite density of unconditional cooperators with unbiased, random mobility? Once mobile, cooperators can move away from the supporting clusters of other cooperators, becoming targets for exploitation, while defectors, on their turn, can exploit cooperators and move away to avoid retaliation. Moreover, mobility helps to increase the random mixing and it is known that when every agent interacts randomly with the whole population, defection prevails. With such considerations, it came as a surprise [46] that even a blind pattern of mobility, without anticipating the future neighborhood and without considering the accumulated payoff, may have a positive effect in the amount of cooperation, and not be a limiting factor for it.

Many previous works have considered the role of diffusion in the problem of cooperation [47, 48, 49, 50, 51, 52, 53, 54, 36, 55, 56]. The present work, which complements the discussion of Ref. [46], differs from all these in several important aspects: we consider non random interactions on a two-dimensional structure, mobility traits do not evolve and individual movements are Brownian, non contingent, and not under the control of the agents, both strategies

CP913, *Nonequilibrium Statistical Mechanics and Nonlinear Physics, XV Conference*
edited by O. Descalzi, O. A. Rosso, and H. A. Larrondo

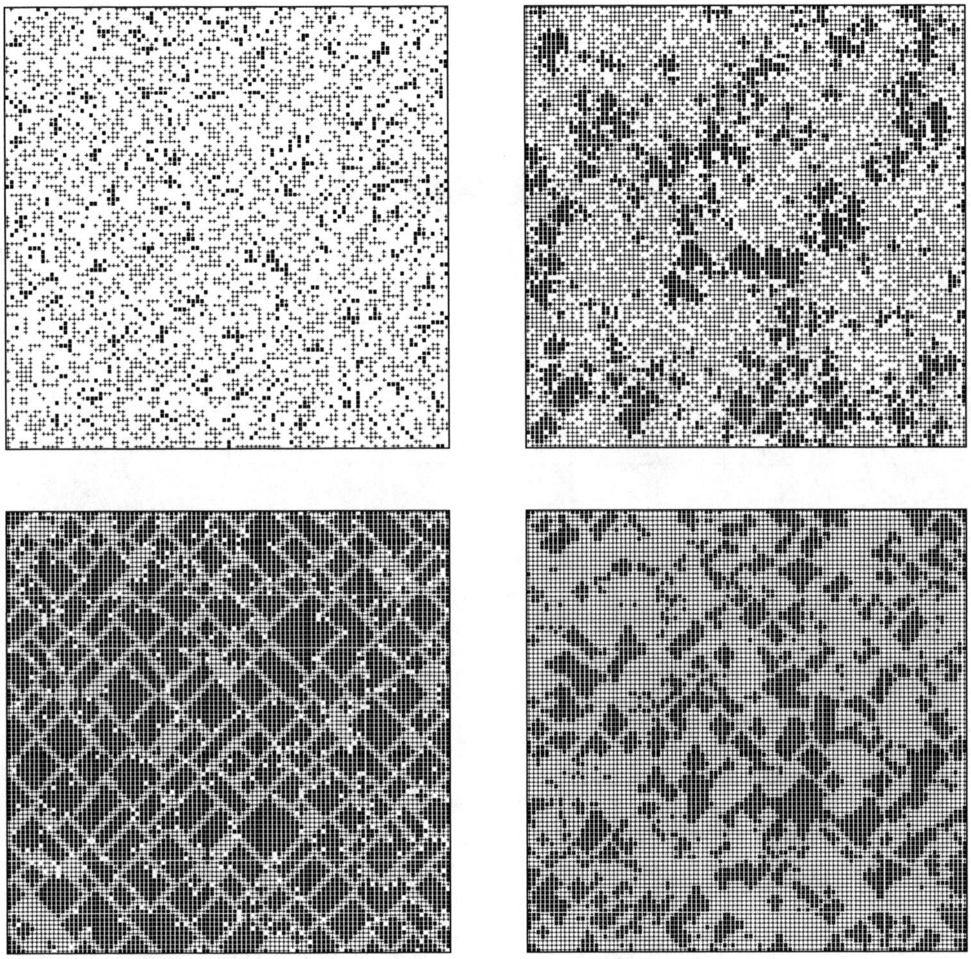

FIGURE 1. Snapshots of typical configurations once the asymptotic state has been reached for a system with no mobility ($m = 0$) and different densities, from top to bottom, left to right: $\rho = 0.4, 0.8, 0.962$ (this corresponds to the optimal density for cooperation) and 1. All simulations are performed on a square lattice of side length $L = 100$, $b = 1.4$, and cooperators (defectors) are represented by squares (crosses). Vacant sites are painted white. In the top left panel, cooperators may survive either because they are isolated or belong to a weakly predated cluster. The remaining panels present high density situations, well above the percolation threshold (a spanning cluster is present), with quite different cluster structures in each case. Notice that once the system percolates, the spanning cluster is formed by defectors, even when they are a minority (like in the bottom left panel).

considered are simple, unconditional and non retaliating, with no memory of previous steps. In other words, we are considering a basic scenario, whose importance is to settle the minimal conditions under which cooperative behavior might be sustained once the requirement of strong viscosity for spatial reciprocity is relaxed. In particular, we show that in the framework of the spatial Prisoner's Dilemma (PD) game under the above conditions, cooperative behavior, besides being robust, can even be diffusion-enhanced.

In any round of the PD game, each of two players either cooperates (C) or defects (D). Here we only consider pure strategies, although more complex rules, with memory of previous encounters, have been devised [2]. The payoff depends on the mutual choice and is given by the matrix whose elements are: a reward R (punishment P) if both cooperate (defect), S (sucker's payoff) and T (temptation) if one cooperates and the other defects, respectively. Moreover, these quantities should satisfy the inequalities $T > R > P > S$ and $2R > T + S$. Here we will take a simplified, one parameter version [19]: $R = 1, P = S = 0$ and $T = b > 1$. In a random mating, infinite population defecting will be the most rewarding strategy, independently of the opponent's choice. On a network of contacts, each individual

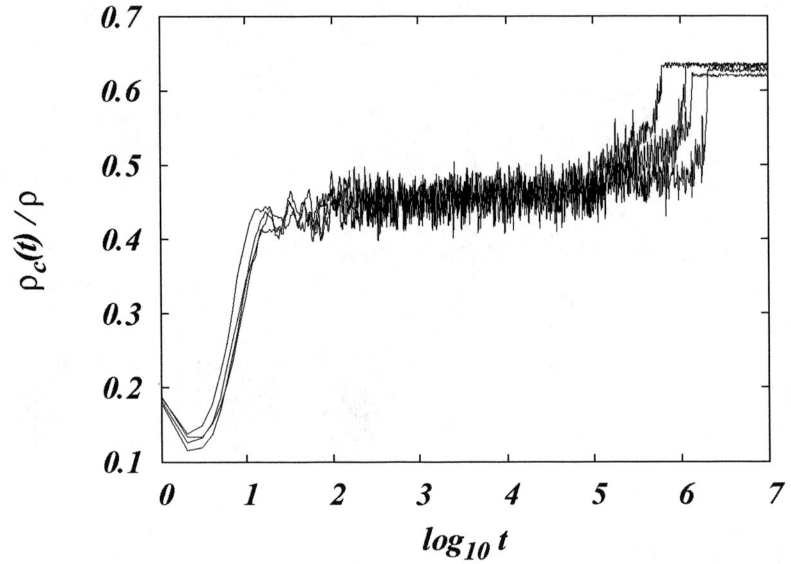

FIGURE 2. Time evolution of cooperator density at the point where their frequency in the population is maximum. Due to the presence of quenched disorder, it evolves very slowly and after many decades it jumps to the equilibrium value. Different lines represent several samples.

combats with its closest neighbors (if any) and accumulates the corresponding payoff, that will be compared, during the offspring step, with the payoff accumulated by the neighbors.

We addressed in an earlier work [31] the question of the robustness of cooperation in spatial games in the presence of heterogeneous environments. By introducing quenched disorder in the lattice (random dilution) each individual would sense a locally varying social environment as the number of neighbors becomes site dependent, and optimal cooperation is achieved for weak disorder as the defects (or inaccessible regions) act as pinning fields, keeping the clusters of cooperators more protected from invasions. In this sense, an irregular landscape may enhance cooperation by introducing natural defenses against invasions of defectors. In fig. 1 we present several snapshots for the case where there are no mobile agents and different lattice occupations. When the density is low, the clusters are disconnected and the fate of each isolated cluster depends on the initial distribution of cooperators and defectors: cooperation is guaranteed by geographical isolation. As density increases, the system percolates. Interestingly, the spanning cluster is always built by defectors, while cooperators tend to organize in more compact groups, what is essential for their stability. Cooperation is maximized for weak disorder ($\rho \simeq 0.962$) and it is the pinning effect of the empty sites that drives the enhancement in the amount of cooperators. Close to this value, the cluster geometry is also dramatically changed as can be observed in the figure. Moreover, the presence of disorder also has a remarkable effect in the way the system evolves, as can be seen in figure 2, where the time evolution of five samples whose density is very close to the critical occupation is shown. After a fast transient there is a very slow increase of the number of cooperators that lasts several decades, followed by an abrupt, avalanche-like jump to the stationary state value. Such slow dynamics is a general feature of systems with quenched disorder and the average jumping time has a very fast increase as the lattice occupation approaches $\rho \simeq 0.962$.

When the vacant sites are no longer held fixed, they may become occupied by a neighbor agent with a probability m that measures the populational viscosity. Only random, unbiased diffusion has been considered in Ref. [46], although extensions to contingent rules are also being studied. As before, individuals combat with their closest neighbors and accumulate a given payoff. Then, they may either move or try to generate their offspring, following the player with the greatest payoff in the neighborhood. The detailed outcome of the game will depend on the precise implementation of the dynamics, in particular, the order in which combats, offspring generation and diffusion occur leads to qualitative and quantitative differences in the population. As an example, in fig. 3 the asymptotic state for a density far above the percolation threshold is shown for two dynamics where the order of the diffusive and offspring steps are reversed (Combat-Offspring-Diffusion, COD, and Combat-Diffusion-Offspring, CDO, dynamics). In the former, as the name says, each step consists of combats followed by the generation of offspring done in parallel, and then diffusion, while

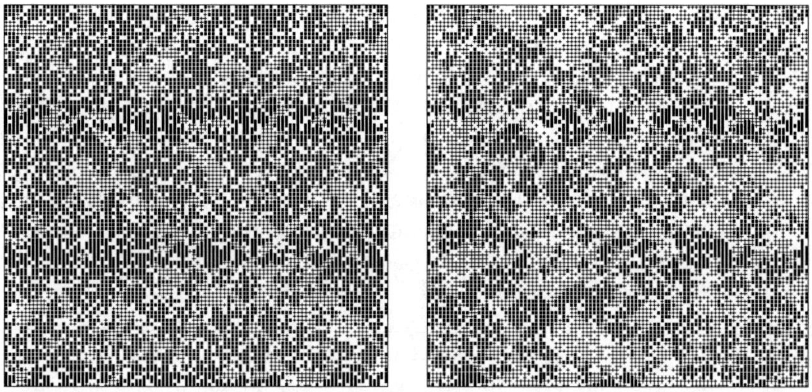

FIGURE 3. Snapshot of typical configurations after the asymptotic state has been achieved for a system with $\rho = 0.8$ and $m = 1$, for CDO (left) and COD dynamics (right). Empty sites are left blank, while cooperators and defectors are represented by squares and crosses, respectively. Compare with the corresponding snapshot in Fig. 1 where $m = 0$.

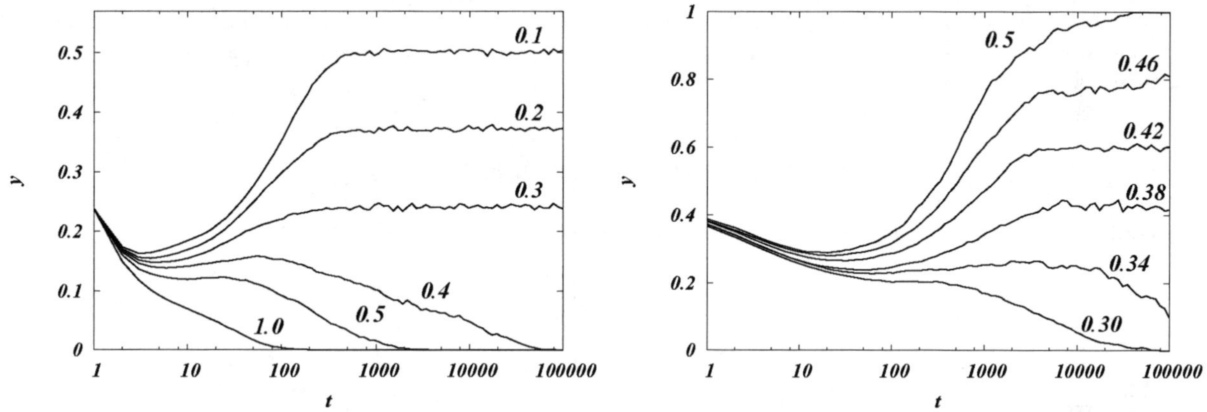

FIGURE 4. Average fraction of cooperators ρ_c/ρ as a function of time for several values of m for the COD dynamics at $\rho = 0.7$ (left) and $\rho = 0.24$ for the CDO dynamics (right). In all cases there is an initial decrease in the cooperator density since cooperators are not yet coordinated and only form small groups, being easily predated. Depending on the mobility, at later times the existing cooperator clusters may either disappear or grow.

in the later, the diffusion and offspring steps are reversed. During the diffusive step, each agent makes an attempt to jump to an empty neighboring site, what is accepted with a probability m.

Fig. 4, with the temporal evolution of the average cooperator density ρ_c, exemplifies the rich behavior presented by the model once mobility is introduced. The ultimate fate of cooperation depends on the population viscosity m, its total density and probably on the initial state: while in the COD dynamics (Fig. 4, left) the asymptotic density of cooperators decreases as m increases, in the CDO case (Fig. 4, right), on the contrary, ρ_c may increase with m for some values of ρ. The detailed dependence on the mobility and density of the asymptotic number of cooperators was studied in Ref. [46], and it was shown that for broad conditions, cooperation is enhanced when compared with the viscous case of $m = 0$. This effect is even stronger in the CDO case: whenever $m \neq 0$, it is good for the cooperators to move away from their partner, whatever its strategy, *after* the combats and *before* the payoff comparison of the offspring step, what favors the CDO dynamics. In this way, an exploited cooperator can still leave descendants if, after diffusing, it is not neighbored by a high payoff defector. Another important effect is that mobility facilitates the onset of cooperation: as shown in Ref. [46], starting with an initial patch with only two cooperators in a sea of defectors, the probability that cooperation overtakes the whole population is an increasing function of the mobility for a wide range of densities. In comparison, this has a very small probability of happening when $m \simeq 0$.

A possible realization of such diffusion scenario may occur in organisms with extracellular metabolism, as is the case of some yeast cells [15]. Sugar is processed outside the cell by a secreted enzyme called invertase, creating a common resource for all surrounding cells. This offers the opportunity for defection as some cells may not have the cost of producing the enzyme but yet benefit from that produced by others. To what extent the cooperative behavior observed in such simple organisms is a sole effect of the underlying spatial structure or whether there is an enhancement factor due to diffusion is an open and interesting question. Moreover, in systems that present polarized motion (chemotaxis), depending on the concentration of cooperators, a gradient of nutrients may be present and both cooperators and defectors can migrate towards (away) high cooperator (defector) density regions. Thus, a possible rule in this case can be: "cooperators attract–defectors repel", but many others can be devised. Such non random rules may also develop cooperative swarming or foraging, relevant for evolving higher levels of biological organization, and is already found to occur in bacteria [13].

Here we presented results for a simple spatial game where the patched environment allows explicit, although random, movement of agents whose strategies are pure, non retaliating (unconditional). The diffusion is Brownian, not relying on any type of explicit, genotypic or phenotypic assortment. Whether a given strategy is able or not to invade another population would strongly depend on how viscous a population is, the global density and the chosen dynamics. However, some universal conclusions can be stated. First of all, cooperation is possible under the above conditions, thus enlarging the limits for cooperative behavior. Second, for a broad range of the parameters (density, viscosity, etc), cooperation is enhanced in respect to the viscous case. Third, a rule like always-move, regardless of the opponent strategy, may increase the capability of cooperators to invade and overtake a population of defectors. In this sense, mobility may have a fundamental role in the problem of the onset of cooperation, and the evolution of a simple mobility as the one studied here, is a first step towards the emergence of more complex foraging behavior or even coherent, cooperative herding behavior.

ACKNOWLEDGMENTS

This work was partially supported by the Brazilian agencies CAPES, CNPq and FAPERGS.

REFERENCES

1. J. Maynard Smith, *Evolution and the Theory of Games*, Cambridge University Press, Cambridge, UK, 1982.
2. R. Axelrod, *The Evolution of Cooperation*, BasicBooks, New York, 1984.
3. L. A. Dugatkin, and H. K. Reeve, *Game Theory and Animal Behavior*, Oxford University Press, New York, 1998.
4. C. Hauert, and G. Szabó, *Am. J. Phys.* **73**, 405–414 (2005).
5. G. Szabó, and G. Fath, Evolutionary games on graphs (2006), cond-mat/0607344.
6. M. A. Nowak, *Evolutionary Dynamics: Exploring the Equations of Life*, Belknap Press, 2006.
7. M. A. Nowak, *Science* **314**, 1560–1563 (2006).
8. P. E. Turner, and L. Chao, *Nature* **398**, 441–443 (1999).
9. B. J. Crespi, *Trends Ecol. Evol.* **16**, 178–183 (2001).
10. M. Vulic, and R. Kolter, *Genetics* **158**, 519–526 (2001).
11. T. Frick, and S. Schuster, *Naturwissenschaften* **90**, 327–331 (2003).
12. P. B. Rainey, and K. Rainey, *Nature* **425**, 72–74 (2003).
13. G. J. Velicer, and Y. N. Yu, *Nature* **425**, 75–78 (2003).
14. A. S. Griffin, S. A. West, and A. Buckling, *Nature* **430**, 1024–1027 (2004).
15. D. Greig, and M. Travisano, *Proc. R. Soc. London B (Suppl.)* **271**, S25–S26 (2004).
16. D. M. Wolf, V. V. Vazirani, and A. P. Arkin, *J. Theor. Biol.* **234**, 255–262 (2005).
17. F. Fiegna, Y.-T. N. Yu, S. V. Kadam, and G. J. Velicer, *Nature* **441**, 310–314 (2006).
18. N. J. Mehdiabadi, C. N. Jack, T. T. Farnham, T. G. Platt, S. E. Kalla, G. Shaulsky, D. C. Queller, and J. E. Strassmann, *Nature* **442**, 881–882 (2006).
19. M. A. Nowak, and R. M. May, *Nature* **246**, 15–18 (1992).
20. M. Nowak, and R. May, *Int. J. of Bifurcation and Chaos* **3**, 35–78 (1993).
21. M. A. Nowak, S. Bonhoeffer, and R. M. May, *Int. J. of Bifurcation and Chaos* **4**, 33–56 (1994).
22. M. A. Nowak, S. Bonhoeffer, and R. M. May, *Proc. Natl. Acad. Sci. USA* **91**, 4877–4881 (1994).
23. K. Lindgren, and M. G. Nordahl, *Physica D* **75**, 292–309 (1994).
24. P. Grim, *J. Teor. Biol.* **173**, 353–359 (1995).
25. T. Killingback, and M. Doebeli, *Proc. R. Soc. Lond. B* **263**, 1135–1144 (1996).
26. M. Nakamaru, H. Matsuda, and Y. Iwasa, *J. Theor. Biol.* **184**, 65–81 (1997).

27. G. Szabó, and C. Töke, *Phys. Rev. E* **58**, 69–73 (1998).
28. K. Brauchli, T. Killingback, and M. Doebeli, *J. Theor. Biol.* **200**, 405–417 (1999).
29. T. Killingback, M. Doebeli, and N. Knowlton, *Proc. R. Soc. B* **266**, 1723–1728 (1999).
30. G. Szabó, T. Antal, P. Szabó, and M. Droz, *Phys. Rev. E* **62**, 1095–1103 (2000).
31. M. H. Vainstein, and J. J. Arenzon, *Phys. Rev. E* **64**, 051905 (2001).
32. G. Abramson, and M. Kuperman, *Phys. Rev. E* **63**, 030901 (2001).
33. C. Hauert, *Int. J. Bif. Chaos* **12**, 1531–1548 (2002).
34. B. Kim, A. Trusina, P. Holme, P. Minnhagen, J. Chung, and M. Choi, *Phys. Rev. E* **66**, 021907 (2002).
35. J. Miekisz, *J. Phys. A: Math. Gen.* **37**, 9891–9906 (2004).
36. C. A. Aktipis, *J. Theor. Biol.* **231**, 249–260 (2004).
37. C. Hauert, and M. Doebeli, *Nature* **428**, 643–646 (2004).
38. H. Fort, and S. Viola, *J. Stat. Mech.: Teor. Exp.* **01**, P01010 (2005).
39. F. C. Santos, and J. M. Pacheco, *Phys. Rev. Lett.* **95**, 098104 (2005).
40. O. Durán, and R. Mulet, *Physica D* **208**, 257–265 (2005).
41. V. M. Eguiluz, M. G. Zimmermann, C. J. Cela-Conde, and M. San Miguel, *Am. J. Sociol.* **110**, 977–1008 (2005).
42. R. O. S. Soares, and A. S. Martinez, *Physica A* **369**, 823–829 (2006).
43. F. C. Santos, J. M. Pacheco, and T. Lenaerts, *Proc. Natl. Acad. Sci.* **103**, 3490–3494 (2006).
44. R. A. Hammond, and R. Axelrod, *Theor. Pop. Biol.* **69**, 333–338 (2006).
45. C. Hauert, *J. Theor. Biol.* **240**, 627–636 (2006).
46. M. H. Vainstein, A. T. C. Silva, and J. J. Arenzon, *J. Theor. Biol.* **244**, 722–728 (2007).
47. L. A. Dugatkin, and D. S. Wilson, *American Naturalist* **138**, 687–701 (1991).
48. D. S. Wilson, G. B. Pollock, and L. A. Dugatkin, *Evol. Ecol.* **6**, 331–341 (1992).
49. P. D. Taylor, *Evol. Ecol.* **6**, 352–356 (1992).
50. M. Enquist, and O. Leimar, *Anim. Behav.* **45**, 747–757 (1993).
51. R. Ferrière, and R. E. Michod, *Proc. R. Soc. B* **259**, 77–83 (1995).
52. R. Ferrière, and R. E. Michod, *Am. Nat.* **147**, 692–717 (1996).
53. M. van Baalen, and D. A. Rand, *J. theor. Biol.* **193**, 631–648 (1998).
54. J. C. Koella, *Proc. R. Soc. B* **267**, 1979–1985 (2000).
55. I. M. Hamilton, and M. Taborsky, *Proc. R. Soc. B* **272**, 2259–2267 (2005).
56. J.-F. Le Galliard, F. Ferrière, and U. Dieckmann, *Am. Nat.* **165**, 206–224 (2005).

Jensen-Shannon Divergence: A Multipurpose Distance for Statistical and Quantum Mechanics

Pedro W. Lamberti, Ana P. Majtey, Marcos Madrid and María E. Pereyra

Facultad de Matemática, Astronomía y Física
Universidad Nacional de Córdoba
Ciudad Universitaria, 5000 Córdoba, Argentina

Abstract. Many problems of statistical and quantum mechanics can be established in terms of a distance; in the first case the distance is usually defined between probability distributions; in the second one, between quantum states. The present work is devoted to review the main properties of a distance known as the Jensen-Shannon divergence (JSD) in its classical and quantum version. We present two examples of application of this distance: in the first one we use it as a quantifiers of the stochastic resonance phenomenon in ion channels; in the second one we use the JSD to propose a geometrical view of entanglement for two qubits states.

Keywords: Metrics for probability spaces; Distances in Quantum Theory; Jensen-Shannon Divergence
PACS: 02.50.-r - 03.65.-w, 89.70.+c

INTRODUCTION

Geometry and physics have lived together in close association for centuries. The relationship has had its ups and downs, but for the last century it has blossomed in a remarkable manner. This is a particularly noticeable after the formulation of the General Relativity theory by A. Einstein and of the Quantum Mechanics by Schrödinger, Dirac, Heisenberg et al. In the first case, the theory is established in terms of a four dimensional manifold, the space-time, with geometrical properties given by the Universe's matter-energy distribution; in the second one the states of a physical system are represented as vectors in a Hilbert space [1, 2].

As was pointed out by Gauss and Riemann in their pioneer works, one of the basic geometrical notion is the concept of distance between points of a given space (differentiable manifold, in modern terminology). According to their conceptions, the distance function (metric) determines all the geometrical properties of the space.

To be precise, let us recall that a distance over an arbitrary set X (a topological space, a Hilbert space, etc.), is a non-negative function D from the cartesian product $X \times X$ into the real numbers set that verifies:

$$D(x,y) = 0 \; iff \; x=y \tag{1}$$
$$D(x,y) = D(y,x) \; \forall \; x,y \in X \; (symmetric) \tag{2}$$

If the function D verifies the *triangle inequality*

$$D(x,z) \leq D(x,y) + D(y,z), \forall x,y,z \in X \tag{3}$$

it is said that D is a metric for the space X.

When distances between (discrete) probability distributions are investigated, the set X is given by a subset of $\Re^N$:

$$X = M_+^1(N) \equiv \{(p_1,\ldots,p_N), p_i \in \Re, p_i \geq 0, \sum_i p_i = 1\}.$$

In the probability theory context, several distances between two probability distributions are found in the literature. Even though the most frequently used are the Minkowskian metrics

$$d_\nu(P,P') \equiv \left(\sum_j |p_j - p'_j|^\nu\right)^{\frac{1}{\nu}} \quad \nu \geq 1 \tag{4}$$

CP913, *Nonequilibrium Statistical Mechanics and Nonlinear Physics, XV Conference*
edited by O. Descalzi, O. A. Rosso, and H. A. Larrondo

several important results are expressed in terms of other metrics [3]. In the expression (4) p_j and p'_j are the event j's probabilities of occurrence for each of the probability distributions P and P', respectively. The case $\nu = 2$ corresponds to the euclidean one.

A frequently used quantity to compare two probability distributions, arose in information theory, is the Kullback-Leibler divergence [4]:

$$K(P,P') \equiv \sum_j p_j \log\left(\frac{p_j}{p'_j}\right), \tag{5}$$

which does not verify properties (2) and (3) of the metric definition. More important even, this "distance" is not always well defined. Throughout this paper, log represents the logarithm in basis two.

Another example is provided by the Wootters's distance, which has the expression:

$$d_W(P,P') = \arccos\left(\sum_j \sqrt{p_j p'_j}\right) \tag{6}$$

This distance was introduced by Wootters in 1981 in the context of studying the problem of distinguishability between quantum states [5]. Roughly speaking, eq. (6) gives the greatest number of distinguishable probability distributions between P and P'. The Wootters's distance is symmetric, but it does not verify the triangle inequality (3).

In spite of these previous examples are very relevant, they do not exhaust the list of all the distances introduced in statistical and probabilistic studies (for a more complete list, see for example, reference [6]). At this point a question naturally emerges: why do we need distances? In answering this question, it should be observed that in several problems of statistical and quantum mechanics we need to compare "objects" (for example symbolic sequences, signals or classical and quantum states.).

Applications of distances range from the development of error detecting and error correcting codes [7], hypothesis testing in statistics [8], and philogenetic studies [9, 10], on the one hand, and to compare the amount of information carried by quantum states [11], on the other. In the realm of statistical physics many relevant problems are stated in terms of distances between probability distributions: to set up the notion of stability [12] and to define a complexity measure [13], just to mention two particularly interesting examples. Several outstanding results arose in classical and quantum information theory can be established in terms of distances too [14], [15].

During last years our research interest was focused in investigating an alternative to the previously mentioned distances: the Jensen Shannon divergence (JSD). Recently this quantity has been exhaustively applied in several problems of interest in theoretical physics and related topics.

Next section is devoted to review the main properties of the classical and quantum versions of this distance.

JENSEN-SHANNON DIVERGENCE

Classical JSD

The JSD was introduced in 1987 by C. Rao and generalized by J. Lin in 1991 as an alternative to the Kullback-Leibler divergence [16, 17]. To overcome the already mentioned limitations of this divergence, they proposed the quantity

$$\begin{aligned} JS(P,P') &\equiv \frac{1}{2}\left[K(P,\frac{P+P'}{2}) + K(P',\frac{P+P'}{2})\right] \\ &= -\frac{1}{2}\left[\sum_j p_j \log\left(\frac{2p_j}{p_j+p'_j}\right) + \sum_j p'_j \log\left(\frac{2p'_j}{p_j+p'_j}\right)\right], \end{aligned} \tag{7}$$

to evaluate the distance between the probability distributions P and P'. In terms of the Shannon entropy $H[P] = -\sum_j p_j \log p_j$, expression (7) can be rewritten in the form:

$$JS(P,P') = H\left(\frac{P+P'}{2}\right) - \frac{1}{2}H(P) - \frac{1}{2}H(P') \tag{8}$$

By using logarithm in basis two, this quantity has units of bits.

The JSD is positive definite, symmetric, bounded and, unlike the Kullback-Leibler divergence, always well defined (in the sense that it can be evaluated even when $p_j \neq 0$ and $p'_j = 0$ for some j). More important yet, its square root verifies the triangle inequality (3) (but JS does not!!). Therefore, it can be concluded that $\sqrt{JS(P,P')}$ is a true metric for the probability distribution space [18].

The JSD can be interpreted in the frame of information theory [19] and in a Bayesian probabilistic sense [17]. Incidentally, it is worth mentioning that for P close to P' the JSD coincides (up to numerical factors) with the Wootters and the Fubini-Study distances and, for the continuous case, is related to the Fisher information measure [22].

Lin has proposed a generalization of (7) to several probability distributions [17]. Let $P^{(k)}$, $k = 1..M$, a set of M probability distributions. The JSD among these probabilities distributions is given by:

$$JS[P^{(1)},..,P^{(M)}] = H[\sum_k \pi^{(k)} P^{(k)}] - \sum_k^M \pi^{(k)} H[P^{(k)}], \tag{9}$$

where the numbers $\pi^{(k)}$, $k = 1..M$, $\sum_k \pi^{(k)} = 1$, are weights properly chosen that allow to assign different relevance to each probability distribution.

The JSD has been extensively applied in different contexts, for example, as a tool for segmenting symbolic sequences [19, 20], for analyzing literary texts and musical score [21], as the disequilibrium for defining a non-triviality measure for chaotic maps [23], to study EEG records [24], as a complexity measure for genomic sequences [25], to compare electron distributions in atoms [26], to define an alternative distinguishability criterium between pure quantum states [22] and as a measure of discernibility between wave functions [27].

Quantum JSD

Distances between quantum states play a central role in quantum information theory (see, for example, chapter 9 of reference [11]). Trace distance, fidelity and the Bures's metric are frequently used as measures of distance between quantum states. Let us recall that a quantum state is mathematically represented by a density operator acting over the Hilbert space of the system, that is, by an hermitian operator, positive definite and with trace equal to one .

An extensively referenced distance between quantum states, is the relative entropy, which is the natural generalization of the Kullback-Leibler divergence, eq. (5), to the context of quantum theory [28, 29]:

$$K(\rho\|\sigma) = \mathrm{Tr}(\rho \log \rho - \rho \log \sigma) \tag{10}$$

with ρ and σ density operators, and where Tr stands for the trace operator. This measure has several inconveniences: it is not symmetric, not bounded and not always well defined (for example, $K(\rho\|\sigma)$ is not defined when σ represents a pure state). To be precise it should be said that in order that $K(\rho\|\sigma) < \infty$ it is necessary that $ker(\sigma) \subset ker(\rho)$, where ker denotes the kernel of the operator (the kernel of an operator is the subspace spanned by eigenvectors corresponding to null eigenvalues).

Following the steps that lead to the definition of the JSD from the Kullback-Leibler divergence (first line of eq. (7)), we can introduce the quantum JSD in the form [30]:

$$JS(\rho\|\sigma) \equiv \frac{1}{2}\left[K(\rho\|\frac{\rho+\sigma}{2}) + K(\sigma\|\frac{\rho+\sigma}{2})\right] \tag{11}$$

which, in terms of the von-Neumann entropy, $H_N(\rho) = -\mathrm{Tr}(\rho \log \rho)$, can be rewritten in the form

$$JS(\rho\|\sigma) = H_N\left(\frac{\rho+\sigma}{2}\right) - \frac{1}{2}H_N(\rho) - \frac{1}{2}H_N(\sigma) \tag{12}$$

This measure is positive definite, symmetric, bounded, and always well defined; in particular the condition over the kernels of the operators ρ and σ can be removed. It has all the adequate dynamical properties for a distance between quantum states . When the quantum JSD is generalized as a distance among several quantum states, according to the recipe (9), the resulting quantity has a very noticeable interpretation: it gives the upper bound of the mutual information that can be gathered between two quantum computing operators (Holevo bound) [30].

TWO EXAMPLES OF APPLICATION OF THE JSD

To conclude this paper, we present two applications of the JSD; in the first one we use the classical JSD as a quantifier of the stochastic resonance phenomenon in ion channels; in the second one we use the quantum JSD to give a geometrical characterization of entanglement for two qubits states.

Example 1

Stochastic resonance (SR) characterizes a cooperative phenomenon wherein the addition of a small amount of noise to an input driving signal can optimally amplify the output response. Generically this phenomenon has three basic ingredients: i) a nonlinear stochastic classical or quantum system; ii) certain kind of meta-stability, such as a potential barrier or a fixed threshold; and iii) a source of noise [31]. SR effect plays a prominent role in such diverse scientific areas as sensory biology, signal information and detection, or in conventional physical and chemical nonlinear noisy systems. Several quantifiers of this phenomenon have been proposed [32, 33] including some arose in information theory [34, 35].

In a very remarkable experiment, Bezrukov and Vodyanoy observed the SR phenomenon in (artificial) alameticyn -voltage gated- ion channels [36]. Motivated in this experimental result Goychuk and Hänggi have advanced in a characterization of the SR in ion channels by means of some information theory originated quantities [34]. According to their approach a ion channel plays the role of a transducing device that transmits an input information, a voltage, to the signal-modulated ion current response: $V \mapsto I$. The input is a time-dependent voltage which has three components:

$$V(t) = V_0 + V_s(t) + V_n(t),$$

where V_0 is a constant bias voltage; $V_s(t)$ is some time-dependent unbiased signal and $V_n(t)$ represents a stationary Gaussian Markovian noise of amplitude σ. Here the signal voltage is assumed of the form $V_s(t) = A\cos(\Omega t)$, $A \ll V_0$ (weak signal).

The gating dynamics of a single ion channel is governed by the equations verified by the time-dependent probabilities for the channel to be open or closed, $P_o(t)$ and $P_c(t)$, respectively. On his part, the probabilities $P_o(t)$ and $P_c(t)$ can be expressed in terms of the joint multi-time probability distributions $Q_s^{c,(o)}(t, \tau_s, \ldots \tau_1)$ for switches between open and closed states to occur at time $\tau_1, \tau_2, \ldots, \tau_s$ and to end up a time t in either the open state "o" or closed state "c", respectively (see reference [34] for a complete description of these probability distributions):

$$P_o(t) = Q_0^o(t) + \sum_{s=1}^{\infty} \int_0^t d\tau_s \int_0^{\tau_s} d\tau_{s-1} \ldots \times \int_0^{\tau_2} d\tau_1 Q_s^o(t, \tau_s, \ldots \tau_1) \tag{13}$$

and a similar expression is valid for the probability of the closed conformation, $P_c(t)$.

The basic idea in quantifying the SR phenomenon by means of the JSD, is to compare the corresponding probability distributions when the signal $V_s(t)$ is present and absent (an upper index "(0)" in a quantity indicates that no voltage signal is applied):

$$JS(I|V_s;T) = \sum_{\alpha=o,c} JS^{\alpha}(I|V_s;T) \tag{14}$$

where

$$JS^{\alpha}(I|V_s;T) = \frac{1}{2}\left[K(Q^{\alpha}, \frac{Q^{\alpha}+Q^{(0)\alpha}}{2}) + K(Q^{(0)\alpha}, \frac{Q^{\alpha}+Q^{(0)\alpha}}{2}\right] \tag{15}$$

with

$$K(Q^{\alpha}, \frac{Q^{\alpha}+Q^{(0)\alpha}}{2}) = Q_0^{\alpha} \log \frac{2Q_0^{\alpha}}{Q_0^{\alpha}+Q_0^{(0)\alpha}} + \sum_{s=1}^{\infty} \int_0^T d\tau_s \int_0^{\tau_s} d\tau_{s-1} \times \ldots$$
$$\times \int_0^{\tau_2} d\tau_1 Q_s^{\alpha}(T, \tau_s, \ldots, \tau_1) \log \frac{2Q_s^{\alpha}(T, \tau_s, \ldots \tau_1)}{Q_s^{\alpha}(T, \tau_s, \ldots \tau_1) + Q_s^{(0)\alpha}(T, \tau_s, \ldots, \tau_1)} \tag{16}$$

and an analogous expression for the second term in (15). All these quantities can be evaluated up to second order in A. In figure 1 we plotted the average over a period ($\frac{2\pi}{\Omega}$) of the JSD for different values of the bias voltage. All the

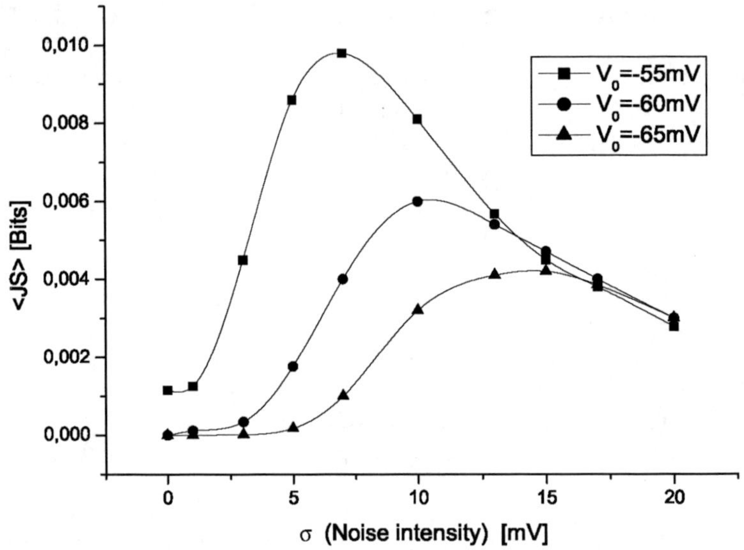

FIGURE 1. Time averaged JSD as a function of noise intensity for different bias voltages.

data required to do this plot has been taken from reference [34], and correspond to a potassium selective *Shaker IR* ion channel. The main result here is that the JSD reveal the typical SR dependence on the noise intensity. By comparing our results with those obtained by Goychuk and Hänggi, we can conclude that the JSD behaves, in this example, like the rate of information gain for the transduction process.

Example 2

Entanglement is the basic resource required to implement quantum information processes [11]. One of the most active issues in quantum information theory is the search of measures that allow to quantify the degree of entanglement of a system [37]. As shown by Vedral and collaborators, a measure of entanglement for bipartite systems can be built from a distance defined over the corresponding Hilbert space, whenever it verifies certain conditions [15]. Remarkably, the quantum JSD verifies these conditions [30]. In order to explore the JSD as a quantifier of entanglement for bipartite systems, we will evaluate it between a Werner state and a pure Bell state. These are states for a system of two 1/2-spin particles.

The Bell basis is expressed in terms of the canonical basis in the form: $|\Phi^{\pm}\rangle = \frac{1}{\sqrt{2}}[|++\rangle \pm |--\rangle]$ and $|\Psi^{\pm}\rangle = \frac{1}{\sqrt{2}}[|+-\rangle \pm |-+\rangle]$. Let us consider the density operator ρ_W corresponding to a Werner state:

$$\rho_W = F|\Psi^-\rangle\langle\Psi^-| + \frac{1-F}{3}\left(|\Psi^+\rangle\langle\Psi^+| + |\Phi^+\rangle\langle\Phi^+| + |\Phi^-\rangle\langle\Phi^-|\right)$$

where F is the fidelity of ρ_W with respect to the pure state $\sigma = |\Psi^-\rangle\langle\Psi^-|$. The Werner state is separable (unentangled) if $F \leq 1/2$; otherwise it is entangled.

The JSD between ρ_W and σ is:

$$d_{JS}(F) \equiv JS[\rho_W \| \sigma] = \frac{1}{2}\left[F \log F - (1+F)\log(\frac{1+F}{2})\right] + \frac{1-F}{2} \tag{17}$$

The JSD is a monotonically decreasing function of the fidelity F. Therefore, it is possible to invert the relationship (17) and express F as a function of the distance d_{JS}:

$$F = F(d_{JS}) \tag{18}$$

Thus the condition for the Werner state ρ_W of being entangled can be rewritten in terms of the distance d_{JS} in the form:

$$0 \leq d_{JS} \leq -\frac{3}{4}\log(\frac{3}{4}) = 0.311.$$

We can interpret this last inequality, from a geometrical point of view, as defining an hyper-sphere centered around the state σ in the 2-qubits space [38]. Outside this sphere only separable Werner states are found. A more detailed study of this geometrical view will be presented in a future work.

ACKNOWLEDGMENTS

The authors are grateful to Secretaria de Ciencia y Técnica de la Universidad Nacional de Córdoba, Argentina, for financial assistance. AM, MP are fellowship holders of CONICET and PWL is a member of CONICET. This work was partially supported by a Grant from CONICET. PWL is very grateful to the organizers of Medyfinol 2006 for their warm hospitality during his stay in Mar del Plata.

REFERENCES

1. R.M. Wald, *General Relativity*, The University of Chicago Press, Chicago, 1984.
2. P.A.M. Dirac, *Quantum Mechanics*, Oxford University Press, 1958.
3. A.N. Shyriaev, *Probability Theory*, Springer, 1996.
4. S. Kullback and R.A. Leibler, *Ann. Math. Stat.* **22**, 79-86 (1951).
5. W. Wootters, *Phys. Rev D* **23**, 357-362 (1981).
6. A. Majtey, PhD Thesis, Universidad Nacional de Córdoba, Argentina (in spanish).
7. R.V. Hamming, *Bell Sys. Tech. Journal* **29**, 147-160 (1950).
8. P.M. Robinson, *Rev. Econ. Stud* **88**, 437-453 (1991).
9. M. Li, J.H. Badger, X. Chen, S. Kwong, P. Kearney and H. Zhang, *Bioinformatics* **17**, 149-154 (2001).
10. M. Li, X. Chen, X. Li, B. Ma and P. M. N. Vitányi, *IEEE Trans. Inf. Theory* **50**, 3250-3264 (2004).
11. M. Nielsen and I. Chuang, *Quantum Computation and Quantum Information*, Cambridge University Press, Cambridge, 2000.
12. B. Lesche, *J. Stat. Phys.* **27**, 419-423 (1982).
13. R. Lopez-Ruiz, H. Mancini and X. Calbet, *Phys. Lett. A* **209**, 321-326 (1995).
14. A. Daffertshofer, A. R. Plastino and A. Plastino, *Phys. Rev. Lett.* **88**, 210601 (2002).
15. V. Vedral, M.B. Plenio, M.A. Rippin and P.L. Knight, *Phys. Rev. Lett* **78**, 2275-2278 (1997).
16. C.R. Rao, *Differential Geometry in Statistical Inference, IMS-Lectures Notes* **10** 217-225 (1987)
17. J. Lin, *IEEE Trans. Inf. Theory* **37**, 145-151 (1991).
18. J. Schindelin and D. Endres, *IEEE Trans. Inform. Theory* **49**, 1858-1860 (2003).
19. I. Grosse, P. Bernaola-Galvan, P. Carpena, R. Roman Roldan, J. Oliver & H.E. Stanley, *Phys. Rev. E* **65**, 041905 (2002).
20. P.W. Lamberti and A.Majtey, *Physica A*, **329**, 81-86 (2003).
21. D. Zanette, *Complex Systems*, To appear (2007).
22. A. Majtey, P.W. Lamberti, M.T. Martin and A. Plastino, *Eur. Phys. J. D* **32**, 413-418 (2005).
23. P.W. Lamberti, M.T. Martin, A. Plastino and O. Rosso, *Physica A* **334**, 119-126 (2004).
24. M.E. Pereyra, P.W. Lamberti and O. Rosso, in *Physica A* **379**, 122-132 (2007).
25. R. Román-Roldán, P. Bernaola-Galván and J. Oliver, Phys. Rev. Lett. **80**, 1344-1347 (1998).
26. K. Ch. Chatzisavvas, Ch. C. Moustakidis and C.P. Panos, *J. Chem. Phys.* **123**, 174111 (2005).
27. M. Casas, P.W. Lamberti, A. Plastino, A.R. Plastino and G. Roston, *Il Nuovo Cimento* **120B**, 521-532 (2005).
28. G. Lindblad *Comm. Math. Phys.* **33**, 305-322 (1973).
29. V. Vedral, *Rev. Mod. Phys.* **74**, 197-234 (2002).
30. A. Majtey, P.W. Lamberti and D. Prato, *Phys. Rev. A* **72** 052310 (2005)
31. L. Gammaitoni, P. Hänggi, P. Jung, and F. Marchesoni, *Rev. Mod. Phys.* **70**, 223-288 (1998).
32. B. McNamara and K. Wiesenfeld, *Phys. Rev. A* **39**, 4854-4869 (1989).
33. P. Jung and P. Hänggi, *Phys. Rev. A* **44**, 8032-8042 (1991).
34. I. Goychuk and P. Hänggi, *Phys. Rev E* **61**, 4272-4280 (2000).
35. C. Tessone, A Plastino and H. Wio, *Physica A* **326**, 37-54 (2003).
36. S.M. Bezrukov and I. Vodyanoy, *Nature* **378**, 362-364 (1995).
37. C.H. Bennett, H.J. Bernstein, S. Popescu and B. Schumacher, *Phys. Rev. A* **53**, 2046-2052 (1996).
38. A. Borras, A.R. Plastino, M. Casas and A. Plastino, *Advaces in Mathematical Physics* To appear (2007).

A General Approach to Describe Nonlinear Semiquantum Motion

C.M. Sarris*,†, M.P. Sassano* and A.N.Proto**,‡

*Laboratorio de Sistemas Complejos, FI-UBA
†Catedra de Fisica, CBC, Universidad de Buenos Aires
**Laboratorio de Sistemas Complejos, Facultad de Ingenieria Universidad de Buenos Aires Av. Paseo Colon 850 (1063) Buenos Aires, Argentina
‡Comision de Investigaciones Cientificas, Buenos Aires State

Abstract.
We describe a general method to study the nonlinear semiquantum systems: as the dynamics of the system develops in a semiquantum space V spanned by the mean values of a Complete Set of Noncommuting observables ($CSNCO$) and the conjugate classical variables, X, P, it is possible to define a positive definite metric for the quantum manifold (QM). Through the components of the second rank covariant metric tensor, a close connection between this metric and the uncertainty principle can be established, meanwhile, the classical manifold (CM) of the system, is preserved as a symplectic one. We present an example of an entangled nonlinear semiquantum system where different dynamics take place.

Keywords: Semiclassical theories in quantum mechanics,Quantum statistical mechanics, Nonlinear dynamics
PACS: 03.65 Sq, 05.30-d, 05.45-a

INTRODUCTION

Nonlinear semiquantum systems constitute a rich area as it is possible, through them, to describe many phenomena (see [1] and references therein). We present, here, a general method to describe this kind of systems which enables us to define the regions where the system exhibits a true semiquantum dynamics (or not) without recoursing to any limiting process (like $\hbar \rightarrow 0$, $t \rightarrow \infty$, or $m \longrightarrow \infty$). To achieve our purpose we define a $V - semiquatum space$ in which the dynamics of the system takes place. This space is spanned by the quantum and classical degrees of freedom of the system. Our approach leads to the same equations as the well-known "semiquantum theory" and, while it preserves the symplectic dynamics in the classical manifold, the quantum one is endowed with a positive definite metric which is closely connected with the generalized uncertainty principle (GUP) of ref [6]. We exemplify our method using the particular case of the Hamiltonian $\hat{H} = B\ \hat{\sigma}_z + C\ q\ \hat{\sigma}_x + \frac{p^2}{2m} + \frac{q^4}{4}$, where neither the quantum subsystem nor the classical one are chaotic by itself. For this system the previous main conclusion [3] was that 'semiquantum chaos' is a 'mathematical artifact', as chaos disappears when $m \longrightarrow \infty$. We do not characterize here, for sake of shortness, the nature of the motion which appears when both, the quantum and classical systems are entangled, however we can conclude: i) similar dynamics can be obtained independently of the classical mass value; ii) under certain conditions the chaotic dynamics is a 'mathematical artifact', as it is not the outcome of an entangled semiquantum system. iii) however, when the adequate conditions are selected, patterns that resemble chaotic motion emerge in the region of full entanglement between the quantum and classical degrees of freedom.

QUANTUM DYNAMICS AND MEP

The Maximum Entropy Principle (MEP) approach provides with a definite prescription in order to find the set of relevant operators, RO, associated to the dynamics of problem at hand, as well as the density matrix. The $RO - set$ is defined through the well-known closure condition [4]:

$$[\hat{H}(t), \hat{O}_j] = i\hbar \sum_{r=0}^{q} g_{rj}(t)\, \hat{O}_r \qquad j, = 1, \ldots, q \tag{1}$$

CP913, *Nonequilibrium Statistical Mechanics and Nonlinear Physics, XV Conference*
edited by O. Descalzi, O. A. Rosso, and H. A. Larrondo

where $g_{rj}(t)$ are the elements of a $q \times q$ matrix, $G(t)$, which may depend upon time if $\hat{H}(t)$ is time-dependent. This condition defines: a) A Complete Set of Non-Commuting Operators ($CSNCO$) $\{\hat{O}_j\}$, RO, which close a partial Lie algebra under commutation with the Hamiltonian and are linearly independent, b) The density matrix, $\hat{\rho}(t)$, of the system which is constructed as $\hat{\rho}(t) = \exp\left(-\lambda_0 \hat{I} - \sum_{j=1}^{q} \lambda_j(t)\hat{O}_j\right)$ where the $q+1$ Lagrange multipliers λ_i are determined in order to fulfill $Tr(\hat{\rho}) = 1, \langle \hat{O}_j \rangle = Tr(\hat{\rho}\hat{O}_j)$, c) This density operator $\hat{\rho}(t)$ maximizes the entropy $S = \lambda_0 + \sum_{j=1}^{q} \lambda_j(t) \langle \hat{O}_j \rangle (t)$ (S is a constant of motion, as $\hat{\rho}$ follows the Liouville equation). The generalized Ehrenfest´s theorem gives the temporal evolution of the mean values of the RO ($CSNCO$) :

$$\frac{d\langle \hat{O}_j \rangle}{dt} = -\sum_{r=1}^{q} g_{rj}(t) \langle \hat{O}_r \rangle (t), \quad j = 1, ..., q \tag{2}$$

The RO defined through (1) are linearly independent, and so are their mean values. It allows us to define a $q-$dimensional generalized phase space
$V = span\{\langle \hat{O}_1 \rangle (t), ..., \langle \hat{O}_q \rangle (t)\}$ in which the dynamics of the system takes place. On this space we can define a positive definite metric function [5] $\bullet : \mathbb{V} \times \mathbb{V} \longrightarrow \mathbb{R}/$

$$\langle \hat{O}_i \rangle (t) \bullet \langle \hat{O}_j \rangle (t) = \frac{1}{2} \langle \hat{O}_i \hat{O}_j + \hat{O}_j \hat{O}_i \rangle - \langle \hat{O}_i \rangle \ \langle \hat{O}_j \rangle = K_{ij}(t) = K_{ji}(t), \tag{3}$$

which, as well, defines the components $K_{ij}(t) = K_{ji}(t)$ of the second-rank covariant metric tensor, $K(t)$, in terms of the quantum correlation coefficients between the operators which belong to RO and fulfill (1). Depending on the Lie algebra selected to fulfill (1), it is possible to find some invariants [5], from the metric tensor, which are very helpful to deal with the dynamics of the system. The metric (3) is closely related to the generalized uncertainty principle (GUP) of ref. [6]:

$$I^H = \sum_{i=1,j=1}^{q} \left\{ (\Delta \hat{O}_i)^2 (\Delta \hat{O}_j)^2 - [\langle \hat{L}_{ij} \rangle - \langle \hat{O}_i \rangle \langle \hat{O}_j \rangle]^2 \right\} \geq \tag{4}$$

$$\geq -\frac{1}{4} \sum_{i=1,j=1}^{q} \langle [\hat{O}_i, \hat{O}_j] \rangle ; \ i < j. \tag{5}$$

$\hat{L}_{ij} = \frac{1}{2} \langle \hat{O}_i \hat{O}_j + \hat{O}_j \hat{O}_i \rangle$, given that it can be obtained from the metric tensor,$K(t)$, as the summation over its principal minors of order 2.

THE NONLINEAR SEMIQUANTUM CASE

Semiquantum non-linear systems can be described by Hamiltonians of the form $\hat{H} = \sum_{j=1}^{q} a_j(X, P)\hat{O}_j + f(X, P)$, where $a_j(X, P)$ and $f(X, P)$ are functions of the canonically conjugate classical variables X and P and the $\hat{O}_j$´s are the $q-$relevant quantum operators which fulfill (1) and compose the $CSNCO$. Notice that, for a semiquantum system, the coefficients $g_{ij}(X;P)$ belonging to the matrix $G(X;P)$ are now functions of the canonically conjugate classical variables X and P because, now, the Hamiltonian $\hat{H}$ represents a system where classical and quantum degrees of freedom coexist and are coupled. So, the dynamics of the system develops in a semiquantum generalized phase space defined by:

$$\mathbb{V}_{SQ} = span\{\langle \hat{O}_1 \rangle (t), ..., \langle \hat{O}_q \rangle (t), X, P\}, \tag{6}$$

where $\langle \hat{O}_1 \rangle (t), ..., \langle \hat{O}_q \rangle (t)$ are the mean values of the q relevant quantum operators and define the quantum manifold (QM) of the $V_{SQ} - space$; X and P define the classical manifold (CM). So, this semiquantum generalized phase space is $(q+2)-$dimensional. In our approach, the energy is taken to coincide with the quantum expectation value of the Hamiltonian: $\langle \hat{H} \rangle = Tr(\hat{\rho}(t)\hat{H})$ and the temporal evolution of the classical variables are:

$$\frac{dX}{dt} = \{X, \langle \hat{H} \rangle\},$$

$$\frac{dP}{dt} = \{P, \langle \hat{H} \rangle\}, \tag{7}$$

($\{,\}$ indicates Poisson brackets) in order the *CM* remains as a symplectic one. In fact, it is very easy to verify that, making use of Eqs.(7), the dynamics in the *CM* results:

$$\begin{pmatrix} \dot{X} \\ \dot{P} \end{pmatrix} = \begin{pmatrix} 0 & 1 \\ -1 & 0 \end{pmatrix} \begin{pmatrix} \frac{\partial \langle \hat{H} \rangle}{\partial X} \\ \frac{\partial \langle \hat{H} \rangle}{\partial P} \end{pmatrix}, \tag{8}$$

which, in turn, depends on the mean values of the quantum operators which belong to the *RO*. Concerning the dynamical evolution of the quantum degrees of freedom's mean values, they are obtained straightforwardly from Eqs. (2). Notice that although the matrix $G(X,P)$ arises from (1), its nature is semiquantum because of the classical degrees of freedom act like they were parameters in the quantum commutation operation. At the same time, the classical degrees of freedom, are semiquantum as well, because it get involved the mean value of the Hamiltonian operator $\hat{H}$. *So, the interplay between quantal and classical variables acquires special relevance through the matrix* $G(X,P)$. On the other side, if we now define the metric (3) on the quantum manifold $QM = span\{\langle \hat{O}_1 \rangle (t), ..., \langle \hat{O}_q \rangle (t)\}$ of the system, we see that, due to the entanglement between the quantum and classical variables, the dynamical invariants that we can get from the metric tensor will be affected by the dynamical evolution of the classical variables X and P as well as the *GUP* (5) itself. So, the Lie algebra selected to fulfill (1) will give rise to some dynamical semiquantum invariants, through which, it is possible to deal with the semiquantum dynamics. Particularly, it is quite interesting the case in which the orthogonal group $O(q,\mathbb{R})$ rules on the *QM* given that this group preserves the positive definite metrics as it is well-known.

EXAMPLE

We consider the Hamiltonian [3]:

$$\hat{H} = B\,\hat{\sigma}_z + C\,q\,\hat{\sigma}_x + \frac{p^2}{2m} + \frac{q^4}{4}, \tag{9}$$

q and p are the canonically conjugate classical variables and $\hat{\sigma}_i$ are spin operators. By considering the Lie algebra $\{\hat{S}_x, \hat{S}_y, \hat{S}_z\}$ $(\hat{S}_i = \frac{\hbar}{2}\hat{\sigma}_i)$ as the *CSNCO* , the semiquantum generalized phase space is 5−dimensional and is defined by $\mathbb{V}_{SQ} = span\{\langle \hat{S}_x \rangle, \langle \hat{S}_y \rangle, \langle \hat{S}_z \rangle, q, p\}$. Eq.(1) leads to the following G matrix ($SU(2)$ Lie Algebra):

$$G = \begin{pmatrix} 0 & -2B & 0 \\ 2B & 0 & -2qC \\ 0 & 2qC & 0 \end{pmatrix} \tag{10}$$

This G matrix, exhibits the influence of the classical degrees of freedom *and it is an antisymmetric one.* Making use of Eqs. (2) and (7) we obtain the semiquantum equations of motion:

$$\begin{aligned} \frac{d\langle \hat{\sigma}_x \rangle}{dt} &= -2B\langle \hat{\sigma}_y \rangle, \\ \frac{d\langle \hat{\sigma}_y \rangle}{dt} &= 2B\langle \hat{\sigma}_x \rangle - 2Cq\langle \hat{\sigma}_z \rangle \\ \frac{d\langle \hat{\sigma}_z \rangle}{dt} &= 2Cq\langle \hat{\sigma}_y \rangle, \\ \frac{dq}{dt} &= \frac{p}{m}, \\ \frac{dp}{dt} &= -C\langle \hat{\sigma}_x \rangle - q^3, \end{aligned} \tag{11}$$

which clearly exhibit the entanglement between the classical and quantum degrees of freedom as well. The antisymmetry of matrix $G(X,P)$ guarantees that the $O(3,\mathbb{R})$ group rules the dynamics in the *QM* [5]. The consequence of this fact is that the metric (3) defined on the $QM = span\{\langle \hat{S}_x \rangle, \langle \hat{S}_y \rangle, \langle \hat{S}_z \rangle\}$ enables us to obtain three dynamical semiquantum invariants from the metric tensor, $K(t)$, (the summation over the principal minors of order 1, 2 and 3 respectively which are always positive because of the positive definite requirement of the metric), and they are:

$$\begin{aligned} Tr\left[K(t)\right] &= K_{xx}+K_{yy}+K_{zz} = \frac{\hbar^2}{4}\left(3-\langle\hat{\sigma}\rangle^2\right), \\ I^H &= \sum_{\substack{j,k=1 \\ j<k}}^{3}\left(K_{jj}\,K_{kk}-K_{jk}^2\right) = \frac{\hbar^4}{16}\left(3-2\langle\hat{\sigma}\rangle^2\right), \\ \det\left[K(t)\right] &= K_{xx}K_{yy}K_{zz}+2K_{zy}K_{xz}K_{yz}- \\ &\quad -K_{xx}K_{yz}^2-K_{yy}K_{xz}^2-K_{zz}K_{xy}^2 \\ &= \frac{\hbar^2}{4}\left(1-\langle\hat{\sigma}\rangle^2\right) \end{aligned} \tag{12}$$

with $\langle\hat{\sigma}\rangle^2 = \langle\hat{\sigma}_x\rangle^2+\langle\hat{\sigma}_y\rangle^2+\langle\hat{\sigma}_z\rangle^2$. Particularly, the invariant I^H , is the left hand side of the generalized uncertainty principle (5) so, from I^H, it follows straightforwardly the well-known relation: $\langle\hat{\sigma}\rangle^2 \leq 1$. The numerical simulations have been done using adimensional variables and parameters, and setting $\hbar = 1$. From them we conclude: a) meantime $\det\left[K(t)\right] < \frac{I^H}{\langle\hat{H}(t)\rangle} < I^H$ both, the quantum and classical systems are fully entanglemented, *independently of the classical mass value.* Poincare surfaces resemble chaotic motion appear, although the detailed analysis of the emergent dynamics is out of the scope of the present contribution. b) If $\frac{I^H}{\langle\hat{H}(t)\rangle} > I^H$ or $\det\left[K(t)\right] < 0$ chaotic dynamics exists but it can be classified as a 'mathematical artifact', as the entanglement between both systems violates the *GUP* (5).

CONCLUSIONS

a) If it is possible to define a proper positive definite metricon the *QM* of a nonlinear semiquantum system, it is possible to establish a close connection between this metric and the uncertainty principle through the summation over the principal minors of order 2 belonging to covariant metric tensor, b) depending on the Lie algebra selected to conform the relevant set of quantum operators, different semiquantum invariants emerge, c) although these invariants of motion arise from the quantal part of the system, they *rule the whole nonlinear semiquantum dynamic,* d) the present outlined approach leads to the same motion equations as the well-known "semiquantum theory" and takes advantage of the metric defined on *QM* through the *invariants* that can be obtained from the metric tensor.

REFERENCES

1. A.M. Kowalski, A. Plastino, and A.N. Proto *Physica A* **236,** 429-444 (1997)
2. C.M. Sarris, F. Caram and A.N. Proto, *Phys. A* **331,** 125-139 (2004)
3. L.E. Ballentine, *Phys. Rev. E* **63**, 1-7 (2001)
4. Y. Alhassid and R.D. Levine, *Phys. Rev. A* **18**, 89-99 (1978)
5. C.M. Sarris, and A.N. Proto, *Physica A* **377** , 33–42 (2007)
6. C.M. Sarris, F. Caram and A.N. Proto, *Phys. Lett. A* **324**, 1-8 (2004)

Relativistic Stochastic Processes

F. Debbasch and C. Chevalier

Université Pierre et Marie Curie-Paris6, UMR 8112, ERGA-LERMA,
3 rue Galilée, 94200 Ivry, France

Abstract. We review all recent contributions to the literature on stochastic processes. In particular, the Relativistic Ornstein-Uhlenbeck Process is presented in detail, as is the intrinsic Brownian motion studied by Franchi and Le Jan. The Relativistic Brownian Motion of Dunkel and Hänggi is also reviewed, together with a model introduced by Oron and Horwitz. We finally suggest some possible future developements of the current research.

Keywords: Brownian Motion, Diffusion, Relativistic Statistical Physics, Lorentzian Geometry
PACS: 02.50.Ey,05.10.Gg,05.40.Jc,04.90.+e

INTRODUCTION

It is probably fair to say that Stochastic Process Theory originated with Einstein's 1905 study on Brownian motion [1]. The theory has since developed into a full grown branch of Mathematics and its current applications include Physics, Chemistry, Biology and Economy [2, 3]. As far as Physics is concerned, one had to wait until the 70's to see what started as an attempt to describe non quantum Galilean diffusions being extended to include Galilean quantum processes [2, 4]; but the wait for a relativistic extension was even longer, since the first paper dealing with a relativistic stochastic process of clear physical interpretation was only published in 1997 [5]. Various relativistic stochastic processes have since then been considered by several authors; it is the ambition of this contribution to propose of short and self-contained review of this growing literature. The material is organized as follows. Section 2 is devoted to the Relativistic Ornstein-Uhlenbeck Process (ROUP) constructed by F. Debbasch, K. Mallick and J.P. Rivet in 1997. Section 2.1 presents the process in flat space-time. Section 2.2 deals with the so-called hydrodynamical limit of this process. Section 2.3 extends the process to a family of curved space-time diffusions and addresses a recent work on cosmological diffusions. A general relativistic H-Theorem is also mentioned. Section 3 is devoted to the other existing relativistic processes. Section 3.1 deals with a process recently considered by J. Franchi and Y. Le Jan [6]; this process can certainly not be interpreted as a standard physical diffusion and its only possible physical interpretation [7] is discussed in Section 3.1. Section 3.2 deals with a process [8] introduced by J. Dunkel and P. Hänggi in 2005 and which is best considered as an alternative to the ROUP. The more recent numerical and analytical work done by these authors on binary collisions and non Markovian processes is also reviewed. Section 3.3 discusses the relativistic Brownian motion developed by O. Oron and L.P. Horwitz [9]. Finally, Section 4 lists a few interesting directions in which reasearch on relativistic stochastic processes could develop in the next few years.

THE RELATIVISTIC ORNSTEIN-UHLENBECK PROCESS

The ROUP in flat space-time

The ROUP can be interpreted as describing the relativistic diffusion of a point particle of mass m in a fluid characterized by a 4-velocity field U. This process has been originally introduced [5] in the special case where the space-time is flat and where U is point-independent. It is then particularly convenient to choose as reference frame the global rest-frame $\mathscr{R}$ of the fluid in which the particle diffuses. The stochastic equations defining the process read in this frame:

$$\begin{aligned} d\mathbf{r} &= \frac{\mathbf{p}}{m\gamma(\mathbf{p})}\,dt \\ d\mathbf{p} &= -\alpha\,\frac{\mathbf{p}}{\gamma(\mathbf{p})}\,dt + \sqrt{2D}d\mathbf{B}_t \end{aligned} \tag{1}$$

CP913, *Nonequilibrium Statistical Mechanics and Nonlinear Physics, XV Conference*
edited by O. Descalzi, O. A. Rosso, and H. A. Larrondo

Here, $\mathbf{r}$ and $\mathbf{p}$ are the position and momentum of the diffusing particle and the expression of the Lorentz factor as a function of $\mathbf{p}$ reads $\gamma(\mathbf{p}) = \sqrt{1+\mathbf{p}^2/m^2c^2}$; the positive coefficient α determines the amplitude of the frictional force experienced by the particle while D fixes the amplitude of the noise term ($\mathbf{B}_t$ is the usual 3-D Brownian motion).

The forward Kolmogorov equation obeyed by the time-dependent phase-space density $\Pi(t,\mathbf{r},\mathbf{p})$[1] of the diffusing particle is then [5]:

$$\partial_t \Pi + \partial_{\mathbf{r}} \cdot \left(\frac{\mathbf{p}}{m\gamma(\mathbf{p})} \Pi \right) - \alpha\, \partial_{\mathbf{p}} \cdot \left(\frac{\mathbf{p}}{\gamma(\mathbf{p})} \Pi \right) = D\Delta_{\mathbf{p}}\Pi. \tag{2}$$

This equation admits as solution the equilibrium Jüttner distribution:

$$\Pi_J(\mathbf{p}) = \frac{1}{4\pi}\frac{1}{Q^2 K_2(1/Q^2)} \exp\left(-\frac{\gamma(\mathbf{p})}{Q^2} \right) \tag{3}$$

where K_2 is the second order modified Hankel function and $Q^2m^2c^2 = D/\alpha$. The distribution Π_J is normalized to unity with respect to the measure d^3p. The temperature T of the equilibrium is defined by $k_BT/mc^2 = Q^2$. One thus has:

$$\alpha = \frac{D}{mk_BT} \tag{4}$$

which is a *bona fide* special relativistic fluctuation-dissipation theorem. It has furthermore been confirmed by direct numerical simulation [5] that an arbitrary physically reasonable phase-space density does relax in time towards the Jüttner distribution (3).

The equations governing the special relativistic Ornstein-Uhlenbeck process can naturally be written down in an arbitrary inertial frame. This will not be discussed here; the interested reader is referred to the original publications [10, 11] and to Section 2.3 of the present contribution, which contains an explicit presentation of the process in an arbitrary, not necessarily inertial frame. Let us also mention that the stochastic equations (1) can be rewritten using the proper-time s of the particle as time-variable; the corresponding Ito process is described in [12].

Hydrodynamical limit in flat space-time

By definition, the hydrodynamical limit of an arbitrary transport phenomenon corresponds to situations where all quantities characterizing the system vary on 'large' scale only, both in time and in space. For the special relativistic Ornstein-Uhlenbeck process described in the previous section, the hydrodynamical limit [13] addresses situations in which the time and space variation scales of the distribution function Π in $\mathscr{R}$ are respectively much larger than the characteristic time $\tau = 1/\alpha$ and the mean free path $\lambda = \tau\sqrt{k_BT/m}$ (see equation (4) above for the definition of T in terms of α, D and m). Technically, the hydrodynamical limit is studied by performing a Chapman-Enskog expansion around the local equilibrium distribution $\Pi_0 = (1/4\pi Q^2 K_2(1/Q^2)) \times n(t,\mathbf{r}) \exp(-\gamma(\mathbf{p})/Q^2)$ where $n(t,\mathbf{r})$ is the spatial density of the diffusing particle in $\mathscr{R}$; the hydrodynamical equation verified by n then appears a necessary condition for the expansion to exist.

Performing a Chapman-Enskog expansion on the ROUP [13] leads at all orders to $\partial_t n = \frac{\lambda^2}{\tau}\Delta n = \chi \Delta n$, which is of course the standard diffusion equation. This result is quite surprising. Indeed, the dynamics of the ROUP is relativistic and, hence causal in Einstein's sense *i.e.* it does not allow faster than light transport. On the other hand, the diffusion equation is parabolic (as opposed to hyperbolic) and does imply faster than light transport.

This apparent contradiction can be resolved by remembering that, by definition of the hydrodynamical limit, the diffusion equation describes the correct behaviour of the density n associated with the ROUP only in situations where the characteristic variation scales of this density are much larger than $1/\alpha$ and λ. Applying these restrictions to the Gaussian Green function $G(t,\mathbf{r})$ of the diffusion equation leads to $1/\alpha \ll t$ and $| \mathbf{r}/t | \ll \lambda/\tau$; the second inequality implies [13] $| \mathbf{r}/t | \ll c$. Since G represents the time-evolution of a density profile initially concentrated on the point $\mathbf{r} = 0$, the last inequality essentially states that the diffusion equation is only valid well inside the light cone originating at $\mathbf{r} = 0$, *i.e.* for effective diffusion velocities much smaller than c. Thus, the diffusion equation describes correctly the time-evolution of the density n associated with the ROUP only at points of space-time where it does not contradict Einstein's causality [13].

---

[1] This density is defined with respect to the Lebesgue measure $d^3r d^3p$

This situation parallels exactly some challenging properties of hydrodynamical limit of the relativistic Boltzman equation. Indeed, performing a Chapman-Enskog expansion on the solution of the relativistic Boltzman equation delivers the so-called first order hydrodynamical theories which, like the diffusion equation, permit faster than light transport [14]. This has long been recognized as troublesome [15], but it seemed to elude any reasonable interpretation. Studying the hydrodynamical limit of the ROUP has furnished the answer: effective equations governing the dynamics of macroscopic fields in the hydrodynamical limit formally permit faster than light transport, but such equations are only physically valid in restricted domains of space-time where faster than light transport does not occur and where their solutions therefore do not contradict Einstein's causality. Thus, it seems the only way to describe a relativistic continuous medium through local equations which do not permit faster than light transport is to work within the framework of statistical physics (typically, stochastic processes or traditional kinetic theory).

Curved space time generalizations

The special relativistic Ornstein-Uhlenbeck process is susceptible of an infinity of curved space-time generalizations. Those considered so far [16, 17] are fully characterized by the velocity field U of the fluid in which the particle diffuses, by a friction coefficient α and a noise coefficient D; both α and D are allowed to depend on position and momentum.

All calculations can be made manifestly covariant by introducing an unphysical distribution function f defined over an 8-D phase-space $\mathscr{S} = \{(x,q)\}$, where q represents a covariant, possibly off-shell particle momentum. Given an arbitrary coordinate system, the distribution f is related to the physical distribution function Π to be used in this system by an equation of the form:

$$\Pi(t,\mathbf{r},\mathbf{p}) = \int f(x,p)\,\delta\left(p_0 - \gamma(\mathbf{p})\right) dp_0, \tag{5}$$

where the δ distribution enforces a mass-shell restriction. Integrating Π against $d^3q/\sqrt{-\det g}$ delivers the spatial particle density $n(t,\mathbf{r})$, which can in turn be integrated against $\sqrt{-\det g}\,d^3r$. Further details over the measures against which Π can be integrated can be found in [16].

A curved space-time generalization of the process discussed in the previous section is then fully defined by a transport equation obeyed by f. The most natural equation [16, 17] reads:

$$\mathscr{D}_\mu(g^{\mu\nu}(x)p_\nu f) + \frac{\partial}{\partial p_\mu}(mc\,F_{d\mu} f) + K^{\mu}{}_{\rho}{}^{\beta}{}_{\nu}\,\partial_{p_\rho}\left(D(x,p)\,\frac{p_\mu p_\beta}{p_\alpha U^\alpha}\,\partial_{p_\nu} f\right) = 0, \tag{6}$$

where $\mathscr{D}_\mu = \left(\nabla_\mu + \Gamma^\alpha_{\mu\nu} p_\alpha \frac{\partial}{\partial p_\nu}\right)$ represents a covariant derivative at momentum *covariantly* constant (here, the Γ's are the usual Christoffel symbols of the metric g) [15]. The deterministic force F_d is defined by

$$F_{d\,\mu} = -\lambda^\nu_\mu p_\nu\,\frac{g^{\mu\nu}p_\mu p_\nu}{m^2c^2} + \lambda^{\alpha\beta}\,\frac{p_\alpha p_\beta}{m^2c^2}\,p_\mu \tag{7}$$

and the friction tensor λ reads

$$\lambda^{\mu\nu} = \frac{\alpha(x,p)(mc)^2}{(p_\mu U^\mu)^2}\,\mathscr{P}^{\mu\nu} \tag{8}$$

where $\mathscr{P}$ is the projector onto the orthogonal to U: $\mathscr{P}_{\mu\nu} = g_{\mu\nu} - U_\mu U_\nu$. Finally, the tensor K is defined in terms of $\mathscr{P}$ and U by:

$$K^{\mu\rho\beta\nu} = U^\mu U^\beta \mathscr{P}^{\rho\nu} - U^\mu U^\nu \mathscr{P}^{\rho\beta} + U^\rho U^\nu \mathscr{P}^{\mu\beta} - U^\rho U^\beta \mathscr{P}^{\mu\nu}. \tag{9}$$

This class of processes can be used to model the diffusion of a particle in a fluid comoving with the expansion of a spatially flat Friedmann-Robertson-Walker universe. The time-evolution of the thermodynamical state of the matter in the universe can be approximately characterized by a time-dependent temperature $T(t)$. It is then natural to wonder if some processes defined by an equation of the form (6) admit the Jüttner distribution of temperature $T(t)$ as possible momentum distribution. A positive answer to this question has been given in [18, 17] for all possible choices of $a(t)$ and $T(t)$ and the coefficients α and D of the corresponding processes obey a generalized fluctuation dissipation theorem.

Let us also mention that the processes defined by (6) verify a remarkable H-theorem [19][2]; this theorem, which states that the 4-divergences of certain conditional entropy 4-currents associated with (6) are always non-positive, is valid in any general relativistic space-time, including those containing closed time-like curves[3].

OTHER RELATIVISTIC PROCESSES

'Intrinsic' Brownian Motion

This process was introduced [6] by J. Franchi and Y. Le Jan in 2004 and is based on ideas originally developed by R.M. Dudley [21] in 1965. To keep the discussion as simple as possible, let us focus on the flat space-time version of this process. A physically intuitive definition of the process is given by the following equation of motion:

$$d\mathbf{p}_* = \sqrt{2\mathscr{D}}d\mathbf{B}_s, \tag{10}$$

where s stands for the proper-time along a stochastic time-like curve and $d\mathbf{p}_*/ds$ designates what physicists would call the proper acceleration along that curve. Note that equation (10) looks like a standard stochastic differential equation but is *not*; indeed, by definition, the proper momentum $\mathbf{p}_*$ along a time-like curve vanishes for all values of s. Thus, to be integrated as a usual stochastic differential equation, (10) has first to be converted into an equation fixing the differential of the momentum $\mathbf{p}$ associated with the same world-line, but in another, s-independent inertial frame.

Though quite natural from a mathematical point of view, this process is not susceptible of any simple physical interpretation. Indeed, usual diffusions of a point particle are physically generated by the interaction of this particle with a surrounding continuous medium; the stochastic equation (10) does not however contain any information pertaining to such a medium [4]; it is therefore far from obvious what kind a physics would induce a diffusion like (10). Dowker et al. [7] have suggested that (10) may be interpreted as an effective diffusion induced by quantum fluctuations of the space-time structure itself and this seems to be the only possible interpretation of (10). Note also that, contrary to the ROUP, there is no inertial frame in which the process (10) admits an invariant measure in momentum space. There is *a fortiori* no fluctuation-dissipation theorem for (10).

Let us finally mention that the curved space-time version of the process has been studied in detail by Franchi and Le Jan in the particular case where the background space-time is the Kruskal extension of the Schwarzschild metric [6]; diffusions in the Gödel universe have been considered by Franchi in [22] and the asymptotic behaviour of the Franchi-Le Jan process in Minkovski space-time has been recently fully characterized by I. Bailleul and Y. Le Jan [23].

The Relativistic Brownian Motion of Dunkel and Hänggi

Dunkel and Hänggi's process can be viewed as an attempt to modify Franchi and Le Jan's process to obtain a new model in which the special relativistic diffusion of a particle is driven by its interaction with a surrounding fluid. As such, Dunkel and Hänggi's process presents itself as a potential alternative to the special relativistic Ornstein-Uhlenbeck process (see Section 2.1); the common points and differences between the two processes will be discussed at the end of the current Section.

As J. Franchi and Y. Le Jan, J. Dunkel and P. Hänggi start their construction in the proper rest frame of the diffusing particle, but they naturally extend it to all inertial frames. Dunkel and Hänggi's process can be defined by the following stochastic equation, valid in the (global) rest frame of the fluid in which the particule diffuses [8, 24]:

$$dp^i = -\nu p^i\, dt + \sqrt{2D}\mathscr{L}^i_j(\mathbf{p})dB^j_t. \tag{11}$$

---

[2] This reference only envisages processes with constant α and D but the given proof can be straightforwardly extended to include all processes in the class defined by (6)

[3] Standard examples of such space-times are the Gödel universe and the Kerr black hole [20]

[4] Quite generally, the diffusion process introduced by Franchi and Le Jan involves only the metric of space-time. Hence the qualifier 'intrinsic' chosen by the present author to characterize this process

where ν is a positive friction coefficient. Here, the tensorial multiplicative noise $\mathscr{L}^i_j(\mathbf{p})$ has to be understood, not in the Ito nor in the Stratonovich sense, but in the so-called Hänggi-Klimontovich manner [5] if one wants the forward Kolmogorov equation associated with (11) to admit Jüttner distribution (3) as an invariant measure in **p**-space.

The process defined by (11) presents important similarities with the special relativistic Ornstein-Uhlenbeck process (see Section 2). Both share the same invariant measure in **p**-space and both obey simple fluctuation-dissipation theorems. The main difference between both processes lies in the choice of noise. The noise used in the special relativistic Ornstein-Uhlenbeck process is a Gaussian white noise in the proper frame of the fluid which surrounds the diffusing particle; on the other hand, the noise in (11) is a Gaussian white noise in the proper frame of the diffusing particle. A detailed study of how this difference influences the relaxation towards the invariant measure in **p**-space is still lacking. In particular, the hydrodynamical limit has not been worked out on Dunkel and Hänggi's process yet. No general relativistic extension has been considered either.

Dunkel and Hänggi have recently extended their work on relativistic Brownian motions in various directions. They have indeed introduced and studied numerically binary collisions models [25] which suggest that fully realistic relativistic proceses might very well have to be non Markovian in both momentum and position spaces. This work has been followed by another one [26] where the authors propose two simple non Markovian models of relativistic diffusions in position space only. These two models are however not related to non Markovian diffusions in momentum space, but rather introduced as simple models satisfying certain list of minimal physical requirements.

Dunkel and Hänggi have also interpreted the results of numerical simulations of binary collisions [27] to suggest that the Jüttner distribution is *not* the correct equilibrium distribution of dilute special relativistic perfect gases; the alternative distribution considered in [27] has been further supported by Dunkel, Talkner and Hänggi through a theoretical maximum entropy argument [28]. It is our opinion that the Jüttner distribution *is* correct and that the argumentation developed in [27, 28] is not adequate because it relies in a crucial way on non covariant equations; actually, the numerical simulations presented in [27] *do* support the Jüttner distribution if they are properly interpreted through covariant definitions of averaged quantities; and so does the argument presented in [28]. This point, which will be fully explored in a forthcoming publication, cannot be elaborated upon here; let us simply note that the Jüttner distribution is the dilute gas limit of the Bose-Einstein (BE) and the Fermi-Dirac (FD) distributions [29, 15]. The alternative distribution considered in [27, 28] thus conflicts with both the BE and FD statistics, which are direct consequences of quantum field theory [30, 31, 32]. Thus, the alternative distribution conflicts with what is currently the most precise model of matter, confirmed by countless experiments. Let us also remark that the applicability of BE statistics to relativistic particles is itself experimentally confirmed by all observations of black body radiation [33].

The Relativistic Brownian Motion of Oron and Horwitz

O. Oron and L.P. Horwitz suggest [9] that the diffusion of a special relativistic point mass under the action of an external force $\beta(x)$ should be described by the following generalization of Smoluchowski equation:

$$\partial_\tau \rho = -\partial_\mu (\beta^\mu \rho) + D \partial_\mu \partial^\mu \rho, \tag{12}$$

where D is a positive coefficient and ρ is a function of both x and a certain 'universal' time τ. Various possible microscopic justifications for (12) are elaborated upon in [9].

Equation (12) as it stands presents various problems. First, the physical meaning of τ remains unclear, not to mention the fact that introducing a 'universal' time in a special relativistic framework might seem unnatural; also unclear is the measure against which ρ should be integrated. Furthermore, equation (12) does not contain any explicit reference to a fluid surrounding the diffusing particle; thus, like Franchi and Le Jan's process, the physical interpretation of Oron and Horwitz's model is at best problematical.

A final point should be mentioned. Even if equation (12) is certainly an original contribution of O. Oron and L.P. Horwitz, the idea to describe transport phenomena by hyperbolic equations instead of parabolic ones is not that recent; it has been probably first proposed by C. Cattaneo [34] and found its most developed expression to date in the construction of the so-called Extended Thermodynamics theories [35, 36]. These theories present serious physical problems in their own right (see [15, 35] and the discussion at the end of Section 2.2 above) but they certainly deserve attention. They fall outside the context of this review because they are not based on stochastic processes; a thorough

[5] This amounts to evaluating the multiplicative noise at the end of the stochastic jump

comparison of the diffusion models constructed according to the principles of Extended Thermodynamics and the Relativistic Brownian Motion of Oron and Horwitz should nevertheless prove to be very rewarding.

CONCLUSION

The last ten years have witnessed an important activity around relativistic diffusion processes (in chronological order: [5, 13, 10, 11, 16, 9, 6, 19, 8, 24, 18, 27, 28, 25, 26, 22, 17, 37]). The model most developed today is certainly the Relativistic Ornstein-Uhlenbeck process. A priority for future research seems to better our understanding of the other models of relativistic diffusion discussed in this review and to systematically compare their physical predictions to those of the ROUP. It has also been remarked that curvature, even in a purely Galilean context, can profoundly affect the phenomenology of diffusion [38, 39, 37]; general relativistic diffusions should therefore become the objects of renewed scrutiny . Finally, describing stochastically the evolution of both Riemanian and Lorentzian manifolds is a problem certainly worthy of all attention.

REFERENCES

1. A. Einstein, *Investigations on the Theory of Brownian Motion*, Reprint of the 1st English edition (1926), Dover, New-York, 1956.
2. N. van Kampen, *Stochastic Processes in Physics and Chemistry*, North-Holland, Amsterdam, 1992.
3. B. Øksendal, *Stochastic Differential Equations*, 5th edn, Springer-Verlag, 1998.
4. C. Gardiner, and P. Zoller, *Quantum Noise*, 2nd. enlarged edn, Springer-Verlag, New-York, 2000.
5. F. Debbasch, K. Mallick, and J.P. Rivet, *J. Stat. Phys.* **88**, 945-966 (1997).
6. J. Franchi, and Y. Le Jan, *arXiv math.PR/0410485* (2004).
7. F. Dowker, J. Henson, and R. Sorkin, *Mod. Phys. lett. A* **19**, 1829–1840 (2004).
8. J. Dunkel, and P. Hänggi, *Phys. Rev. E* **71**, 016124 (2005).
9. O. Oron, and L. Horwitz, *arXiv:math-ph/0312003 v1* (2003).
10. C. Barbachoux, F. Debbasch, and J.P. Rivet, *Eur. Phys. J. B* **19**, 37-47 (2001).
11. C. Barbachoux, F. Debbasch, and J.P. Rivet, *Eur. Phys. J. B* **23**, 487-496 (2001).
12. C. Barbachoux, *Etude de quelques propriétés du processus d'Ornstein-Uhlenbeck relativiste. Thèse.*, Université Paris 6, 2000.
13. F. Debbasch, and J.P. Rivet, *J. Stat. Phys.* **90**, 1179-1199 (1998).
14. W. Hiscock, and L. Lindblom, *Phys. Rev. D* **31**, 725-733 (1985).
15. W. Israel, "Covariant Fluid Mechanics and Thermodynamics: An Introduction," in *Relativistic Fluid Dynamics*, edited by A. Anile, and Y. Choquet-Bruhat, vol. 1385 of *Lecture Notes in Mathematics*, Springer-Verlag, Berlin, 1987, pp. 152-210.
16. F. Debbasch, *J. Math. Phys.* **45**, 2744-2760 (2004).
17. C. Chevalier, and F. Debbasch, *J. Math. Phys.* **48**, 023304 (12 pages) (2007).
18. C. Chevalier, and F. Debbasch, "Stochastic relativistic processes and fluctuation-dissipation theorems," in *Proceedings of the ISTP17, Toyama, 2006*, 2006.
19. M. Rigotti, and F. Debbasch, *J. Math. Phys.* **46**, 103303 (11 pages) (2005).
20. S. Hawking, and G. Ellis, *The large scale structure of space-time*, Cambridge Monographs on Mathematical Physics, Cambridge University Press, Cambridge, 1973.
21. R. Dudley, *Arkiv f. Mat.* **6**, 241–268 (1965).
22. J. Franchi, *arXiv:math.PR/0612020 v1* (2006).
23. I. Bailleul, *Doctoral Thesis*, University Paris 11, 2007.
24. J. Dunkel, and P. Hänggi, *Phys. Rev. E* **72**, 036106 (11 pages) (2005).
25. J. Dunkel, P. Talkner, and P. Hänggi, *arXiv:cond-math/0607082 v4* (2006).
26. J. Dunkel, P. Talkner, and P. Hänggi, *arXiv:cond-math/0608023 v2* (2006).
27. J. Dunkel, and P. Hänggi, *Phys. A* **374**, 559–572 (2007).
28. J. Dunkel, P. Talkner, and P. Hänggi, *arXiv:cond-math/0610045 v2* (2006).
29. S. de Groot, W. van Leeuwen, and C. van Weert, *Relativistic Kinetic Theory. Principles and Applications*, North-Holland, 1978.
30. R. Feynman, *Statistical Mechanics. A set of Lectures*, Addison-Wesley Publishing Company, inc., 1972.
31. C. Itzikson, and J. Zuber, *Quantum Field Theory*, Mc Graw and Hill, 1980.
32. R. Haag, *Local Quantum Physics*, Texts and Monographs in Physics, Springer-Verlag, Berlin, 1992.
33. F. Reif, *Fundamentals of Statistical and Thermal Physics*, McGraw-Hill, Auckland, 1965.
34. C. Cattaneo, *Atti Sem. Mat. Fis. Univ. Modena* **3** (1948).
35. I. Müller, and T. Ruggeri, *Extended Thermodynamics*, vol. 37 of *Springer Tracts in Natural Philosophy*, Springer-Verlag, New-York, 1993.
36. D. Jou, J. Casas-Vázquez, and G. Lebon, *Extended Irreversible Thermodynamics*, 2nd edn, Springer-Verlag, Berlin, 1996.

37. C. Chevalier, and F. Debbasch, "Diffusion Processes on Manifolds," in *Submitted to the proceedings of the Kurtzfest, Madison, 2006*, 2007.
38. F. Debbasch, and M. Moreau, *Phys. A* **343**, 81–104 (2004).
39. C. Chevalier, and F. Debbasch, *EPL* **77**, 20005–20009 (2007).

Synchrony and Critical Behavior: Equilibrium Universality in Nonequilibrium Stochastic Oscillators

Kevin Wood*, C. Van den Broeck†, R. Kawai** and Katja Lindenberg‡

*Department of Chemistry and Biochemistry, Department of Physics, and Institute for Nonlinear Science, University of California San Diego, 9500 Gilman Drive, La Jolla, CA 92093-0340, USA
†Hasselt University, Diepenbeek, B-3590, Belgium
**Department of Physics, University of Alabama at Birmingham, Birmingham, AL 35294, USA
‡Department of Chemistry and Biochemistry and Institute for Nonlinear Science, University of California San Diego, 9500 Gilman Drive, La Jolla, CA 92093-0340, USA

Abstract. We review our work on a discrete model of stochastic, phase-coupled oscillators that is sufficiently simple to be characterized in complete detail, lending insight into the universal critical behavior of the corresponding nonequilibrium phase transition to macroscopic synchrony. In the mean-field limit, the model exhibits a supercritical Hopf bifurcation and global oscillatory behavior as coupling eclipses a critical value. The simplicity of our model allows us to perform the first detailed characterization of stochastic phase coupled oscillators in the locally coupled regime, where the model undergoes a continuous phase transition which remarkably displays signatures of the XY equilibrium universality class, verifying recent analytical predictions. Finally, we examine the effects of spatial disorder and provide analytical and numerical evidence that such disorder does not destroy the capacity for synchronization.

Keywords: stochastic phase coupled oscillators, synchronization, critical behavior, universality
PACS: 64.60.Ht, 05.45Xt, 89.75.-k

Time and periodicity play a critical role in a host of physical, biological, and chemical systems [1, 2]. In particular, a vast number of systems consist of noisy individual entities characterized by periodic (oscillatory) dynamics that give rise to a potentially complex competition between individual dynamics on a microscopic scale and large scale cooperative behavior. Originating with the pioneering work of Kuramoto [3, 4], the scientific literature is replete with studies of such synchronization [1, 2]. The typical approach involves systems of coupled, nonlinear differential equations. As such, work has been traditionally limited to relatively small, deterministic systems in the mean field limit [5, 6, 7, 8, 9].

While providing a mature understanding of the synchronization dynamics in coupled oscillators, the numerical and analytical complexities of these familiar models have prohibited a complete characterization in the large system limit. Therefore, the analogy between the cooperative behavior of these oscillators with phase transitions and critical phenomena from statistical mechanics [10, 11] has not been fully exploited. Specifically, the onset of synchronization when viewed on a long wavelength scale is characterized by a macroscopic change in an inherently nonequilibrium symmetry: for systems below the synchronization threshold, the large-scale dynamics appears time-translationally invariant, while synchronized systems oscillate collectively, breaking this time-based symmetry.

With these possible analogies in mind, and with an eye towards critical behavior, we develop a simple, discrete model of noisy phase-coupled oscillators [12, 13, 14] which displays the desired macroscopic behavior while remaining sufficiently simple to be characterized as a phase transition. Owing to the well-established notion of universality, which renders microscopic specifics essentially irrelevant for study of the typically long-wavelength behavior occurring near criticality, our model of the phase transition in question must preserve the macroscopic symmetry-breaking while incorporating the relevant physical details, namely, stochasticity and microscopic periodicity.

Our starting point is a three-state unit [12, 13, 14] governed by transition rates g (Fig. 1). Loosely speaking, we interpret the state designation as a generalized (discrete) phase, and the transitions between states, which we construct to be unidirectional, as a phase change and thus an oscillation of sorts. The probability of going from the current state i to state $i+1$ in an infinitesimal time dt is gdt, with i=1,2,3 modulo 3. For an isolated unit, the transition rate is simply a constant (g) that sets the oscillator's intrinsic frequency; for many coupled units, we will allow the transition rate to depend on the neighboring units in the spatial grid, thereby coupling neighboring phases.

CP913, *Nonequilibrium Statistical Mechanics and Nonlinear Physics, XV Conference*
edited by O. Descalzi, O. A. Rosso, and H. A. Larrondo

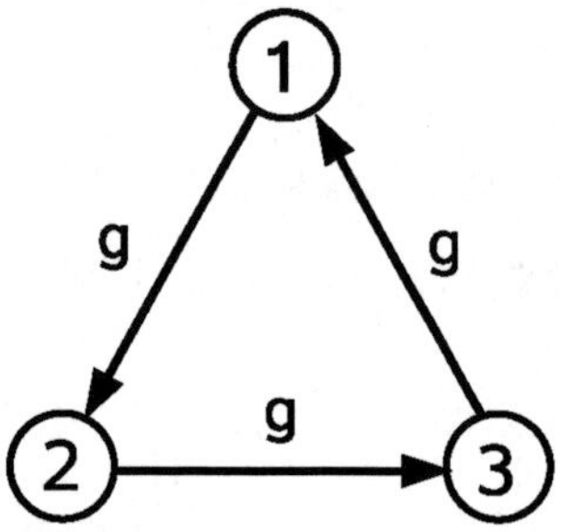

FIGURE 1. Three-state unit with generic transition rates g.

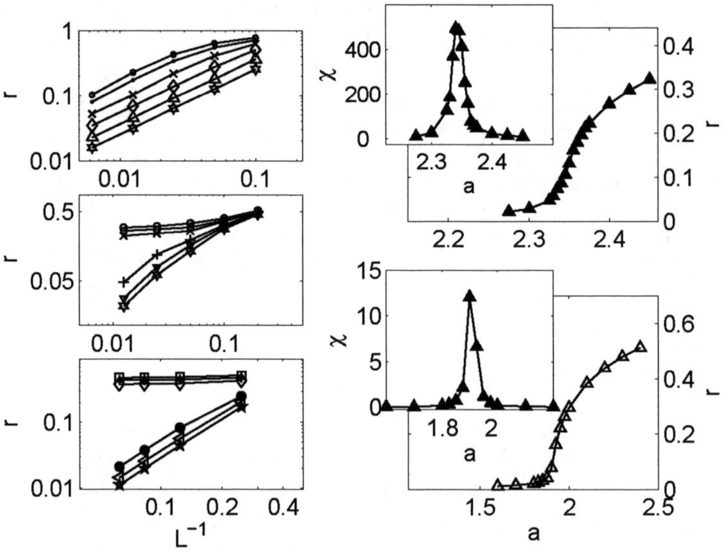

FIGURE 2. Dimensionality (d) dependent phase transition. Left column shows r vs L on a log-log scale for $d = 2$, 3, 4, top to bottom. For $d = 2$, the curves start from $a = 2{,}0$ (top) with increments of 0.5. For $d = 3$, the curves start from $a = 2.275$ (top) with increments of 0.025 (with the exception of $a = 2.350$, not shown). For the $d = 4$ case, the curves are $a = 1.6, 1.7, 1.8, 2.0, 2.1, 2.2$, top to bottom. In the limit of infinite system size, a synchronous phase emerges for $d > 3$ for sufficiently large a. The right column shows r vs a and χ vs a for $d = 3$, $L = 80$ (top) and $d = 4$, $L = 16$, bottom. Fluctuations peak near the critical point, giving an estimation of $a_c = 2.345 \pm 0.005$ and $a_c = 1.900 \pm 0.025$ for $d = 3$ and $d = 4$, respectively.

For an isolated unit we write the linear evolution equation $\partial P(t)/\partial t = MP(t)$, where the components $P_i(t)$ of the column vector $P(t) = (P_1(t)\ P_2(t)\ P_3(t))^T$ are the probabilities of being in state i at time t, and

$$M = \begin{pmatrix} -g & 0 & g \\ g & -g & 0 \\ 0 & g & -g \end{pmatrix}. \tag{1}$$

The system reaches a steady state for $P_1^* = P_2^* = P_3^* = 1/3$. The transitions $i \to i+1$ (with $i+1 \equiv 1$ when $i = 3$) occur with a rough periodicity determined by g; that is, the time evolution of our simple model qualitatively resembles that of the discretized phase of a generic noisy oscillator.

We implement microscopic coupling by allowing the transition probability of a given unit to depend on the states of the unit's nearest neighbors in a spatial grid. Specifically, we choose a function which compares the phase at a given site with its neighbors, and adjusts the phase at the given site so as to facilitate phase coherence. We settle on the following form of the transition rate from state i to state j:

$$g_{ij} = g \exp\left[\frac{a(N_j - N_i)}{2d}\right] \delta_{j,i+1}. \tag{2}$$

Here the constant a is the coupling parameter, g is a parameter related to the intrinsic frequency of each oscillator, and δ is the Kronecker delta. N_k is the number of nearest neighbors in state k, and $2d$ is the total number of nearest neighbors in d dimensional cubic lattices. While this choice is by no means unique and these rates are somewhat distorted by their independence of the number of nearest neighbors in state $i-1$, the form (2) does lead to synchronization and allows for fast numerical simulation of large lattices near the critical regime [12, 13].

To verify the emergence of global synchrony, we first consider a mean field version of the model. In the large N limit with all-to-all coupling we write

$$g_{ij} = g\exp\left[a(P_j - P_i)\right]\delta_{j,i+1}. \tag{3}$$

Note that in the mean field limit g_{ij} does not depend on the location of the unit within the lattice. Also, there is an inherent assumption that we can replace N_k/N with P_k. With this simplification we arrive at a nonlinear equation for the mean field probability, $\partial P(t)/\partial t = M[P(t)]P(t)$, with

$$M[P(t)] = \begin{pmatrix} -g_{12} & 0 & g_{31} \\ g_{12} & -g_{23} & 0 \\ 0 & g_{23} & -g_{31} \end{pmatrix}. \tag{4}$$

Normalization allows us to eliminate P_3 and obtain a closed set of equations for P_1 and P_2. We can further characterize the mean field solutions by linearizing about the fixed point $(P_1^*, P_2^*) = (1/3, 1/3)$. The complex conjugate eigenvalues of the Jacobian evaluated at the fixed point, $\lambda_\pm = g(2a - 3 \pm i\sqrt{3})/2$, cross the imaginary axis at $a = 1.5$, indicative of a Hopf bifurcation at this value, which following a more detailed analysis [12, 13] can be shown to be supercritical. Hence, as a increases, the mean field undergoes a qualitative change from disorder to global oscillations, and the desired breaking of time translational symmetry emerges. Numerical solutions confirm this behavior, yielding results that agree with simulations of an all-to-all coupling array [12, 13].

We now proceed with a detailed numerical characterization of the locally coupled case. We performed simulations of the locally coupled model in continuous time on d-dimensional cubic lattices with periodic boundary conditions. Time steps were 10 to 100 times smaller than the fastest local average transition rate, i.e., $dt \ll e^{-a}$ (we set g=1). We find that much smaller time steps lead to essentially the same results. Starting from random initial conditions, all simulations were run until an apparent steady state was reached, and statistics are based on 100 independent trials.

Following other works on phase synchronization, we introduce the order parameter [3, 4]

$$r = \langle R \rangle, \quad R \equiv \frac{1}{N}\left|\sum_{j=1}^{N} e^{i\phi_j}\right|. \tag{5}$$

Here ϕ_j is the discrete phase $2\pi(k\text{-}1)/3$ for state $k \in \{1,2,3\}$ at site j. The brackets represent an average over time in the steady state and over all independent trials. Nonzero r in the thermodynamic limit indicates synchrony. We also calculate the generalized susceptibility $\chi = L^d[\langle R^2\rangle - \langle R\rangle^2]$.

As shown in Fig. 2, the model undergoes a dimension-dependent transition marked by characteristics of a typical phase transition, including a macroscopic change in r in the infinite system limit, a peak in fluctuations at the critical point, and a diverging spatial correlation length [12, 13] (not shown). Specifically, in $d = 2$ we do not see the emergence of a synchronous phase. Instead, we observe intermittent oscillations (for very large values of a) that decrease drastically with increasing system size. In fact, $r \to 0$ in the thermodynamic limit, even for very large values of a. We conclude that the phase transition to synchrony cannot occur for $d = 2$.

In contrast to the d=2 case, which serves as the lower critical dimension, a clear thermodynamic-like phase transition occurs in three dimensions. Figure 2 shows explicitly that for $a < a_c$ when $d = 3$, $r \to 0$ as system size is increased, and a disordered phase persists in the thermodynamic limit. For $a > a_c$, the order parameter approaches a finite value as the system size increases. In addition, Fig. 2 shows the behavior of the order parameter as a is increased for the largest system studied (L=80); the upper left inset shows the peak in χ at $a = 2.345 \pm 0.005$, thus providing an estimate of the critical point a_c. While the accuracy of our current estimation of the critical point is modest, it nonetheless suffices to determine the universality class of the transition. Similarly, for $d = 4$ we estimate the transition coupling to be $a_c = 1.900 \pm 0.025$ (Fig. 2), and for $d = 5$ we see a transition to synchrony at $a_c = 1.750 \pm 0.015$ [12, 13] (not shown).

To further characterize this transition, we use finite size scaling analysis by assuming the standard scaling

$$r = L^{-\frac{\beta}{\nu}} F[(a - a_c)L^{\frac{1}{\nu}}]. \tag{6}$$

Here $F(x)$ is a scaling function that approaches a constant as $x \to 0$. To test our numerical data against different universality classes we choose the appropriate critical exponents for each, recognizing that there are variations in the

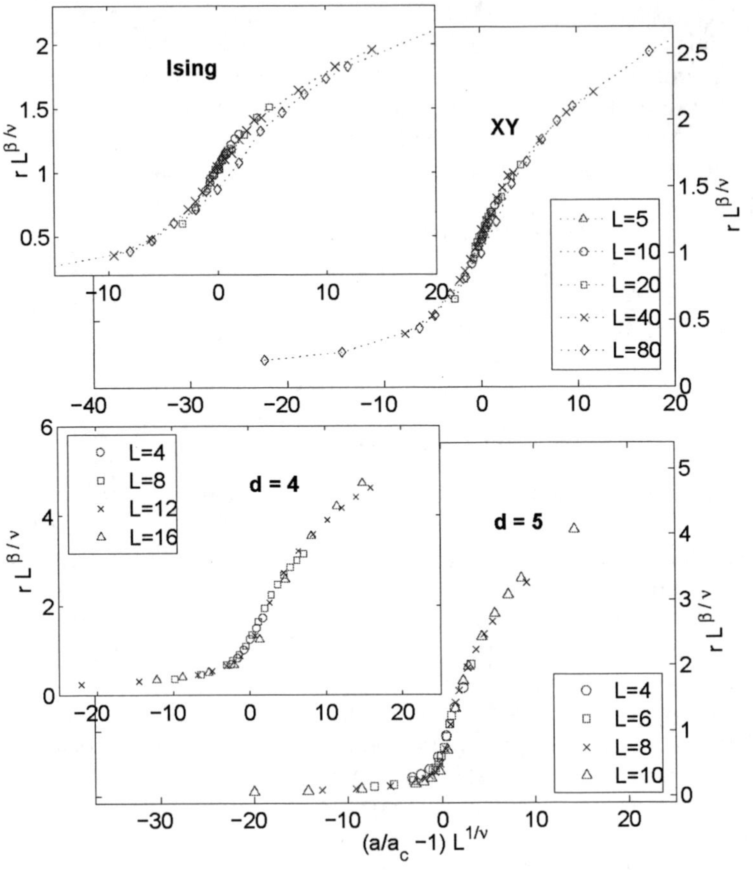

FIGURE 3. Finite size scaling analysis for $d=3$, $d=4$, $d=5$: Data collapse using ansatz (6) with mean field exponents.

reported values of these exponents [15]. For the XY universality class we use the exponents β=0.34 and ν=0.66 [17]. For the Ising exponents we use β=0.31 and ν=0.64 [11]. In Fig. 3, we see quite convincingly a collapse when exponents from the XY class are used, verifying recent analytic predictions [18, 19]. For comparison, we also show the data collapse with $d=3$ Ising exponents (note the scale differences). Because we expect $d=4$ to be the upper critical dimension in accordance with XY/Ising behavior, we anticipate a slight breakdown of the scaling relation (6). Nevertheless, as shown in Fig. 4, the data collapse is very good with the mean field exponents. As such, our simulations suggest that $d=4$ serves as the upper critical dimension. The case $d=5$, where the data collapse with the mean field exponents is excellent, is also shown in Fig. 3, and further supports the claim that d_{uc}=4.

Having observed convincingly equilibrium-like critical behavior in systems of inherently nonequilibrium oscillators, we now turn to the question of spatial disorder. That is, when units are no longer identical (g not equal for all units), is the capacity for synchronization destroyed? To address this question, we consider first a dichotomously disordered population consisting of two subpopulations of oscillators, each characterized by a different frequency parameter $g=\gamma_1$ and $g=\gamma_2$. We assume a modified form of the inter-unit coupling given by

$$g_{ij} = g \exp\left[\frac{a(N_j - N_{i-1})}{n}\right] \delta_{j,i+1}, \tag{7}$$

where n is the number of oscillators to which unit ν is coupled, and N_k is the number of units among the n that are in state k. This form maintains a closer macroscopic analogy with coupled oscillators far above the frequency threshold [16], and while more time consuming for in depth studies of the critical regime, it proves more suitable for the current disorder studies.

In the mean field limit, our dichotomously disordered array can be characterized by a six dimensional equation for P_{i,γ_1} and P_{i,γ_2}, $i \in 1,2,3$, where, for example, P_{1,γ_1} represents the probability for a unit with transition rate parameter

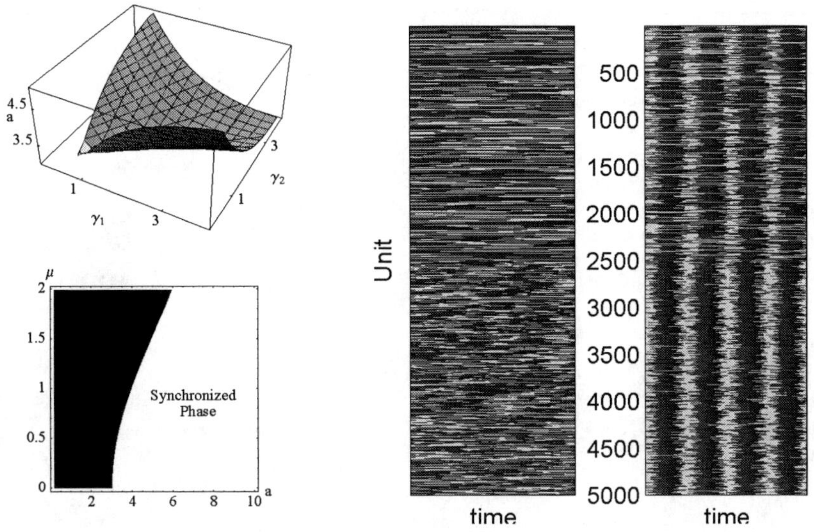

(a) **(b)**

FIGURE 4. (**a.**) Upper panel: The critical surface a_c is shown for a dichotomously disordered system. Specifically, the boundary given by the contour $\mathrm{Re}\lambda_+ = 0$ is plotted in (γ_1, γ_2, a) space. This contour indicates the critical point, where the Hopf bifurcation occurs and the disordered solution becomes unstable. The region above the contour represents the synchronized phase. Lower panel: Stability boundary in terms of relative width parameter. (**b.**) Time evolution snapshots of a dichotomously disordered system with $\gamma_1 = 0.5$ and $\gamma_2 = 1.5$ are shown for $a = 3.5 < a_c$ (left) and $a = 4.1 > a_c$ (right). Each color represents the state of the oscillator.

γ_1 to be in state 1. Following normalization, the set reduces to four equations which can be linearized about the nonsynchronous fixed point $(1/3, 1/3, 1/3, 1/3)$. The eigenvalues of the relevant Jacobian are given by:

$$\begin{aligned} \frac{\mathrm{Re}\lambda_\pm}{\gamma_1+\gamma_2} &= \frac{1}{8}\left[a - 6 \pm B(a,\mu)\cos\left(C(a,\mu)\right)\right], \\ \frac{\mathrm{Im}\lambda_\pm}{\gamma_1+\gamma_2} &= \frac{1}{8}\left[\sqrt{3}(a+2) \pm B(a,\mu)\sin\left(C(a,\mu)\right)\right], \end{aligned} \tag{8}$$

where

$$\begin{aligned} B(a,\mu) &\equiv \sqrt{2}\left[a^4 - 6a^2\mu^2 + 3\mu^4(a^2+3)\right]^{1/4}, \\ C(a,\mu) &\equiv \frac{1}{2}\tan^{-1}\left(\frac{-\sqrt{3}(a^2-(a+3)\mu^2)}{a^2+3(a-1)\mu^2}\right). \end{aligned} \tag{9}$$

In fact, one pair of eigenvalues crosses the imaginary axis (Hopf bifurcation) at a critical value $a = a_c$, but the other pair shows no qualitative change as a is varied. Aside from an overall factor $(\gamma_1 + \gamma_2)$, Eqs. (8) depend only on the relative width variable

$$\mu \equiv \frac{2(\gamma_1 - \gamma_2)}{(\gamma_1 + \gamma_2)} \tag{10}$$

$(-2 \leq \mu \leq 2)$, and in fact the Hopf bifurcation occurs at a single value of $a_c(\mu)$. As shown in Fig. 4, the surface a_c separates regions of synchronous and asynchronous phases. A small-μ expansion leads to an estimate of a_c to $O(\mu^2)$,

$$a_c \approx \frac{1}{8}\left(12 + 3\mu^2 + \sqrt{3}\sqrt{(12+\mu^2)(4+3\mu^2)}\right), \tag{11}$$

a result that exhibits these trends explicitly. In particular, we note the increase in a_c with increasing μ. Figure 4 also shows snapshots from simulations of a globally coupled, dichotomously disordered population ($N = 5000$ units, or 2500 with $\gamma_1 = 0.5$ and 2500 with $\gamma_2 = 1.5$), confirming the emergence of macroscopic synchrony for sufficiently strong coupling.

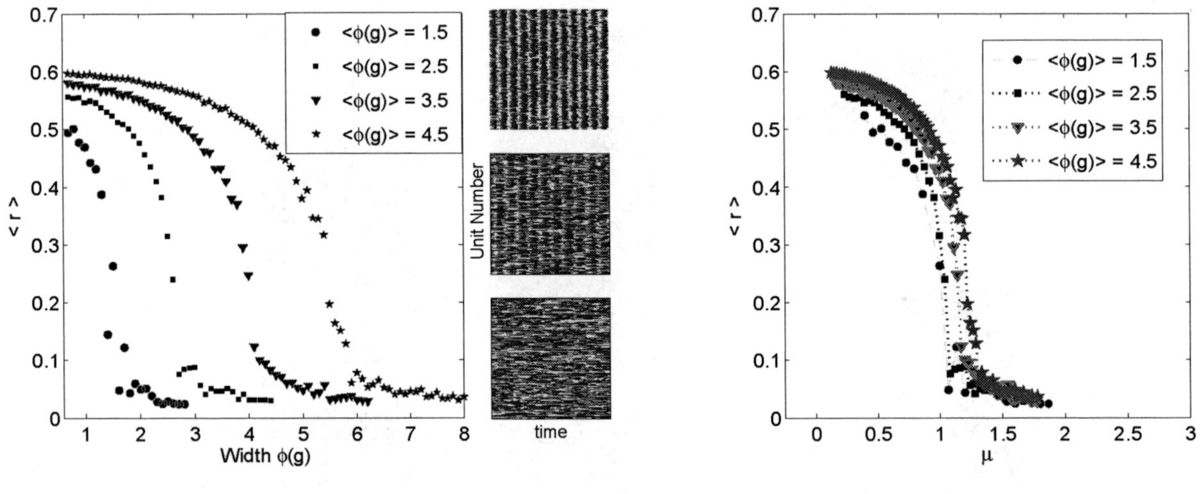

(a) **(b)**

FIGURE 5. (**a.**) As the width of the $\phi(g)$ distribution increases, a critical width is reached beyond which synchronization is destroyed. The coupling is chosen to be $a = 3.2$, and the four curves represent the steady state, time-averaged order parameter for distributions with different means. As the mean of the $\phi(g)$ distribution increases, the transition to disorder occurs at a greater width. The insets at the right show the long-time behavior of an entire population of mean transition rate parameter 3.5 (corresponding to the triangle order parameter data) and widths of 0.6, 4.0, and 6.2. (**b.**) The same data is plotted against μ, the relative width parameter.

Finally, we show that these trends carry over to fully disordered systems, where γ_i is drawn from a uniform distribution characterized by a relative width variable μ. Explicitly, μ is the ratio of the distribution mean to the distribution width. As shown in Fig. 5, synchronization still occurs in these systems and furthermore, the critical coupling a_c is essentially determined by the distribution parameter μ. We find that, analogous to the dichotomously disordered system, synchronization occurs at a *single* value of a_c which depends crucially on μ.

In conclusion, we have demonstrated the remarkable result that a fundamentally nonequilibrium transition, namely, a phase transition that breaks the symmetry of translation in time, is described by an equilibrium universality class. By utilizing a simple discrete model for active noisy oscillators, we have shown compelling numerical and analytical evidence that the emergence of synchronous oscillations in these systems contains signatures of an equilibrium phase transition, including diverging fluctuations at criticality, a macroscopic change in the order parameter, and classic exponents belonging to the XY universality class. In addition, we show that synchronization can and does occur in both dichotomously and fully disordered populations, leading to large-scale cooperativity in spite of the non-identical nature of the constituents.

ACKNOWLEDGMENTS

This work was partially supported by the National Science Foundation under Grant No. PHY-0354937.

REFERENCES

1. S. H. Strogatz, *Nonlinear Dynamics and Chaos*, Westview Press, 1994.
2. A. Pikovsky, M. Rosenblum, and J. Kurths, *Synchronization*, Cambridge Univ. Press, Cambridge, 2001.
3. S. H. Strogatz, *Physica D* **143**, 1–20 (2000).
4. J. A. Acebón *et al.*, *Rev. Mod. Phys.* **77**, 137–185 (2005).
5. H. Sakaguchi, S. Shinomoto, and Y. Kuramoto, *Prog. Theor. Phys.* **77**, 1005–1010 (1987).
6. H. Daido, *Phys. Rev. Lett.* **61**, 231–234 (1988).
7. S. H. Strogatz and R. E. Mirollo, *J. Phys. A* **21**, L699–L706 (1988).
8. S. H. Strogatz and R. E. Mirollo, *Physica D* **31**, 143–168 (1988).

9. H. Hong, H. Park, and M. Choi, *Phys. Rev. E* **71**, 054204 (2004).
10. N. Goldenfeld, *Lectures on Phase Transitions and the Renormalization Group*, Westview Press, 1992.
11. K. Huang, *Statistical Mechanics* Second Edition, Wiley, New York, 1988.
12. K. Wood, C. Van den Broeck, R. Kawai, and K. Lindenberg, *Phys. Rev. Lett.* **96**, 145701 (2006).
13. K. Wood, C. Van den Broeck, R. Kawai, and K. Lindenberg, *Phys. Rev. E* **74**, 031113 (2006).
14. T. Prager, B. Naundorf, and L. Schimansky-Geier, *Physica A* **325**, 176–185 (2003).
15. A. Pelissetto and E. Vicari, *Phys. Rep.* **368**, 549–727 (2002).
16. K. Wood, C. Van den Broeck, R. Kawai, and K. Lindenberg, to appear in Phys. Rev. E (2007).
17. A. P. Gottlob and M. Hasenbusch, *Nucl. Phys. B Suppl.* **30**, 838–841 (1993).
18. T. Risler, J. Prost, F. Jülicher, *Phys. Rev. Lett.* **93**, 175702 (2004).
19. T. Risler, J. Prost, F. Jülicher, *Phys. Rev. E* **72**, 016130 (2005).

Intermittent Search Processes: Chance Against Strategy

M. Moreau, O. Bénichou, C. Loverdo, P.H. Suet and R. Voituriez

Laboratoire de Physique Théorique de la Matière Condensée (UMR 7600)
Université Pierre et Marie Curie, 4 Place Jussieu, 75252 Paris France

Abstract. Intermittent search processes are widely observed in nature and in social activities. Simplified models show that they can be very efficient. Here we give new evidences of this efficiency. We discuss the conditions which can make intermittence favorable to the search, generally due to the imperfection of the detectors. We show that organized strategies, which are possible for searchers with sufficient abilities, in principle lead to shorter search times than random behaviors. Nevertheless, their advantage is not determinant, and random can be practically preferable because of its easier and more general use.

Keywords: intermittence, stochastic search, optimization
PACS: 87.23.-n

INTRODUCTION

Search processes are present at all scales in natural phenomena as well as in human activities. In many circumstances, a "searcher", which may be a molecule, a living organism or a social organization, has to find a given target by displacing its position, or its attention, till it discovers its objective. Search problems have been widely studied as optimization problems [1], often for practical motivations, like the search for submarines in 2nd world war, or in rescue operations. In general, information or signs concerning the location of the target play a predominant role for optimizing the search with respect to a given quantity, which can be, for instance, the energy needed for the search, or the number of targets reached in a given time.

The problem is significantly different in the case of hidden targets, when they can not be distinguished from a large distance and no indication allows one defining any non uniform target distribution probability. Then the searcher has to follow an arbitrary trajectory to prospect the territory until a target enters into the action field of the sensors which can detect it. However, careful scanning is generally a slow process, and it often occur that the searcher alternate phases of slow, detailed inspection with phases of fast displacements, which significantly reduce the perception abilities but rapidly lead to new territories. Although apparently not rational, such an intermittent behaviour is commonly observed for a large class of animals in quest of food [2, 3]. In the recent years, many articles have been published on these problems, mainly to find conditions which allow optimizing the search [4, 5, 6, 7, 8, 9, 10, 11]. In particular, it has been shown that intermittent processes, which alternate slow scanning and fast displacements, can allow for minimizing the first arrival time on the target [4, 5, 6, 7, 8, 9, 10, 12, 13].

Here we will extend these results and propose some general responses to questions which where formerly addressed by means of specific, restricted models: when can intermittent search be justified, instead of a systematic search? Is it possible that random be more efficient than a well determined strategy, and in some sense, that disorder be stronger than order? and eventually: how can the simplified models used for reasoning be useful in the huge complexity of actual situations?

To such general problems we shall not give complete answers, but just suggest some hints. First, we will consider the conditions which are necessary for justifying intermittent search processes and give examples where they are actually realized. Then, we will briefly review the results obtained on simple intermittent models, and mention various extensions which allow us considering more realistic situations. On this basis, and from reasonable extrapolations, we will eventually give some elements for answering the previous questions.

CP913, *Nonequilibrium Statistical Mechanics and Nonlinear Physics, XV Conference*
edited by O. Descalzi, O. A. Rosso, and H. A. Larrondo

SEARCH AND INTERMITTENCE

Hidden Targets, Finite Range Detectors And Random Search

We consider an immobile target located inside a finite region V of some space, assuming that no other information is available on its location, and that it can not be seen from a large distance a. As a consequence, the searcher has to move till the moment when its distance to the target is less than a, the detection range of activation. This motion should be decided arbitrarily, and in many cases it is chosen randomly.

The first search model with random trajectories was proposed by Vishwanathan and coworkers in 1999 and extended later [5]. In this model, based on experimental observations of albatrosses searching for food, the targets have a random distribution with uniform average density in infinite space. The searcher performs truncated Levy flights, consisting in a straight displacements in random directions. The distance l covered in one of these straight flights has a probability density $p(l) \propto l^{-\mu}$ for $l > a$ and $p(l) = 0$ for $l < a$, with $\mu > 1$. A target is found instantaneously if its distance from the searcher is less than the range a of the detectors. At the end of a flight, a new, independent Levy flight is immediately performed. The authors showed that, if the targets are left unchanged or immediately reconstituted after discovery, the average number of targets found during a given time (for constant velocity) is maximum for $\mu = 2$. This value corresponds to the maximum efficiency of the process in the present situation. However, if the targets are destroyed when they are discovered, the maximum efficiency is obtained for an infinite straight displacement, since this behaviour obviously allows to sweep the maximum volume during a given time.

In fact, targets often disappear after their discovery (except if they are distributed in patches), as it occurs for the preys destroyed by the predator or for victims of some accident when they are found by rescuers. Nevertheless, indefinite straight trajectories are not widely observed.

Imperfect Detectors

In fact, actual detectors need some time to analyze the data and do not allow for instantaneous detection, at least for hidden targets which can not be easily found. Thus their performances significantly decrease when the searcher moves rapidly, and detection can even need the searcher to stop completely. Thus efficient scanning must be slow, which may lead the searcher to interrupt it for moving rapidly to unexplored regions. Nevertheless, the relevance of this intermittent behaviour is unclear, and various models have been introduced to study if, and in what conditions, it can be justified [3, 6, 7, 8, 9, 10].

MODELS

We now consider simplified models alternating a slow motion allowing for detection, and fast displacements during which the target is not seen. The slow motion can be represented by a diffusion or a random walk: then the searcher moves relatively slowly, and he can detect a target within the range a of its detector. The fast displacements may be modeled by a ballistic motion, or even by a "teleportation" which randomly relocates the searcher in the search territory. Clearly, these are simplifying assumptions, but they represent typical cases and allow for analytical solutions and explicit answers to our questions. Generalizations will be considered later.

The duration of each phase, scanning and moving, is assumed to be an independent, stochastic variable, generally - but not always - following an exponential law. This is justified if the switching between the regimes is decided independently of searcher situation. In order to characterize the efficiency of the search, we use the mean search time, which is the first time a target is detected, averaged on the position of the target. The mean search time is primordial in most practical circumstances, and it does not depend whether the target is or is not destroyed by the searcher after discovery.

Diffusion And Ballistic Motion

The Example Of Foraging Predators: One Dimensional Model

Numerous observations of animals searching for preys show that they often use an intermittent strategy, or saltatory behaviour. During their fast phases, their velocity is approximately constant and their direction almost fixed: the turning angle between two ballistic phases is usually small. This example inspired the first, one dimensional model of intermittent search [4]. In this model, the searcher moves along an axis $0x$. It can be in two dynamical regimes: a *search regime*, labeled by 1 and modeled by a 1-dim diffusion with diffusion coefficient D, and a *move regime* 2, modeled by a ballistic motion with constant velocity v. The duration Ti of each regime i is an exponential stochastic variable with mean value τ_i. Preys are supposed to be regularly distributed along $0x$. They are found during regime 1 only, as soon as the searcher reaches them. The first arrival time of the searcher on a prey, starting in regime 1 from a given position, obeys the Chapmann-Kolmogorov backward equation [14] and can be computed classically. Its average on initial position yields the average search time S. In the low target density limit, it is found [4] that S scales as the inter-distance of the preys L, whereas it would scale as L^2 for a purely diffusive search. Thus it is clear that intermittence is favourable for the search.

Furthermore, it is shown that there is no global minimum of the mean search time S for finite values of τ_1 and τ_2: in order to minimize S both τ_1 and τ_2 should be as small as possible and satisfy the scaling law $\tau_2/\tau \propto (\tau_2/\tau)^{\nu}$, where $\tau = D/v^2$ is a characteristic time of the process, and the exponent ν depends on the conditions: if $\tau_1 \gg \tau$, $\nu = 2/3$: then $\tau_1 \gg \tau_2$ and the searcher spends more time searching than moving, which seems quite natural. But if $\tau_1 \ll \tau$, $\nu = 3/5$, and $\tau_1 \ll \tau_2$: then, the searcher *spends more time moving than searching* - although it can not find any target while moving! This non intuitive conclusion is not the only surprising one in these studies, but the results are in good agreement with experimental data on various, very different species including fishes, birds insects and lizards [2, 3, 4]. Thus in this example intermittence is an efficient strategy for minimizing the search time.

Memoryless Search In N Dimensions Space

The previous model is clearly limited by its simplifying assumptions. In particular, it should be extended to more general spaces, or in practice, to 2-dim and 3-dim spaces. On the other hand, the previous 1–dim searcher has space memory, since it remembers its velocity from one move phase to the other one. However, such memory can be impossible for a searcher with low abilities (molecules or elementary organisms). We now present a n-dim model for a searcher without such memory [9], which could be used, for instance, for addressing a diffusion limited chemical reaction where a species B reacts on immobile, catalytic sites A: $A + B \rightarrow A + C$.

The searcher now moves in a n-dim finite space V containing a target A. It alternates two regimes: during regime 1, it experiments a n-dim diffusion with coefficient D allowing for immediate detection when the searcher reaches the target; during regime 2, it follows a non-reactive ballistic motion with constant velocity v and arbitrary direction, randomly reoriented at each phase 2. The waiting times in both regimes are again exponential, independent variables with mean waiting times τ_i , $i = 1, 2$. Then the Chapman-Kolmogorov backward equations for the search times $S_1(x)$ and $S_2(x, v)$, starting from position x and velocity v in regimes 1 or 2, cannot be solved explicitly. However, it can be shown that they can be replaced by two diffusion-like equations which can be solved analytically in spherical geometry. For instance, in 2 dimensions, we consider that region V is a circle of radius b, and the target is a concentric circle of radius $a \ll b$. Then the explicit solution allows to compute the mean search time S and to treat the optimization problem analytically. It results that

- if $a < b \ll D/v$ intermittence is not favorable for minimizing the search time, and the best strategy is to stay in the diffusive regime indefinitely,
- if $a \ll D/v \ll b$ intermittence is (moderately) favorable. It allows minimizing the search time for finite values of τ_1 and τ_2 of the mean durations of the diffusion and ballistic regimes such that the diffusion and ballistic lengths are approximately equal. Then the minimum search time S_{min} is approximately half the search time in a purely diffusive regime: $S_{min}/S_{diff} \approx 0.5$.
- if $D/V \ll a \ll b$ there is also an optimal intermittent strategy with now $S_{min}/S_{diff} \ll 1$ if $b/a \gg 1$.

Diffusion And Teleportation

Protein Search For A Specific Site On DNA: One-Dim Model

Genetics events often depend on the interaction of a protein, or restriction enzyme, with DNA: the restriction enzyme cleaves the DNA chain at a specific sequence. This process it is of great importance, since it controls the kinetics of many genetics events. However, its mechanism remains mysterious. In fact, the cleaving rate is very fast, several orders of magnitude faster than if it was driven by diffusion. It has been suggested that this translocation process could alternate phases of diffusion along the chain, and jumps towards other sites of DNA.

Some years ago, we presented [4] a simple stochastic model for the localization of a protein on DNA based on these mechanisms. We assumed that the duration of each regime, 1-dim diffusion and 3-dim excursion, is an exponential, independent stochastic variable. Since the DNA molecule has generally a very intricate form, a small jump in space may correspond to a very large displacement along the DNA strand. Thus we can suppose that after an excursion in the bulk of any duration, the protein is randomly relocated on DNA: this is *teleportation*.

We showed that this model provides an optimal search strategy which can explain the observed fast rate of the process. In fact, the minimum search time is obtained when the mean durations of both phases are approximately equal, another non intuitive result of intermittence that was previously anticipated by simple considerations [15]. Furthermore, we proved [6] that the minimum search time S_{min} , compared to the search time S_{diff} in case of pure diffusion along the chain, is

$$\frac{S_{min}}{S_{diff}} = \frac{6}{L}\sqrt{D\tau} \tag{1}$$

(1). where L is the length of the DNA chain, τ is the average duration of each phase and D is the diffusion coefficient along the chain. Thus for a long DNA strand, the intermittent, 3-dim mediated strategy is extremely efficient and can accelerate the processes significantly. This conclusion and other results of the model agree with experimental observations.

Teleportation in n dimensions

It is possible to generalize the 1-dim model in a general, n-dim space. Then, teleportation, i.e. random relocation of the searcher anywhere in the (finite) accessible region V, independently of the jump duration, may be questioned. Nevertheless, it can be justified in various examples, in particular, if the average scale of the jumps is much larger than the average distance between targets. On the other hand, it is not necessary to assume that the duration of teleportation is an exponential variable, but only that it has a finite average τ_2. Supposing that the diffusive phase 1 has an exponential duration with average τ_1, it can be proved [16] that the mean search time is

$$S = \frac{1-\alpha-\langle\tilde{j}_1(1/\tau_1)\rangle_V}{\alpha+\langle\tilde{j}_1(1/\tau_1)\rangle_V}(\tau_1+\tau_2) \tag{2}$$

where α is the ratio $v(A)/v(B)$ of the volumes of A and B, and $\langle\tilde{j}_1(1/\tau_1)\rangle_V$ is the Laplace transform of the search time in regime 1 averaged on region V. If V and A are two concentric spheres with respective radius b and a, this quantity can be calculated exactly [16]. Then it is seen that in order to minimize the search time, τ_2 should be as small as possible, which is obvious; as for τ_1, two situations are possible, according to the value of τ_2 :

- if $\tau_2 > \tau_{2c}$ (τ_{2c} being a critical value that can be computed explicitly) S decreases continuously with τ_1 . Then the shortest search time is obtained for a purely diffusive search, and intermittence is not favourable.
- if, on the contrary, for $\tau_2 < \tau_{2c}$, the search time is minimum for a finite, non trivial value of τ_1: then intermittence is an efficient strategy which can make the search much shorter than pure diffusion.

These conclusions can be generalized to more general situations. It should be remarked that teleportation does not make use of any space memory of the searcher: this point is important for the following discussion.

ORDER, CHANCE AND STRATEGY

The previous models assume the searcher to have more or less developed abilities. We have seen that the technical performances of their detectors play an important role on the behavior to be adopted. On the other hand, in order to follow a determined trajectory the searcher needs to have space memory. It is the case in our first, 1-dim model: the searcher remembers its initial velocity when it begins a new ballistic phase. Similarly, it would need time memory if the durations of the dynamic regimes would obey non-exponential probability distributions, which will be considered later. We now discuss these memory effects.

Memory And Deterministic Exploration

Let us assume that the searcher is able to pre-determine certain trajectory Γ for exploring its territory, and to follow it between two successive scanning phases. Will the search be more efficient than if it randomly re-orients its trajectory after each search phase, as it was assumed in the n-dim ballistic model of previous section ?

In fact, an efficient trajectory Γ should obviously avoid self-crossing, since scanning the same region several times is time-consuming and useless for immobile targets and good detectors. If random is introduced in the "fast" trajectory because the searcher has no memory, the possibility of such self crossings cannot be avoided and the search time necessarily increases. This effect was indeed checked by numerical simulations in the 2-dim ballistic model [10].

On the other hand, multiple scanning of the same spot could have interest if we consider a new kind of imperfect detectors, i.e. detectors which can miss a target actually present inside their range. Then it might be useful to return at the same place. Nevertheless, it can be shown [17] that, if the detector has no time memory, the search time is always shorter for a self avoiding trajectory, and thus *a deterministic strategy* (without multiple scanning) during the fast regime *is always more efficient than a stochastic strategy*. This conclusion does not necessarily hold if the detector has time memory and if the probability to find the target increases at each passage.

Memory And Non Exponential Waiting Times

Searchers with time memory can chose non exponential waiting times for one or both regimes. Specific models should then be considered, since the waiting time laws should depend strongly on the example under study. In some cases, for instance teleportation (see previous section), the mean search time only depends on the mean duration of the regime. In general, however, this is not true. As a typical example, we considered the case of deterministic waiting times τ_1 and τ_2 when alternating diffusion and ballistic motion [9]. We have shown numerically that the search time is then slightly shorter than for exponential waiting times with the same mean durations.

This result seems intuitively justified: if the search is shorter for some definite values of τ_1 and τ_2 , it is reasonable to expect that efficient stochastic laws should have these average durations, and, furthermore, that the dispersion around these optimal values should increase the search time, or at least not decrease it.

Order And Disorder In Target Distribution.

A completely ordered target distribution is typically represented by a regular lattice, or in one dimension, by equally spaced targets. We used such a distribution in several works, or studied equivalently one target in a finite space with reflecting boundaries.

On the other hand, a typical disordered distribution is Poisson distribution, where each target is located independently of the other ones: different distributions should be motivated by specific examples. We studied Poisson distribution [18] in the case of alternating diffusion and ballistic motion. Comparing the results for distribution of equal density, both in numerical simulations and in some analytical approximations, we found [18] that the search time is slightly shorter for an ordered distribution.

Thus, once more the spatial memory implied by the regular distribution proves to be favorable to the efficiency of the search, without, however, producing a significant advantage for the searcher.

CONCLUSION

In this paper we have given new evidences, in particular for processes with teleportation, that intermittence can be a very efficient strategy in the search of hidden targets, thus confirming former results. In many circumstances, although not always, it allows decreasing the search time significantly, even when restricting the processes to simple diffusions and ballistic motions. Clearly, intermittence does not exclude using a larger class of processes, such as Levy flights, which have proved to be efficient in various cases: conjugating them with intermittence should presumably give excellent results in certain situations.

It has been shown that the interest of intermittence is in great part due to the imperfection of detectors and depends on the abilities of the searcher. As suggested by intuition, larger abilities can lead to better performances. In particular, space memory allows using a deterministic fast motion, which gives a shorter search time than stochastic fast motion. A similar advantage is obtained by imposing convenient deterministic waiting times to the alternating regimes, in case the searcher is able of time memory. On the other hand, an ordered distribution of the targets allows for a shorter search time than a disordered one - but this may be an inconvenient for the targets, which could, in this case, prefer disorder!

Thus we should conclude that the search is, in principle, more efficient using a relevant strategy than relying on random, which seems better for the sake of order and organization! We have given some quantitative arguments for this assertion which seems quite natural, but it should be pointed out that in all cases which could be treated accurately, the advantage of strategy is in no way determinant, and the search time is of the same order than in convenient random behaviors. In practice, the last remark leads to temper our conclusion: since finding the best strategy should depend on the specific problem to be treated, it is time and energy consuming. It depends on the abilities of the searcher, and it does not yield a definitive advantage. Thus it can be far simpler, and eventually preferable, to use a standard random behavior, if such standard behavior is available for the kind of search to be addressed. This can be the basis for further applied researches.

REFERENCES

1. J. R. Frost and L. D. Stone, *http.//www.rdc.uscg.gov/reports/2001/cgd1501dpexsum.pdf*
2. W.J. Bell, Searching behaviour, *the behavioural ecology of finding resources*, Chapman and Hall Animal Behaviour Series, 1991
3. W.J. O'Brien, H.I. Browman and B.I. Evans, *American Scientist* **78**, 152–157 (1990)
4. O. Benichou, M. Coppey , M. Moreau, P.-H. Suet and R. Voituriez , *Phys. Rev. Lett.* **94**, 198101–198105 (2005)
5. G.M. Viswanathan, S.V. Buldyrev , S. Havlin et al., *Nature* **401**, 911–914 (1999)
6. M. Coppey, O. Bénichou, R. Voituriez and M. Moreau, *Biophys. J.* **87**, 1640–1649 (2004)
7. O. Bénichou, M. Coppey, M. Moreau, P.-H. Suet and R. Voituriez, *J. Phys.: Condens. Matter* **17**, S4275–S4286 (2005)
8. O. Bénichou, M. Coppey, M. Moreau and R. Voituriez, *Europhys. Lett.* **75**, 349–354 (2006)
9. O. Bénichou, C. Loverdo, M. Moreau and R. Voituriez, *Phys. Rev. E.* **74**, 020102 (2006)
10. O. Bénichou , C. Loverdo, M. Moreau and R. Voituriez, *J. Phys.: Condens. Matter* **19**, 065141 (2007)
11. M. F. Shlesinger, *Nature* **443**, 281–282 (2006)
12. I. Eliazar, T. Koren and J. Klafter, *J. Phys.: Condens. Matter* **19**, 065140 (2007)
13. G. Oshanin, H.S. Wio, K. Lindenberg and S.F. Burlatsky *J. Phys.: Condens. Matter* **19**, 065142 (2007)
14. S.A. Redner, *A guide to first passage time processes*, Cambridge University Press 2001
15. M. Slutsky and L.A. Mirny, *biophys.J.* **87**, 4021–4035 (2004)
16. O. Bénichou, M. Moreau, P. H. Suet and R. Voituriez, Intermittent search and teleportation, *submitted*
17. M. Moreau, O. Bénichou, C.Loverdo and R. Voituriez, Random and strategy in stochastic search processes, *in preparation*
18. M. Moreau, 0. Bénichou, C. Loverdo and R. Voituriez, *EPL* 77 20006 (2007)

Noise-Induced Phase Transitions: Effects of the Noises' Statistics and Spectrum

Roberto R. Deza*, Horacio S. Wio† and Miguel A. Fuentes**,‡

*Departamento de Física, Facultad de Ciencias Exactas y Naturales, Universidad Nacional de Mar del Plata, Deán Funes 3350, 7600 Mar del Plata, Argentina.
†Instituto de Física de Cantabria (Universidad de Cantabria and CSIC), Av. de los Castros s/n, E-39005 Santander, Spain.
**Centro Atómico Bariloche (Comisión Nacional de Energía Atómica), Av. Exequiel Bustillo 9500, 8400 S. C. de Bariloche, Argentina.
‡Santa Fe Institute, 1399 Hyde Park Road, Santa Fe, New Mexico 87501, USA.

Abstract. The local, uncorrelated multiplicative noises driving a second-order, purely noise-induced, ordering phase transition (NIPT) were assumed to be Gaussian and white in the model of [Phys. Rev. Lett. **73**, 3395 (1994)]. The potential scientific and technological interest of this phenomenon calls for a study of the effects of the noises' statistics and spectrum. This task is facilitated if these noises are dynamically generated by means of stochastic differential equations (SDE) driven by white noises. One such case is that of Ornstein–Uhlenbeck noises which are stationary, with Gaussian pdf and a variance reduced by the self-correlation time τ, and whose effect on the NIPT phase diagram has been studied some time ago. Another such case is when the stationary pdf is a (colored) Tsallis' *q–Gaussian* which, being a *fat-tail* distribution for $q > 1$ and a *compact-support* one for $q < 1$, allows for a controlled exploration of the effects of the departure from Gaussian statistics. As done before with stochastic resonance and other phenomena, we now exploit this tool to study—within a simple mean-field approximation and with an emphasis on the *order parameter* and the "*susceptibility*"—the combined effect on NIPT of the noises' statistics and spectrum. Even for relatively small τ, it is shown that whereas fat-tail noise distributions ($q > 1$) counteract the effect of self-correlation, compact-support ones ($q < 1$) enhance it. Also, an interesting effect on the susceptibility is seen in the last case.

Keywords: noise-induced phase transition, mean field, order parameter, q–Gaussian
PACS: 05.40.Ca, 05.70.Fh, 05.70.Ln, 47.20.Hw, 47.20.Ky, 64.60.-i

INTRODUCTION

As it is known, the role of temperature in equilibrium phase transitions can equally well be played by any set of spatially uncorrelated Gaussian white noises, regardless of their origin, provided that they have the same mean and variance and act *additively* on the system. Under the influence of *multiplicative* noises, the extended-system correlative of the phenomenon of noise-induced transitions (NIT) [1], namely a purely noise-induced *phase* transition (NIPT), may occur. A comprehensive account of the many ways the phenomenon may take place can be found in [2, 3]. However, for consistency with our previous work [4, 5], we shall restrict here to the 1994 model by Van den Broeck, Parrondo and Toral (VPT) [6, 7], in which they proposed the following mechanism for the NIPT:

1. An initially unimodal pdf gets rapidly destabilized towards a multimodal one.
2. If spatial coupling is strong enough, the new states couple to form ordered domains that might subsequently coarsen and grow.

Aiming at finding a nonequilibrium phase transition arising solely from the *multiplicative* nature of the noise, and characterized (in the limit of an infinite system) by *ergodicity breakdown* (only microstates compatible with the macroscopic broken symmetry should appear) and *multiple* steady state probability distributions $P^{\rm st}(\{x_i\})$, the authors in [6] set up a model that is the lattice version of a scalar reaction–diffusion model submitted to *multiplicative* local noises $\eta_i(t)$. The system's state at (continuous) time t is given by $N = L^d$ stochastic variables $x_i(t)$ defined at the sites $\mathbf{r}_i$ of a hypercubic lattice, and obeying a system of coupled ordinary stochastic differential equations (SDE)

$$\dot{x}_i = f(x_i) + g(x_i)\eta_i + \Delta_i, \quad \text{with} \quad f(x) = -x(1+x^2)^2 \quad \text{and} \quad g(x) = 1+x^2. \tag{1}$$

CP913, *Nonequilibrium Statistical Mechanics and Nonlinear Physics, XV Conference*
edited by O. Descalzi, O. A. Rosso, and H. A. Larrondo

Here $\Delta_i = (D/2d)\sum_{j\in n(i)}(x_j - x_i)$ is a discretization of the Laplacian $\nabla^2 x(\mathbf{r},t)$, D is the lattice version of the diffusion coefficient, and $n(i)$ stands for the set of $2d$ nearest neighbors of $\mathbf{r}_i$. The $\eta_i(t)$ are uncorrelated, Gaussian and white

$$\langle \eta_i(t)\rangle = 0 \quad , \quad \langle \eta_i(t)\eta_j(t')\rangle = \sigma^2 \delta_{ij}\delta(t-t'). \tag{2}$$

As argued in [6], establishing a phase transition rigorously is a difficult task even in equilibrium, where at least the explicit form of the steady-state pdf is known. Hence, one cannot resort to the traditional techniques from equilibrium statistical physics. However, the oldest and simplest ansatz that can reproduce (albeit not always faithfully) ergodicity breakdown, namely Weiss' mean field, can be readily adapted to this nonequilibrium situation.

The method proposed by [6] can be sketched as follows: By integrating the multivariable FPE over all variables except x_i and using the isotropy and translational invariance of the steady-state properties, one gets an *exact* (but implicit) steady-state equation for the *one-site* pdf in terms of the steady-state conditional average $E(y)$ of $y \in n(i)$, given the value of $x \equiv x_i$. To determine the unknown function $E(y)$, they introduce Weiss' mean-field (MF) approximation: to neglect fluctuations in neighboring sites, so that $E(y) = \bar{x}$ independent of y. The value of $\bar{x}$ then follows from the self-consistency condition

$$\bar{x} = m, \quad \text{with} \quad m \equiv \langle x\rangle_{\bar{x}=\text{const}} = \int_{-\infty}^{+\infty} x P^{\text{st}}(x,\bar{x})\,dx = F(\bar{x}). \tag{3}$$

When this nonlinear equation has multiple solutions, there are several corresponding steady state probabilities $P^{\text{st}}(x)$ and the MF approximation predicts a phase transition with ergodicity breakdown (usually accompanied by symmetry breakdown). If, for example, f is *odd* and g *even*, then any realization $\{x_i(t)\}$ is equally probable as $\{-x_i(t)\}$ and one should expect $\langle x\rangle = 0$. However, with the appearance of multiple solutions, this symmetry need not be fulfilled by the separate solutions, and one typically finds "ordered" phases with an order parameter $m \equiv |\langle x\rangle| \neq 0$.

Since $F(\bar{x})$ is odd, $\bar{x} = 0$ is always a root of Eq. (3). Hence, if the phase transition is second-order, its *phase boundary*—which provides rich *qualitative* information—lies where this root becomes unstable, i.e. where $(dF/d\bar{x})|_{\bar{x}=0} = 1$. Now, the results of measurements and numerical simulations are correlations, order parameters and susceptibilities. Although the MF approximation is unable to predict the former ones, it yields predictions of the remaining two that can thus serve as a guide for numerical and (prospective) real experiments. The "susceptibility" we look upon is the MF correlative of the one defined in [6, 7] for numerical simulations:

$$\chi \equiv \frac{1}{\sigma^2}\left[\int_{-\infty}^{+\infty} x^2 P^{\text{st}}(x,\bar{x})\,dx - m^2\right]. \tag{4}$$

Given the exploratory character of this work, we shall perform a still simpler MF approximation consisting in replacing in Eq. (1) $\Delta_i \to \bar{\Delta}_i \equiv D(\bar{x} - x_i)$, where $\bar{x}$ is a parameter that will be determined self-consistently. Since the N SDEs get decoupled, we hereafter consider a generic one: $\dot{x} = f(x) + g(x)\eta + D(\bar{x} - x)$. We postpone the discussion of the numerical implementation of the MF method until we have described the process with q–Gaussian distribution. For the sake of comparison with the results obtained in that case, we illustrate the findings in [6, 7] by plotting in Fig. 1 the order parameter m and the "susceptibility" χ as functions of the coupling D and the white noise intensity σ^2. That the NIPT is a *combined* effect of noise and coupling is evidenced by the existence of a threshold value of D. Note also that the deterministic system is *disordered* however strong the coupling and that, of course, the NIPT is reentrant as a function of σ^2.

Early in the long list of works inspired by [6, 7], two of us were involved in the study of the consequences of the multiplicative noise being Ornstein–Uhlenbeck (OU) [4, 5]. In this case, Eq. (2) is replaced by $\langle \eta_i(t)\rangle = 0$ and $\langle \eta_i(t)\eta_j(t')\rangle = \delta_{ij}(\sigma^2/2\tau)\exp(-|t-t'|/\tau)$, where the $\eta_i(t)$ obey

$$\tau\dot{\eta}_i = -\eta_i + \xi_i, \quad \text{with} \quad \langle \xi_i(t)\xi_j(t')\rangle = \sigma^2\delta_{ij}\delta(t-t'). \tag{5}$$

Although the colored noises $\{\eta_i\}$ make the process $\{x_i\}$ non-Markovian, some approximations (interpolation schemes) render a *Markovian* (i.e. tractable) process, still capturing some of the essential features. Of course, at the price of adding a new unsystematic approximation to the MF one. However, we dispose of a neat control parameter, namely τ, to compare with the white-noise case. Figure 2 is the corresponding plot for $\tau = 0.1$. For further details, see [4, 5].

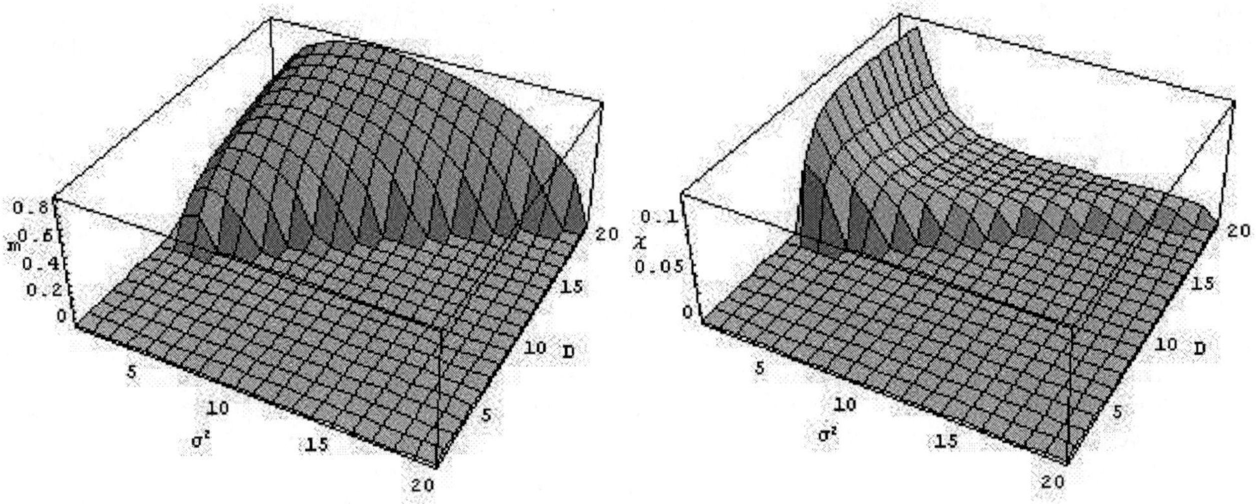

FIGURE 1. Case of white multiplicative noise. a) order parameter and b) susceptibility as functions of coupling D and white-noise intensity σ^2.

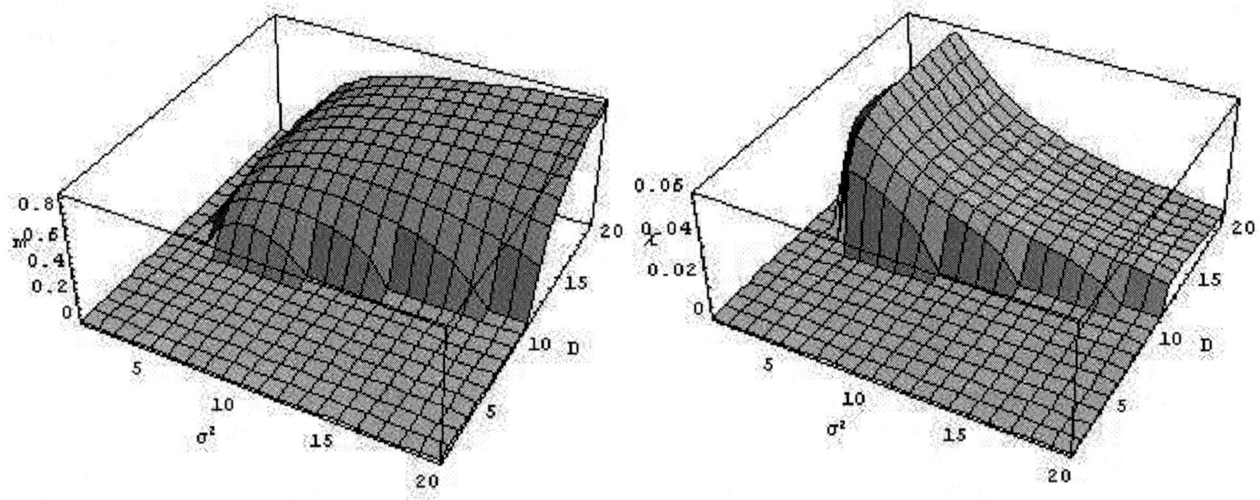

FIGURE 2. Case of OU multiplicative noise, with $\tau = 0.1$. a) order parameter and b) susceptibility as functions of coupling D and white-noise intensity σ^2.

COMBINED EFFECT OF SPECTRUM AND STATISTICS

One possible generalization of Eq. (5) is

$$\tau\dot{\eta} = -\frac{d}{d\eta}V_q(\eta) + \xi(t), \quad \text{with} \quad V_q(\eta) = \left[\frac{\sigma^2}{2\tau(q-1)}\right] \ln\left[1 + \frac{\tau(q-1)}{\sigma^2}\eta^2\right], \tag{6}$$

proposed some time ago as model for correlated diffusion [8]. As it occurred previously with the OU noise, this generalization provides a device to explore statistics effects by varying just one parameter (namely q, at constant τ and σ^2; note that the proper control parameter here is $\tau(q-1)/\sigma^2$, contrarily to the OU case in which it was τ/σ^2).

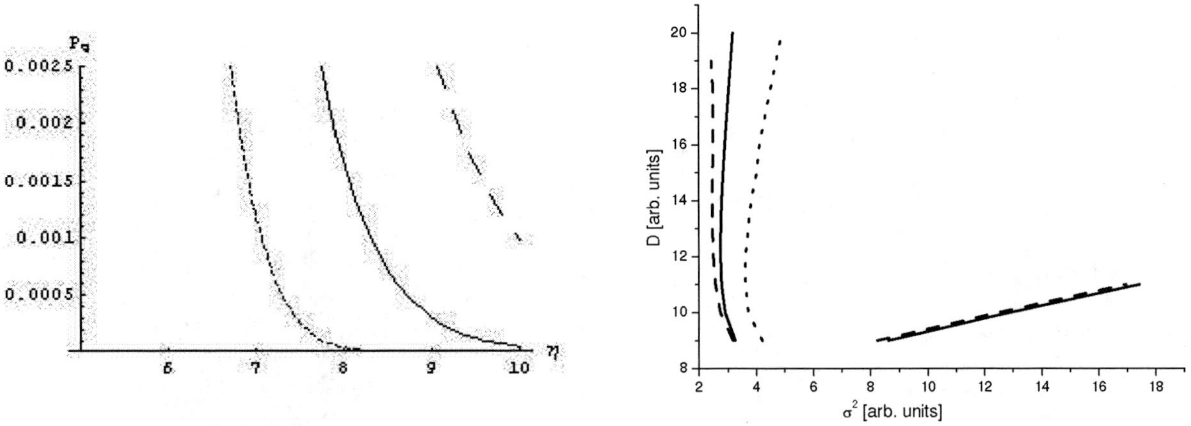

FIGURE 3. a) Detail of the unnormalized $P_q^{\rm st}(\eta)$ near its cutoff for $q = 0.9$, and b) phase boundaries in (σ^2,D) plane, for $\tau = 0.1$. Dotted line: $q = 0.9$; solid line: $q = 1.0$. dashed line: $q = 1.1$.

The stationary properties of the noise η, including the time-correlation function, have been studied in [11] so here we summarize the main results. The stationary probability distribution is given by

$$P_q^{\rm st}(\eta) = \frac{1}{Z_q}\left[1+\frac{\tau}{\sigma^2}(q-1)\eta^2\right]^{\frac{1}{1-q}}, \tag{7}$$

where Z_q is the normalization factor. This distribution can be normalized only for $q < 3$. The first moment $\langle\eta\rangle = 0$ is always equal to zero, and the second moment

$$\langle\eta^2\rangle = \frac{\sigma^2}{\tau(5-3q)} \tag{8}$$

is finite only for $q < 5/3$, being larger than $\frac{\sigma^2}{2\tau}$ for $q > 1$. For $q < 1$ the distribution has a cut-off, and it is only defined for $|\eta| < \eta_c \equiv \sqrt{\frac{\sigma^2}{\tau(1-q)}}$ (Fig. 3a). Finally, the correlation time τ_q of the stationary regime of the process $\eta(t)$ diverges near $q = 5/3$ and it can be approximated over the whole range of values of q as $\tau_q \approx 2\tau/(5-3q)$. Clearly, when $q \to 1$ we recover the limit of η being a Gaussian colored noise, namely the Ornstein–Uhlenbeck process $\xi_{OU}(t)$, with correlations $\langle\xi_{OU}(t)\xi_{OU}(t')\rangle = \frac{\sigma^2}{2\tau}\exp-|t-t'|/\tau$ and probability distribution $P^{\rm st}(\xi_{OU}) = Z^{-1}\exp-\frac{\tau}{\sigma^2}\xi_{OU}^2$. This process gives rise to interesting phenomena when it drives different kinds of nonlinear systems [9, 10, 11, 12, 13, 14, 15, 16? , 17, 18].

The expression for $P^{\rm st}(x,\bar{x})$ arises from a consistent Markovian approximation based on phase-space functional integration (details to be published elsewhere). The numerical implementation of the MF method compromises precision and speed: infinite integrals like those of Eqs. (3) and (4) [unless $q < 1$, see below] are performed by means of a 160 pt. Gauss–Hermite algorithm, whereas the finite integrals in the exponent of the stationary pdf are performed by means of a 96 pt. Gauss–Legendre algorithm. To determine the phase boundary, the Newton–Raphson algorithm is used (unless the corresponding function is badly conditioned, in which case a succession of finer sweeps is resorted to). In the $q < 1$ case, when the integrals over x in Eqs. (3) and (4) are bounded because of the bound in η, it would be faster to solve for the x-bounds using the Newton–Raphson algorithm and then applying the 96 pt. Gauss–Legendre algorithm. However, the function seems not to be well conditioned for this algorithm. Hence we resort to naive integration inside a `while` loop.

The integrals performed in the aforementioned way seem to be precise enough, except for large D/σ^2 where unphysical ordered states appear. For that reason, we have limited the exploration of the (σ^2,D) plane to $(\sigma^2 > 2)$ (except in Fig. 1, where the analytical expression for $P^{\rm st}(x,\bar{x})$ is used). Figures 4 and 5 exhibit the order parameter and susceptibility results respectively for $q = 1.1$ and $q = 0.9$ in the $\tau = 0.1$ case, showing that fat-tail noise distributions

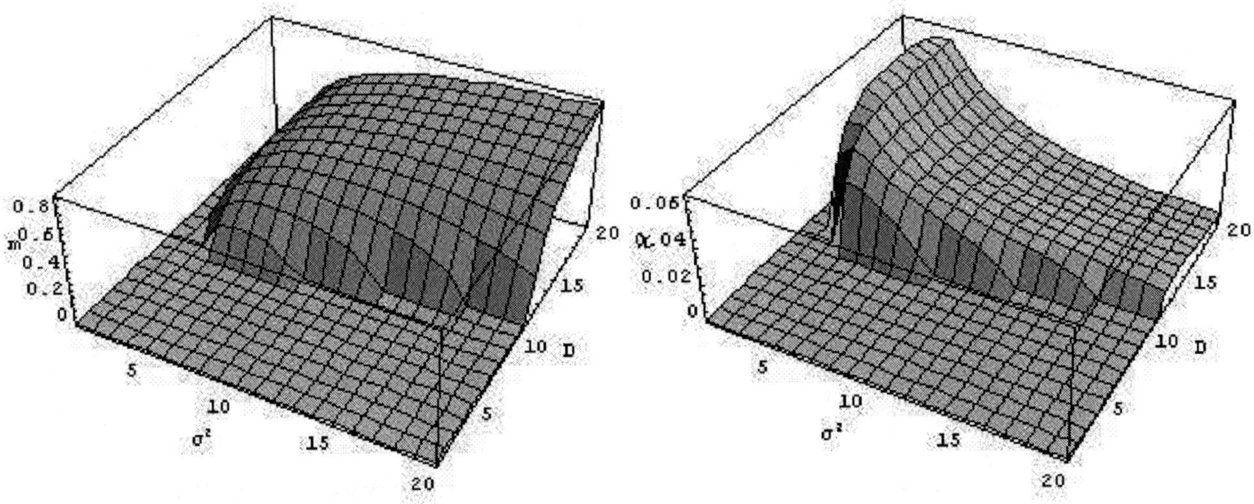

FIGURE 4. Case of non-Gaussian multiplicative noise, with $\tau = 0.1$ and $q = 1.1$. a) order parameter and b) susceptibility as functions of coupling D and white-noise intensity σ^2.

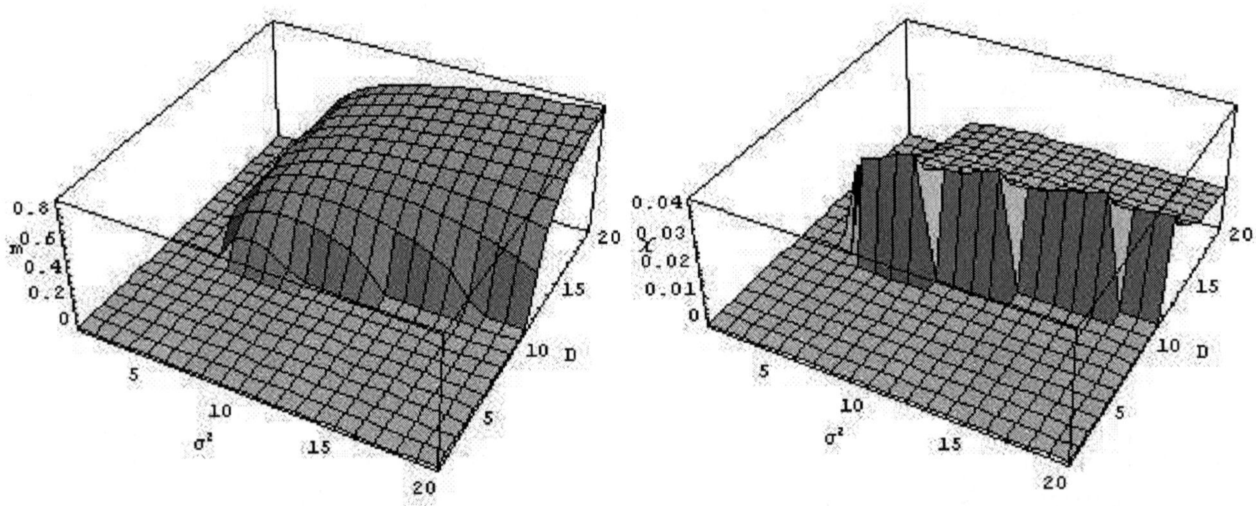

FIGURE 5. Case of non-Gaussian multiplicative noise, with $\tau = 0.1$ and $q = 0.9$ a) order parameter and b) susceptibility as functions of coupling D and white-noise intensity σ^2.

($q > 1$) *counteract* the effect self-correlation (namely, they *advance* the ordering boundary as σ^2 is increased at constant D), and compact-support ones ($q < 1$) *enhance* it (they *retard* the ordering boundary). Particular interest rises the effect of ($q < 1$) multiplicative noises on the susceptibility: as seen in Fig. 5b, it shifts from being larger on the *ordering* boundary to being larger on the *disordering* boundary.

DISCUSSION

As warned earlier, the character of this work is exploratory, and the value of its results is to be orientative of what to expect with numerical integration of the system of SDEs, as well as with more refined MF ansätze and/or interpolation schemes. In order not to run into the "forbidden" σ^2 region, and given the *ordering* effect of $q > 1$ multiplicative noises, we have chosen to limit our exploration to $\approx 10\%$ around $q = 1$, just in order to discover trends. But in principle, our consistent Markovian approximation allows to explore the whole meaningful range of q values.

ACKNOWLEDGMENTS

HSW thanks the European Commission for the award of a *Marie Curie Chair* at the Universidad de Cantabria, Spain. RD thanks support by ANPCyT, CONICET and UNMdP.

REFERENCES

1. W. Horsthemke and R. Lefever, *Noise-Induced Transitions*, Springer, Berlin, 1984.
2. J. García-Ojalvo and J. M. Sancho, *Noise in Spatially Extended Systems*, Springer, New York, 1999.
3. J. M. Sancho and J. García-Ojalvo, in *Stochastic Processes in Physics, Chemistry, and Biology*, edited by J. A. Freund and T. Pöschel, Lecture Notes in Physics 557, Springer, Berlin, 2000, pp. 235–247.
4. S. E. Mangioni et al., *Phys. Rev. Lett.* **79**, 2389–2393 (1997).
5. S. E. Mangioni et al., *Phys. Rev. E* **61**, 223–232 (2000).
6. C. Van den Broeck et al., *Phys. Rev. Lett.* **73**, 3395–3398 (1994).
7. C. Van den Broeck et al., *Phys. Rev. E* **55**, 4084–4094 (1997).
8. L. Borland, *Phys. Lett. A* **245**, 67–72 (1998).
9. M. A. Fuentes et al., *Physica A* **295**, 114–122 (2001).
10. F. J. Castro et al., *Phys. Rev. E* **64**, 051105 (2001).
11. M. A. Fuentes et al., *Physica A* **303**, 91–104 (2002).
12. J. A. Revelli et al., *Physica D* **168-169**, 165–170 (2002).
13. M. A. Fuentes et al., *Fluct. and Noise Lett.* **3**, L365 (2003).
14. H. S. Wio and R. Toral, *Physica D* **193**, 161–168 (2004).
15. S. Bouzat and H. S. Wio, *Eur. Phys. J. B* **41**, 97–106 (2004).
16. H. S. Wio, in *Nonextensive Entropy. Interdisciplinary Applications*, edited by M. Gell-Mann and C. Tsallis, OUP, Oxford, 2004, pp. 177–194.
17. S. Bouzat and H. S. Wio, *Physica A* **351**, 69–78 (2005).
18. H. S. Wio, *Europhys. News* **36**, 197–201 (2005).

Chaos Synchronisation and Message Extraction in Optical Chaos Communications

K. A. Shore, M. W. Lee, J. Paul, Y. Hong, A. Murakami, I. Pierce and P. S. Spencer

University of Wales, Bangor, School of Electronics, Bangor, LL57 1UT, Wales, UK

Abstract. This paper provides an overview of experimental and theoretical work undertaken in the authors' laboratory and directed at the development of optical chaos communications technology. The specific focus of the paper is on the use of external cavity edge-emitting semiconductor lasers for this purpose.

Keywords: optical chaos, chaos synchronisation, chaos communications
PACS: 05.45.Gg, 05.45.Vx, 42.62.-b, 42.65.Sf, 42.79.Sz

INTRODUCTION

Several approaches have been adopted for the practical implementation of the concept of performing digital communications using synchronised chaotic lasers [1]. Due to their direct compatibility with existing optical fibre communications technology, semiconductor lasers (laser diodes) have attracted the attention of many workers seeking to exploit optical chaos in communications [2]. The specific focus of the present paper is on the use of external cavity edge-emitting semiconductor lasers for this purpose. In this configuration the laser source is subject to optical feedback from an external mirror. An extensive treatment of the dynamical behaviour of external cavity semiconductor lasers is provided in a recent research monograph [3]. The challenge faced in seeking to exploit this behaviour is to show that the chaotic dynamics can be generated in a robust manner which will allow engineering deployment. That has been the thrust of experimental work undertaken within the research group of the present authors and is the theme of the present paper.

The most fundamental requirement for effecting communications using a chaotic carrier is that of chaos synchronisation. The counter-intuitive proposal of Ott, Greboggi and Yorke [4] that such synchronisation could be effected stimulated the pioneering experimental activity of Roy et al [5] and led to subsequent experimental realisations in a number of laser systems and, in particular, the first demonstration of chaos synchronisation in external cavity laser diodes by Sivaprakasam et al. [6]. The achievement of high-quality chaos synchronisation opens the opportunity for message transmission using a chaotic carrier. The simple concept here is that a message added to a chaotic carrier generated in a transmitter laser can be recovered using a receiver laser in chaotic synchrony with the transmitter laser. That concept does translate into a practical demonstration but simultaneously opens quite subtle issues concerning the precise mechanism for message extraction at the receiver.

SYNCHRONISATION

The general concept of synchronisation has a long historical tradition and its modern manifestations have been summarised in a major work [7]. The general context for chaos synchronisation in external cavity semiconductor lasers has been delineated by Sivaprakasam and Masoller [8] to which attention is directed for further references. Here we outline experimental work on chaos synchronisation in external cavity lasers from the perspective of implementing optical chaos communications.

The generic experimental configuration includes a transmitter or master laser and a receiver or slave laser. Unidirectional optical coupling between the transmitter and receiver laser enables synchronisation. Account is taken of the 'time-of-flight' delay between the physically separated lasers. Two options arise in respect of the receiver laser: it may be configured as an external cavity laser and enter the chaotic regime via optical feedback from the external cavity mirror or else it may be a stand alone laser whose dynamics is affected via the optical coupling from the chaotic

CP913, *Nonequilibrium Statistical Mechanics and Nonlinear Physics, XV Conference*
edited by O. Descalzi, O. A. Rosso, and H. A. Larrondo

transmitter laser. The former case is generally termed a closed-loop configuration whilst the latter is referred to as an open loop configuration. In such experiments several regimes of nonlinear dynamics can be accessed. A specific regime which is convenient for careful examination of laser dynamics is so-called low-frequency fluctuations (LFF) where the laser output power exhibits a sudden drop-out followed by a relatively long recovery time. A number of forms of chaos synchronisation have been demonstrated including:

- **Two-mode synchronisation**

Here the aim is to achieve the analogue of wavelength division multiplexed (WDM) operation. Strong optical injection and optical frequency matching have been used to effect an experimental demonstration of two-mode synchronisation using a multi-mode external-cavity chaotic transmitter laser and two single-mode stand-alone receiver lasers [9].

- **Cascaded and contemporaneous synchronisation**

As described below, synchronising a series of chaotic lasers becomes essential for message relay in optical chaos networks and an experimental demonstration of the synchronisation of three chaotic diode lasers was reported in [10].

As may be expected in such an arrangement, time of flight lag times appear in the synchronization. It was shown in [11] that the effect of time-of-flight delays may be nullified. That observation has been supplemented by the observation of so-called contemporaneous synchronization: zero time lag, complete-chaos synchronization [12]. The synchronization of three cascade-coupled lasers has also been used to extract messages in wavelength-multiplexed chaotic communications [13].

- **Synchronisation-switching in chaotic DFB laser diodes**

Two general forms of synchronisation may be identified and are often termed complete and generalised chaos synchronisation. It was shown in [14] that frequency detuning between transmitter and receiver lasers enables switching between complete and generalised chaos synchronisation.

- **VCSELs Synchronisation**

A salient feature of VCSELs is their tendency to exhibit changes in the emission polarisation due to changes e.g. of bias current and operating temperature. Polarisation effects may therefore be expected to be of some importance in VCSEL synchronisation. Spencer et al. [15] investigated theoretically the synchronisation of chaotic VCSELs. Later Ohtsubo et al [16] reported an experimental observation of chaotic synchronisation in mutually coupled stand-alone VCSELs.

Chaos synchronisation was achieved experimentally in unidirectionally coupled external cavity vertical-cavity surface-emitting semiconductor lasers (VCSELs) [17]. Synchronisation was observed when the polarisation of the transmitter is perpendicular to that the free-running receiver (x-polarisation). The transmitter output versus the x-polarized receiver output power shows normal (positive-slope) synchronisation. However, inverse (negative-slope) synchronisation is found to arise between the transmitter output and the y-polarized receiver output power which is consistent with previous observations of anti-phase dynamics of orthogonal polarisations in unstable VCSELs [18,19].

- **Self-pulsating lasers**

While the thrust of work done on chaos synchronisation in diode laser systems has utilised external-cavity lasers an alternative method based on self-pulsating lasers is available. Such devices are commonplace and inexpensive, having been adopted in Compact Disc players. The lasers can be rendered chaotic through direct modulation of their bias current. This offers a possible advantage over external cavity systems through the elimination of the external mirror and hence a potential reduction in the overall size of the transmitter and receiver systems. Theoretical work [20] showed that it is possible to generate chaotic pulse-sequences from narrow-stripe self-pulsating lasers through direct sinusoidal modulation of the drive current. Those predictions have been confirmed by recent experimental results where the possibility of synchronisation between self-pulsating lasers driven chaotic by external modulation has been demonstrated.

- **Optimisation of Synchronisation**

The foregoing sections have emphasised practical demonstrations of chaos synchronisation. Many theoretical studies have also been undertaken of the phenomenon but here attention is drawn to recent theoretical work which highlights the complexity of the process of optimising such chaos synchronisation. The accepted method for determining the quality of the synchronisation attained between the master and slave laser is via the evaluation of the cross-correlation function at two specific times, namely, at zero delay for injection locking synchronisation and at the difference between the time-of-flight and external cavity round trip time for anticipated synchronisation. In [21] it has been demonstrated that the maximum correlation does not always occur at those times expected and in certain special circumstances this can lead to a mis-interpretation of the results. It has also been shown that the precise maximum correlation in the injection-locking synchronisation regime is a function of both injection power and the frequency detuning between the master and slave lasers. These two features have far-reaching consequences for the majority of earlier numerical studies of chaos synchronisation and point to the need to re-evaluate that earlier work.

MESSAGE TRANSMISSION

Having established the capability of effecting chaos synchronisation using external cavity lasers attention is turned to the use of such a chaotic carrier for the transmission of messages. In the cases considered here proof-of-principle of chaotic message transmission utilise simple sinusoidal signals as the message for transmission and reception. Here a number of specific aspects of message transmission capabilities are addressed.

- **Chaos Pass Filtering**

Understanding the physical mechanism which allows message extraction is of central importance and thus merits the investment of significant effort. However, the message amplitude in the receiver is smaller than that in the transmitter and the message can again be recovered by subtracting the output of the receiver from that of the transmitter.

Theoretical work by Murakami et al [22, 23] has employed a simple theory based on driven damped oscillators to clarify the physical background of CPF. The work of Murakami has revealed several fundamental physical aspects of CPF in injection-locked LDs and argues that CPF is in essence a simple phenomenon, in which linear rather than nonlinear effects seem to dominate. If a nonlinear effect is apparent in the receiver response, a large modulation may cause mixing or coupling of periodic and chaotic components. However, no such tendency appears in the numerical results obtained by Murakami. It is considered that such enhancement of linearity comes from the extremely strong optical injection, which may suppress nonlinearity in the injection-locked-laser response [24]. Moreover, it is very interesting to consider that the large periodic signals have no influence on synchronisation of the chaotic carriers. This fact is contrary to a prediction of the conventional complete-chaos-synchronisation theory, in which message signals should be very small as a perturbation to chaotic carriers to enable synchronisation of the carriers [25]. Noticing that the main oscillating frequency of the chaotic signal we are dealing with is around the relaxation oscillation frequency it is valid to consider that the receiver can linearly respond to the most parts of the chaotic driving signal. This agrees well with the experiments in Ref. [26], in which the matching of the relaxation oscillation frequencies of the transmitter and receiver was needed to obtain the optimal synchronisation state. Recent theoretical work undertaken within the authors' laboratory has provided a deeper insight into the CPF process and has identified circumstances when a nonlinear response is apparent [27].

- **Networked Chaotic Optical Communications**

In the preceding discussions a rather simple transmitter-receiver configuration has generally been utilised for message transmission. However, to meet the requirements of network communications systems several functionalities would be advantageous. Specifically, chaos message multiplexing (where several messages have been transmitted using a number of channels); chaos message broadcasting (enabling distribution of messages from a single transmitter to a number of receivers) and message relay (to allow extension of the reach of the chaos communications link) are of great interest. All these functionalities have been demonstrated experimentally and serve to confirm the suitability of chaos communications techniques for realising networking functionality.

In [13] it was shown experimentally that two chaotic optical communication channels can be configured over a single transmission path using two external-cavity laser diodes as transmitter lasers and a single stand-alone receiver laser. The two chaotic communication channels were shown to operate independently. In this case to achieve message extraction the decoder and receiver lasers are wavelength tuned such that they synchronise to the appropriate transmitter and comparisons are made of the signals injected into the decoder laser with its output. Thus, by tuning the decoder and receiver lasers it is possible to select which of the two messages is to be extracted. As shown in [13] good quality message encoding and decoding is observed in both communication channels.

In conventional communications systems, network operations are often performed with the assistance of relay stations which provide the opportunity to enhance the distance over which information can be transmitted. In order to implement such a system, the requirement is to couple three or more chaotic transmitters and receivers in cascade configuration and then successfully extract a message at the terminal receiver. A basic requirement for successful message transmission in such a configuration is the achievement of robust chaos synchronisation between the cascaded transmitters and receivers. As noted above, cascade synchronisation has been recently reported using three diode lasers. Based on such cascade synchronisation the successful relay of a 1-GHz chaotic message using three cascade-coupled distributed feedback (DFB) laser diodes was reported in [28].

In order to implement a chaotic communications network it is highly attractive to have the capability of broadcasting a message from one transmitter to several receivers. The challenge to the successful demonstration of chaos broadcasting is to ensure that such synchronisation can be robustly achieved between the transmitter and the several receivers. The experimental arrangement for the demonstration of GHz message broadcasting at telecommunications wavelengths using a chaotic external-cavity DFB laser transmitter and two DFB laser receivers is provided in [29]. The LFF (Low Frequency Fluctuation) regime was used here for message encoding. The bandwidth of the chaotic outputs is more than 4 GHz which is adequate to encode the 1 GHz message. The technique used for message encoding is that of chaotic modulation where the TL is directly modulated. The 1 GHz message masked by the chaotic signal was transmitted to both the receiver lasers.

SUMMARY

This paper has summarised experimental and theoretical work concerned with the use of external cavity semiconductor lasers in optical chaos communications. The versatility of this configuration has been illustrated by its use in delivering networking functionalities. Further opportunities for development of this activity is indicated in respect of recent experiments performed using chaotic self-pulsing laser diodes.

Having established the framework for networked communications using chaotic external cavity laser diodes attention is given to some of the outstanding challenges which merit further exploration. A strong message here is the opportunity to broaden the application of the techniques to allow alternative technological platforms to benefit from the progress which has been made using laser diode devices. Challenges also arise in engendering a dialogue with workers in other communities - such as classical cryptography - who may be able to assist in further progressing the significant achievements already made in the use of chaotic external cavity laser diodes.

ACKNOWLEDGEMENTS

The work described in this paper has been undertaken in collaboration with a large number of past and present colleagues whose contributions are acknowledged via the relevant paper citations but who fully merit explicit recognition. To this end we express our deep appreciation of the key contributions made to experimental work by Dr S. Sivaprakasam as well theoretical work performed by Dr S. Peters-Flynn. Support for this work was received from the UK EPSRC and the EU.

REFERENCES

1. P. Colet and R. Roy , *Opt. Lett.*, **19**, 2056–2058 (1996).
2. See 'Feature Section on Optical Chaos and Applications in Cryptography', *IEEE J. Quantum Electron.*, **38**(9) (2002).
3. D. M. Kane and K. A. Shore (Editors), *Unlocking Dynamical Diversity*, Wiley, 2005, pp. 1–339.
4. E. Ott, C. Greboggi and J. A. Yorke, *Phys. Rev. Lett.*, **64**, 196–1199 (1990).
5. R. Roy et al., *Phys. Rev. Lett.*, **68**, 1259–1261 (1992).
6. S. Sivaprakasam and K. A. Shore, *Opt. Lett.*, **24**, 466–468 (1999).
7. A. Pikovsky, M. Rosenblum and J. Kurths, *Synchronisation: A Universal Concept in Nonlinear Sciences*, Cambridge University Press, 2001.
8. S. Sivaprakasam and C. Masoller Ottieri, Chapter 6 in [3].
9. M. W. Lee and K. A. Shore, *IEEE J. Lightwave Technol.*, **23**, 1068–1073 (2005).
10. S. Sivaprakasam and K. A. Shore, *Opt. Lett.*, **26**, 253–255 (2001).
11. S. Sivaprakasam and K. A. Shore, *Opt. Lett.*, **28**, 1397–1399 (2003).
12. M. W. Lee, J. Paul, C. Masoller and K. A. Shore, *J. Opt. Soc. Am. B*, **23**, 846–851 (2006).
13. J. Paul, S. Sivaprakasam and K. A. Shore, *J. Opt. Soc. Am. B* **21**, 514–521 (2004).

14. M. W. Lee et al., *IEEE J. Quantum Electron.*, **41**, 302–307 (2005).
15. P. S. Spencer et al., *IEEE J. Quantum. Electron.*, **34**, 1673 (1998).
16. N. Fujiwara, Y. Takiguchi, and J. Ohtsubo, *Opt. Lett.*, **28**, 1677–1679 (2003).
17. Y. Hong et al., *Opt. Lett.*, **29**, 1215–1217 (2004).
18. C. Masoller and N. B. Abraham, *Phys. Rev. A*, **59**, 3021 (1999).
19. S. Bandyopdhyay et al., *Opt. Comm.*, **202**, 145 (2002).
20. R. J. Jones et al., *J. Opt. Soc. Am. B*, **18**, 166–172 (2001).
21. S. Peters-Flynn. et al., *IEEE J. Quantum Electron.*, **42**, 427–434 (2006).
22. A. Murakami and K. A. Shore, *Phys. Rev. A*, **72**, 053810 (2005).
23. A. Murakami and K. A. Shore, *Phys. Rev. A*, **73**, 043804 (2006).
24. A. Murakami et al., *IEEE J. Quantum Electron.*, **39**, 1196 (2003).
25. A. Uchida, Y. Liu and P. Davis, *IEEE J. Quantum Electron.* **39**, 963 (2003).
26. I. Fischer, Y. Liu and P. Davis, *Phys. Rev. A* **62**, 011801(R) (2000).
27. S. J. Lea and P. S. Spencer "Nonlinear Amplitude Response Mechanism of the Chaos Pass Filtering Process in Synchronized Semiconductor Laser Diodes" submitted to Phys. Rev. E., 2006.
28. M. W. Lee and K. A. Shore, *IEEE Photon. Technol. Lett.*, **18**, 169–171 (2006).
29. M. W. Lee and K. A. Shore, *Electron. Lett.*, **40**, 614–615 (2004).

Non Stationary Highly Oscillatory Phenomena Characterization Using the Wavelet Transform

Eduardo Serrano* and Alejandra Figliola†

*Escuela de Ciencia y Tecnología, Universidad Nacional de General San Martín.
M. Irigoyen 3100, San Martín, Pcia. de Buenos Aires, Argentina.
†Instituto del Desarrollo Humano, Universidad Nacional de General Sarmiento.
Juan María Gutierrez 1150, Los Polvorines, Pcia. de Buenos Aires, Argentina.

Abstract. We are interested in to characterize nonlinear oscillatory phenomena distinguished for signals whose frequency increases quickly while its amplitude changes slowly and the regularity of the signal collapses and it shows a singularity when the frequency goes to the infinite.

Keywords: Oscillatory phenomena, Chirps, Gravitational waves, Instantaneous frequency, Littlewood Paley Analysis, Wavelet Transform
PACS: 02.60.-x, 0.30.-w, 42.65.Re

INTRODUCTION

We are interested to characterize non-linear oscillatory phenomena distinguished for signals whose frequency increase quickly while its amplitude changes slowly. In general, the regularity of the signal collapses and it shows a singularity when the frequency tends to the infinite. Such phenomena are very common in nature. As an example, we could mention the sing of the birds, the human words, the gravitatory waves or many others, [5]. Characterization of these phenomena requires refined analytical tools to process this signals, suitable to fit the quick frequency variation and to detect the singularities. Then, ordinary techniques based on Fourier methods are not suitable for this propose. Rather than, the character of the problem requires time-scale or time-frequency methods. In particular, has good capacities to this kind of estimations for this context. Nowadays, Wavelet Transforms play a significant role in pure and applied mathematics. In physics or signal processing applications wavelet analysis can be consider as a time-scale technique. Therefore, for many problems, it performs better than Fourier methods. In particular, it has good capacities to this kind of estimations for the above exposed context, [2]-[4].
However, it is a multifaceted tool. For instance, the Discrete Wavelet Transform (DWT) is correlated with the atomic decomposition of the given signal, using an orthonormal wavelet basis in the context of a multiresolution scheme. On the other hand, when the shift-invariance is needed or large scales phenomena are as important as small scales ones, the correct tool for analyzing is the Continuous (CWT) or the Semi-Discrete Wavelet Transform (SDWT). Recently, Littlewood - Paley analysis seems to have the better performance for analizing turbulent phenomena, fractional Brownian processes or frequency modulated signals and chirps. It can be considered as a particular semi-discrete wavelet transform. This is our choice for the characterization problem and we will give a quick survey in a next section and we refer to [3]-[5] for details.

CHIRPS AND GRAVITATIONAL WAVES

There is a very large class of oscillatory phenomena, but we only consider for this work signals modeling by:

$$s(t) \quad = \quad A(t)\cos(\omega(t)) \tag{1}$$

The phase $\omega(t)$ is monotonous. In general, it has a slow behaviour but the changes become faster when the time tends to some finite limit t_0. The module of the derivate $|\omega'(t)|$ is the *instantaneous frequency*. It enlarges or tends to infinite when t tends to t_0. The amplitude $A(t)$ varies with relative slowness but it presents some singularity at t_0. The conjunction of the singularity of $A(t)$ with highly oscillations give us an isolated *oscilatory singularity* of the signal in

CP913, *Nonequilibrium Statistical Mechanics and Nonlinear Physics, XV Conference*
edited by O. Descalzi, O. A. Rosso, and H. A. Larrondo

t_0. We could exhibit to classic prototypes. For one hand, a *chirp* prototype is defined around t_0 as:

$$s(t) = |t-t_0|^{\alpha} \sin\left(\frac{\omega}{|t-t_0|^{\beta}}\right) \quad (2)$$

when $0 < \alpha < 1$, $\beta > 0$ and $\omega \gg 1$. The instantaneous frequency tends to infinite when t tends to t_0. The amplitude is not differentiable in t_0 being α the *pointwise Hölder exponent* at this point. (see Fig. 1) A second prototype are the *gravitational waves*:

$$s(t) = |t-t_0|^{-\alpha} \cos\left(\omega |t-t_0|^{\beta}\right) \quad (3)$$

with $0 < \alpha, \beta < 1$ and ω is large. Here, instantaneous frequency increases when t tends to t_0 while the amplitud collapses to infinity (see Fig. 2). The general *Estimation Problem* consists in characterize the signal properties, from a given data set, ie to establish the instantaneuos frequency beahviour, the singularities nature or, assuming a model, to estimate the corresponding parameters. Particullarly, in this work, we will estimate the parameters α and β for the gravitational waves [3].

WAVELET TOOLS

Littlewood - Paley Analysis, can be considered as a particular wavelet transform, [1]. Let us give a brief description for this tool. Let φ an appropriate spline function in $L^2(\mathscr{R})$, with compact support and with the Fourier Transform $\widehat{\varphi}$ having a fast decay outside the interval $| w_0 | \leq 2\pi$ and $\widehat{\varphi}(w_0) \approx 1$ on $| w_0 | \leq \pi$, [1], [6]-[7]. We define a *Littlewood-Paley wavelet* as:

$$\psi(t) = 2\varphi(2t) - \varphi(t) \quad (4)$$

The function ψ is an *admissible wavelet* with compact support and oscillating. The Fourier transform $\widehat{\psi}(\omega)$ has a peak in the frequency $\omega_0 \approx 1.32 * \pi$. For any signal s and $\nu \in \mathscr{R}$

$$D_{\nu}s = 2^{\nu}\ \psi(2^{\nu}.) * s \quad (5)$$

is the Wavelet Transform at scale ν. Suppose that the signal is sampled at the points $t_k = h.k$, and let R a positive integer. Fixed R and h, we will define *rational wavelet sub-scales* as follows: Let $j = -1, ..., j_{min}$ and for $r = 0, ..., R-1$:

$$\psi_{j,r}(t) = \frac{2^j}{h(R+r)} \psi\left(\frac{2^j t}{h(R+r)}\right) \quad (6)$$

These wavelets are splines with knots on the sampled points and their support are scaled by the factor $2^j/h(R+r)$, [6]. On the other hand, each sub-scale (j,r) corresponds to the frequency: $\omega_{j,r} = \frac{2^j \omega_0}{h(R+r)}$.
We have large family of time-frecuency atoms and we could compute at the sampling points the convolutions $D_{j,r}(t_k)$ from the given data s_k using an efficient *Algorithm à Trous*, [4].

INSTANTANEOUS FREQUENCY

Let $g(t) = \cos(\omega(t))$ in $[a,b]$, and

$$\omega(t) = \omega(t_0) + \omega'(t_0)(t-t_0) + \frac{1}{2}\omega''(\alpha)(t-t_0).^2 \quad (7)$$

$|\omega'(t_0)| \geq L > 0$ for $t_0 \in (a,b)$ and $\frac{|\omega''(t_0)|}{|\omega'(t_0)|^2} \approx 0$
Then $\phi(t_0) = |\omega'(t_0)|$ is the *instantaneous frequency* in t_0: the *local oscillation pattern* at the point.
For the gravitational waves we have $\omega(t) = \omega_0(1-t)^{\beta}$ then,

$$\frac{|\omega''(t)|}{|\omega'(t)|^2} = \frac{1-\beta}{\beta\omega_0}(1-t)^{-\beta} \quad (8)$$

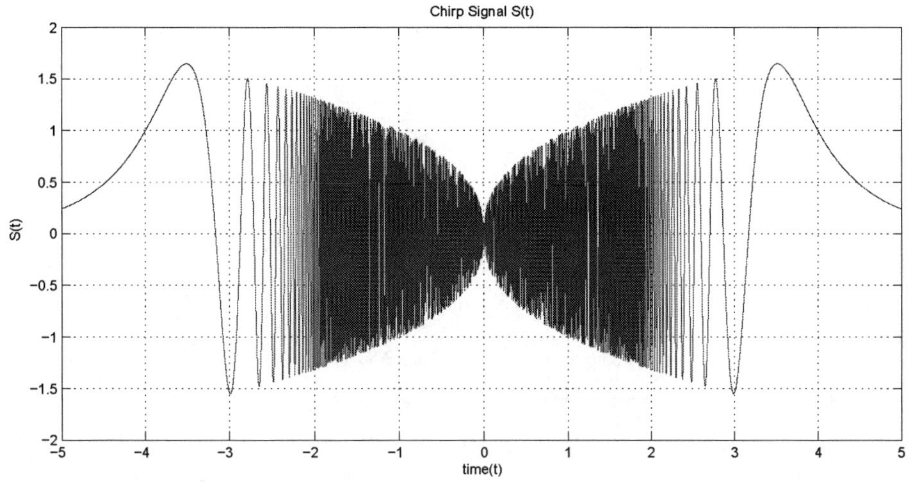

FIGURE 1. Chirp Signal

Note that the instantaneous frequency collapses in the limit $t = 1$ but, is ω_0 is large, it may be characterized still near this singular point. On the other hand, observe that if we analize the signal in the interval $0 < t_1 < t_2 < 1$, where makes sense the relation $\phi(t) = |\omega'(t)| = \beta\omega_0(1-t)^{\beta-1}$ that the sampling rate h must be consistent with the maximum instantaneous frequency $\phi(t_2)$, according to the Nyquist ratio.

Let us discuss the estimate of parameter β under the above given model (3). where α, β and ω are unknown. The main idea is, first to characterize the behaviour of the instantaneous frequency $\phi(t) = |\omega'(t)| = \beta\omega_0(1-t)^{\beta-1}$ in some time interval $0 < t_1 < t_2 < 1$. Then we will estimate β.

But, although the Fourier spectrum is not stationary, the sampling rate h must be consistent with the maximum instantaneous frequency $\phi(t_2)$.

We remark that, according the Nyquist ratio, we must verify the constrain $\phi(t_2) \leq \frac{\pi}{h}$. Then, the question is: since we cannot evaluate *a priori* $\phi(t_2)$, which is the interval where the instantaneous frequency $\phi(t)$ makes sense? We emphatize that a wavelet method will give the chance to overcome this problem.

ESTIMATION METHOD

Choice R and j_{min} and compute the wavelet transform in the interval $[h, 1-h]$, for $j = -1, ..., j_{min}$ and $r = 0, ..., R-1$. Owing the model, in any sub-interval $[t_1, t_2]$ where the instantaneous frequency makes sense, we have the linear relation:

$$\log(\phi(t)) = (\beta - 1) + \log(\beta\omega_0(1-t)) \tag{9}$$

For this reason, the wavelet spectrum, $|D_{j,r}|$ must have a maximum near a point $t_{j,r}$ where, $\phi(t_{j,r}) \approx \omega_{j,r}$

Then, we only must consider the sub-scales (j, r) where such maxima $t_{j,r}$ can be detected and evidence the linear relation:

$$\log(\omega_{j,r}) \approx (\beta - 1) + \log(\beta\omega_0(1-t_{j,r})) \tag{10}$$

In this case, we can estimate β using a linear regression. Observe that we only consider an admissible range of sub-scales (j, r), where the above maximum conditions hold. This automatically give us the time interval for the analysis. The parameter α can be easely estimated from the maximum values $(t_{j,r}; |D_{j,r}(t_{j,r})|)$.

EXAMPLE

We simulate a gravitational wave where $\alpha = -1/4$, $\beta = 5/8$ and $\omega_0 = 1000$. The sampling rate is $h = 4 * 10^{-5}$ (see Fig. 2). Observe that the amplitude goes to infinite while the Fourier Transform has a fast decay, when the frequency

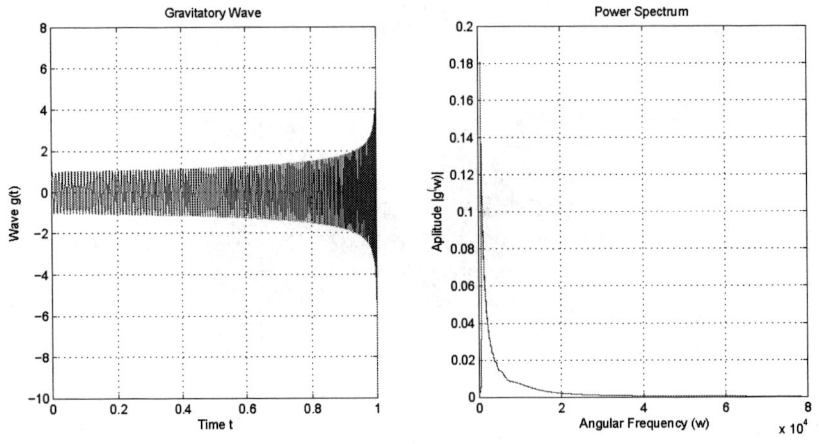

FIGURE 2. Gravitional Wave and its Fourier Spectrum

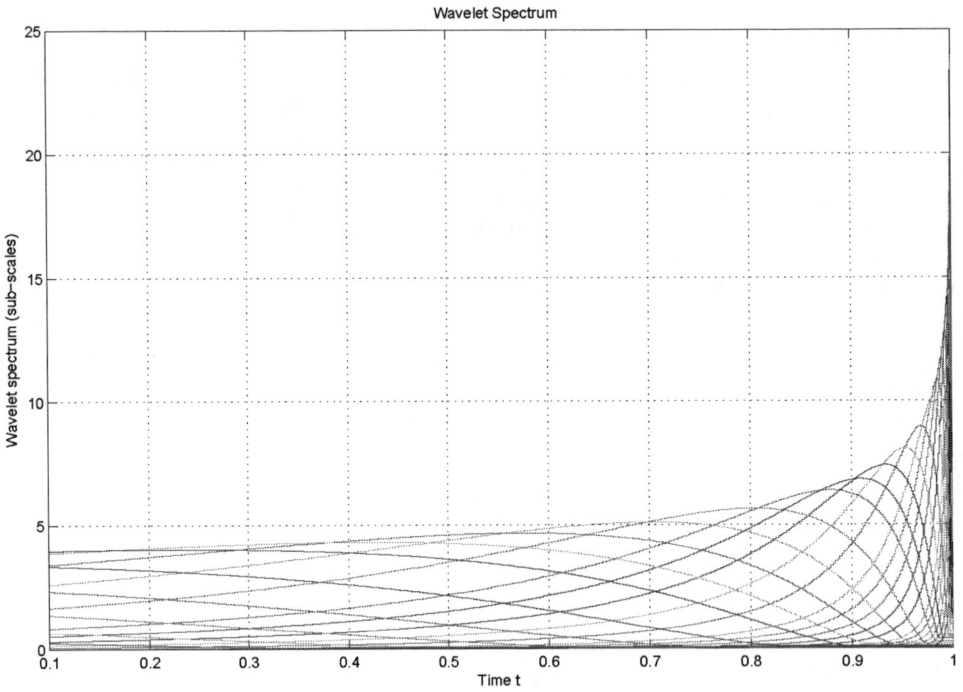

FIGURE 3. Wavelet Spectrum of the signal showed in Fig. 2

increases. We compute the Wavelet Transform for the sub-scales corresponding with $j = -1, \dots -6$, with $R = 5$ and the rate h. The wavelet spectrum for each sub-scale are perfiled in the Fig. 3. We compute the maximum $(t_{j,r}; |D_{j,r}(t_{j,r})|$ and we can observe the admissible range of sub-scales in Fig. 4. The linear fit lead us to estimate $\beta \approx 0.6239$ and the estimation of the parameter $\alpha \approx .2505$ is given from the maximum values. We remark that wavelet method above exposed is a powerfull tool to estimate the parameters α and β under the given model (3), from an appropiate data set. We well here attack a particular problem, however, we emphatize that analogous techniques are possible to use for resolve generalized problems, represented from sampled values, that involved a general clases of chirps and oscillatory singularities.

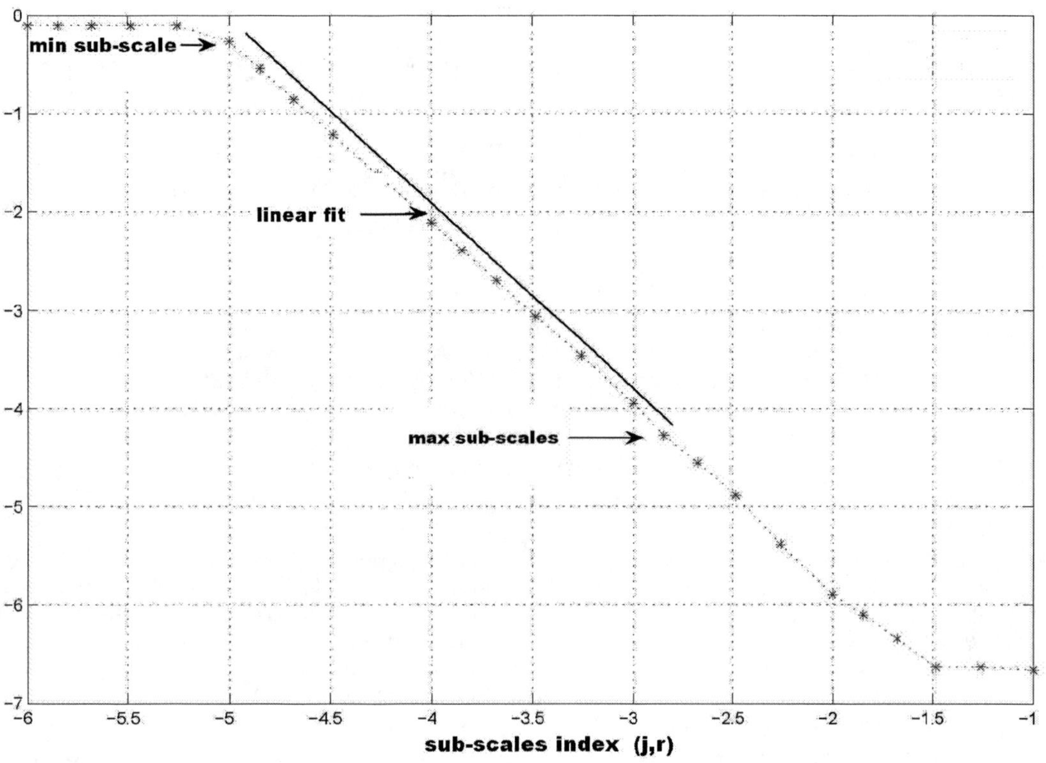

FIGURE 4. Wavelet Maxima: the full line correspond the admissible sub-scales range. The dots give us the maximum values in the wavelet spectrum give in Fig. 3. The value β is estimate from the slope of the full line over the admisible range.

ACKNOWLEDGMENTS

This work was supported by Consejo Nacional de Investigaciones Científicas y Técnicas (CONICET) (PIP 5687/05) and Agencia Nacional de Promoción Cientifica y Tecnológica (PICT 20648/04), Argentina.

REFERENCES

1. Y. Meyer, *Oscillating Patterns in Image Processing and Nonlinear Evolutions Equations*, AMS Eds., New York, 2001, pp. 1-60.
2. T. Blu, *IEEE Signal Processing*, **16** N° 6, 109-135 (1999).
3. I. Daubechies, *Ten Lectures on Wavelets*, SIAM, Philadelphia, 1992, pp. 1-300
4. S. Mallat, *A Wavelet Tour of Signal Processing*, Academic Press, Philadelphia, 1998, pp. 100-520.
5. S. Jaffard, Y. Meyer and R. Ryan *Wavelets, Tools for Science and Technology*, SIAM, Philadelphia, 2001, pp. 1-250.
6. I. J. Schoenberg, *Cardinal Spline Interpolation*, SIAM, Philadelphia, 1973, pp. 1-125.
7. M. Unser, A. Aldroubi and M. Eden, *IEEE Trans. Acoust., Speech and Signal Process*, **41** N° 12, 821-833 (1993).

Evolutionary and Ecological Trees and Networks

Emilio Hernández-García*, E. Alejandro Herrada*, Alejandro F. Rozenfeld†, Claudio J. Tessone*, Víctor M. Eguíluz*, Carlos M. Duarte†, Sophie Arnaud-Haond**,‡ and Ester Serrão**

*Unidad de Física Interdisciplinar-IMEDEA (CSIC-UIB).
Campus Universitat de les Illes Balears, E-07122 Palma de Mallorca, Spain
†Instituto Mediterráneo de Estudios Avanzados IMEDEA (CSIC-UIB).
C/ Miquel Marquès, 21, E-07190 Esporles, Mallorca, Spain
**CCMAR, CIMAR-Laboratório Associado, Universidade do Algarve
Gambelas, 8005-139, Faro, Portugal
‡DEEP/LEP-Laboratoire Environnement Profond
IFREMER Centre de Brest, BP 70 29280, Plouzane, France

Abstract. Evolutionary relationships between species are usually represented in phylogenies, i.e. evolutionary trees, which are a type of networks. The terminal nodes of these trees represent species, which are made of individuals and populations among which gene flow occurs. This flow can also be represented as a network. In this paper we briefly show some properties of these complex networks of evolutionary and ecological relationships. First, we characterize large scale evolutionary relationships in the Tree of Life by a degree distribution. Second, we represent genetic relationships between individuals of a Mediterranean marine plant, Posidonia oceanica, in terms of a Minimum Spanning Tree. Finally, relationships among plant shoots inside populations are represented as networks of genetic similarity.

Keywords: Complex networks, Phylogenies, Tree of Life, Population structure, Genetic similarity, Posidonia oceanica
PACS: 89.75.Hc , 87.23.-n

INTRODUCTION

The study of complex networks, representing interactions among components, has become a central tool in the science of complex systems [1, 2, 3]. Evolutionary relationships between species are usually represented in phylogenies, i.e. evolutionary trees. One branching event represents the evolution of an ancestral species into descendent ones. The whole set of relationships among all known species is conceptually represented as a huge *Tree of Life*. A tree is a network in which there are no cycles, i.e., there is a unique path from one node to another inside the network. This is probably a good approximation to the correct large scale structure of the Tree of Life, but processes such as lateral gene transfer or hybridization would need a richer network structure to be properly represented [4]. If analyzing the Tree of Life at a finer detail, entering the scale appropriate for ecological interactions, we observe that species are composed of different populations, and that those are made of individuals that interchange genes and recombine their genomes in processes such as sexual reproduction. Thus, there are gene flow processes, particularly obvious when looking at the intraspecific level, which add loops to the Tree of Life, and make the whole structure a rather complex object.

In this paper we briefly analyze, at three different scales, some properties of this complex network of evolutionary and ecological gene flow. Our aim is just to provide an overview of the many interesting features of genetic relationships and phylogenies that can be addressed from the point of view of complex systems, with the hope that this will stimulate further and more detailed work. First, we characterize large scale evolutionary relationships, the ones more traditionally represented in a tree topology. To this end we analyze a large scale reconstruction of the Tree of Life and characterize it by its degree distribution. Second, we represent the genetic relationships between individual shoots of a Mediterranean marine plant, Posidonia oceanica, sampled across its entire geographical extent, in terms of a Minimum Spanning Tree that would represent the most parsimonious pattern of gene flow between distant populations. Finally, relationships among shoots in the same population are represented as networks of genetic similarity.

CP913, *Nonequilibrium Statistical Mechanics and Nonlinear Physics, XV Conference*
edited by O. Descalzi, O. A. Rosso, and H. A. Larrondo

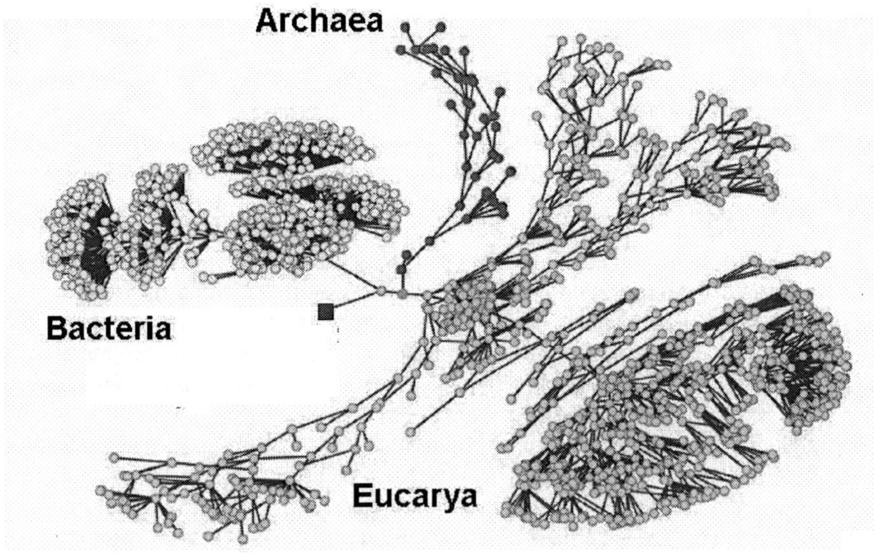

FIGURE 1. A rendering of the first branchings of the "Tree of Life". Many additional subdivisions occur at each of the final branches in this plot, where we can distinguish the three main Domains (Bacteria, Archaea, Eucarya) in which the different organisms are classified.

SCALE-FREE DEGREE DISTRIBUTION IN THE TREE OF LIFE

Since the beginning of the studies on evolutionary biology, one of the most ambitious goals has been the assemble and comprehension of the complete evolutionary history of biodiversity. This assembly has produced a huge phylogenetic tree, baptized with the name "Tree of Life". The data set analyzed here is the reconstruction of the Tree of Life which is available at the database of the *Tree of Life Web Project* (`http://www.tolweb.org/`). It contains about 6×10^5 nodes. A very small portion of it, showing the first ramifications, is shown in Fig. 1.

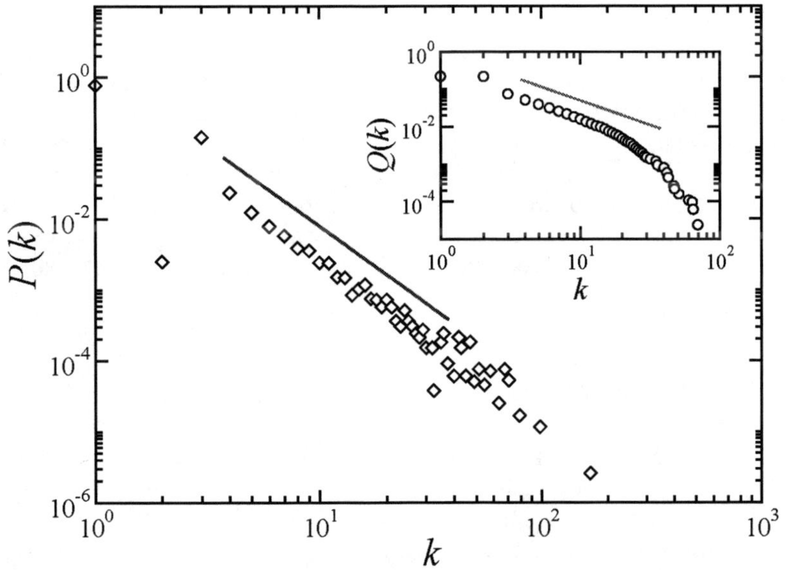

FIGURE 2. Degree distribution in the Tree of Life. The fit is a power law $P(k) \sim k^{-\gamma}$ with the scaling exponent $\gamma \approx 2.33$. The inset shows it in cumulative form: $Q(k) = 1 - \sum_{r=1}^{k} P(r)$.

One of the first quantities to discuss when analyzing complex networks is the degree distribution, $P(k)$, i.e. the proportion of nodes in the network which are connected to k neighbors. For our tree, this function is plotted in Fig. 2 (together with the function $Q(k) = 1 - \sum_{r=1}^{k} P(r)$, giving an accumulated version of the distribution). We see that the probability of finding a node with degree k decays as a power law $P(k) \sim k^{-\gamma}$, with $\gamma \approx 2.33$. Thus, this phylogenetic tree has a broad degree distribution of the scale-free type. The scaling exponent γ characterizing this distribution is consistent with the results obtained by Cartozo *et al.* [5] from a previous taxonomic diversity analysis, where the values obtained were in the range $1.9 - 2.7$.

The use of the degree distribution to characterize a phylogenetic tree needs some qualification: It is generally accepted that a true phylogenetic tree at high enough resolution will contain only binary branchings [6]. This corresponds to $k = 3$ for all internal nodes, and $k = 1$ for the root and the terminal tips, that would constitute about one half of the nodes. The presence of nodes with a very high degree (see Fig. 2) in our data set reveals a large amount of polytomies. This would correspond to successive branchings of several species or groups in an order that can not be resolved with the techniques used in the phylogenetic reconstruction, so that all of them are assigned the same branching point. In this sense, it can be thought that a characteristic such as the degree distribution is quantifying more the limitations of the methodology used to reconstruct the phylogenetic history than the true evolutionary branching itself. The limitation would be clearer if considering the statistics of taxonomic classifications, in which branchings are forced to fit into the restricted set of levels established by taxonomic science (Domain, Kingdom, Phyllum, Class, Order, Family, Genera, Specie). The *Tree of Life Web Project* used here does not use a strict taxonomic classification, since new levels are added as needed. It is likely that the broad distribution of degrees obtained here reflects a property of *radiation* processes during evolutionary history, i.e. events in which many new species appeared during a relatively short period of time [7]. Clearly, further analysis is needed to establish which features of the degree distribution arise from phenomena of biological relevance and which are consequences of the lack of resolution of available reconstruction methods. In any case, it is seen in Fig. 2 that the proportion of nodes with $k = 3$, representing binary branchings, stands up above the background power law behavior. We see also the presence of some nodes with $k = 2$ (they represent taxa which consist of only one subtaxon), although in a very small proportion.

INSIDE ONE LEAF OF THE TREE: GENETIC RELATIONSHIPS IN A PLANT SPECIES

We now consider in more detail the genetic structures present at one of the tips of the above Tree of Life. We focus in a particular species and analyze its internal genetic relationships, first across its whole geographical range, and then at particular locations.

We consider the case of Posidonia oceanica, a marine angiosperm living in meadows submerged between 0 to 40 m in the Mediterranean Sea [8]. It combines clonal reproduction with episodes of sexual reproduction. This plant is experiencing basin-wide decline and is subject to specific protection and conservation measures [8]. Genomic DNA data are available from approximately 40 shoots sampled in each of 37 localities across the Mediterranean. A set of seven microsatellite markers was used to characterize the genotype of each individual. This provides us with a data set [9, 10, 11] consisting on the number of repetitions of the microsatellite motif at each locus of each shoot sampled. In this set, a convenient genetic distance d_{ij} can be defined [10, 11] which measures the degree of dissimilarity among every pair $\{i, j\}$ of sampled shoots. From this distance matrix, several trees and networks can be built and analyzed, as described in the following.

The minimum spanning tree of plant shoots

We can consider the fully connected network linking all pairs of shoots in our data set. Given a connected, undirected graph, a spanning tree of that graph is a subgraph which is a tree, i.e. contains no loops, and connects all the vertices together. A single graph can have many different spanning trees. Our distance matrix d_{ij} assigns a weight to each of the links. We can use this information to assign a total *length* to each of the possible spanning trees by computing the sum of the distances of the edges which are kept in that spanning tree. A minimum spanning tree (MST) is then a spanning tree with the minimum possible total length. In the case in which all edges have a different distance, the MST is unique. But more generally, there could be several MSTs for a particular network and distance matrix. A MST is in fact the minimum-cost subgraph connecting all vertices, since subgraphs containing cycles necessarily have more

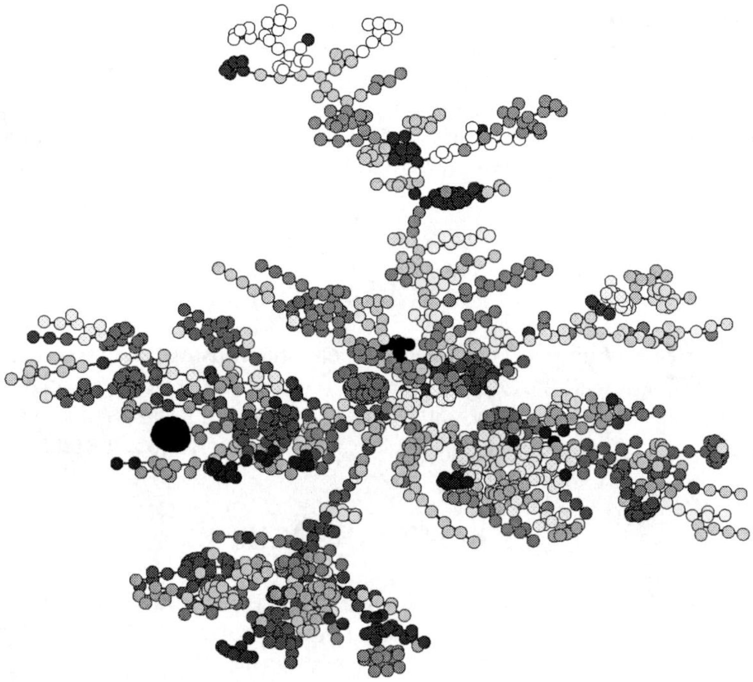

FIGURE 3. Minimum spanning tree containing all sampled ramets of Posidonia oceanica. Each node represents a shoot, with colors (or grey shades) indicating the sampling meadow, and each link has an associated distance. The minimum spanning tree minimizes the total sum of distances.

total weight. This allows to interpret the MST as the main path of gene flow among the plant populations, as it links all the specimens in the genetically shorter way.

Fig. 3 shows the MST associated to the set of shoots and the corresponding distances (it is in fact one of the several equivalent MST with identical total length and similar shape that can be constructed from our distance matrix). The nodes, representing shoots, are colored with the same shade if they have been collected from the same meadow. The upper part of the tree contains shoots from populations in the Eastern and Central Mediterranean, and the lower part from the Northern Spanish coast, in the Western Mediterranean. The central part of the MST is occupied by populations in the Balearic Islands. The tightly packed balls of nodes group together genetically identical shoots, arising from clonal reproduction. Although most of the shoots from the same location appear close together in the tree, there is some scatter, indicating that there has been some migration between different populations. This would need to be represented with additional links introducing cycles and transforming the tree into a more complex network.

Figure 4 shows the cumulative degree distribution characterizing the MST. It follows a power law with exponent -1.95, so that the degree distribution is of the scale-free type, satisfying $P(k) \sim k^{-2.95}$.

Intrapopulation networks

Still at a smaller scale, we can address the genetic relationship among shoots sampled inside the same meadow. It is clear that a tree structure is not appropriate to represent the intense gene mixing that would be provided by sexual reproduction at this scale. Starting from the fully connected network containing links among all pairs of shoots, we can discard the links containing higher distances and retain only those that represent a sufficiently close genetic similarity. This is related to the network construction methods based in correlation thresholds [12, 13]. In [10] we constructed networks of genetic similarity among *genets* (the set of clonally identical shoots) by choosing as the distance threshold the so called *outcrossing distance*. This is the average distance between the genets in a population and its offspring obtained from a simulation of outcrossing, i.e. sexual reproduction among genetically different individuals. The same idea is applied here to obtain networks of shoots. Network links will join shoots which are genetically closer than the typical outcrossing distance for their population. Figure 5 shows two examples of networks obtained in such a way. The network representation visually highlights the main features of the population structure. For example the population in Es Pujols (Fig. 5a) consists of a central core of interconnected shoots to which less central organisms are linked. Campomanes (Fig. 5b), instead, is structured in two main components. The size of these networks is too small to search for scaling properties in the degree distribution, but the analysis in [10] at the genet level reveals that they have the characteristics of small worlds [14], i.e. clustering significatively higher than a random network with the same number of nodes and links, while at the same time keeping the same low diameter values characteristic of random networks.

CONCLUSION

We have presented a brief overview of properties of trees and networks constructed in the context of phylogenetic and genetic relationship at very different scales, from evolutionary to ecological. The visual inspection of these structures reveals interesting clues such as paths of gene flow, patterns of speciation, or population structure. The open challenge is to relate more quantitative topological properties of these complex networks to relevant biological mechanisms.

ACKNOWLEDGMENTS

We acknowledge financial support from the Spanish MEC (Spain) and FEDER through projects CONOCE2 (FIS2004-00953) and SICOFIB (FIS2006-09966), the Portuguese FCT through project NETWORK (POCI/MAR/57342/2004), the BBVA Foundation (Spain), and the European Commission through the NEST-Complexity project EDEN (043251).

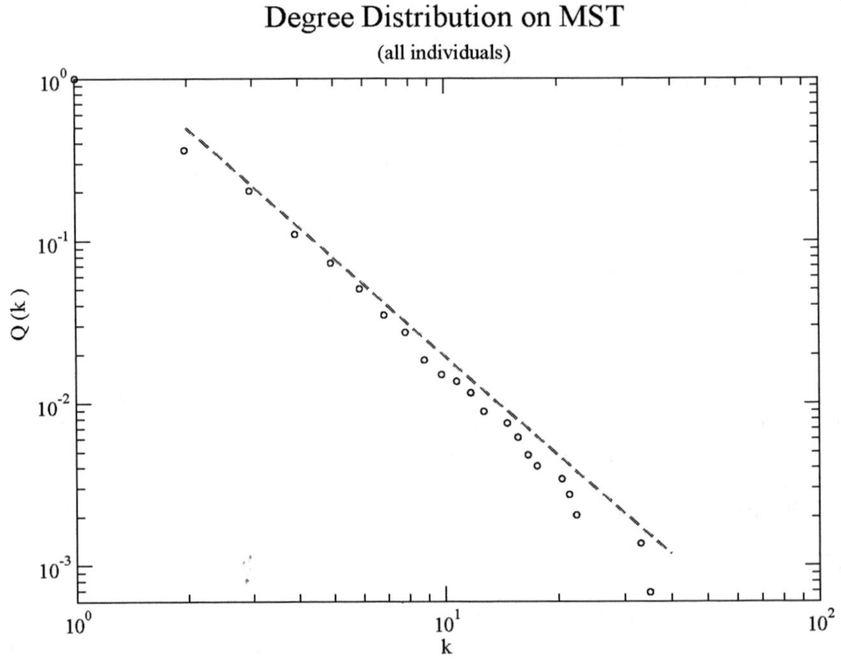

FIGURE 4. Representation of the cumulative degree distribution $Q(k) = 1 - \sum_{r=1}^{k} P(r)$ characterizing the MST. The straight line is a fit to $Q(k) \sim k^{-\gamma+1}$, with $\gamma \approx 2.95$

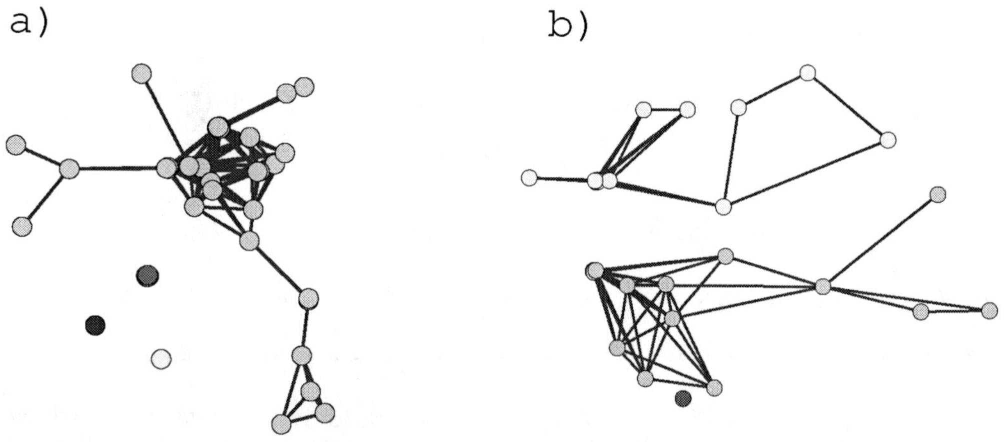

FIGURE 5. Networks of genetic similarity constructed for the shoots sampled at a) Es Pujols (Formentera, Balearic Islands), and b) Campomanes (Mediterranean Spanish coast).

REFERENCES

1. R. Albert and A.-L. Barabási, *Rev. Mod. Phys.* **74**, 47–97 (2002).
2. S. Dorogovtsev and J. Mendes, *Adv. Phys.* **51**, 1079–1187 (2002).
3. S. Boccaletti, V. Latora, Y. Moreno, M. Chavez, and D.-U. Hwang, *Phys. Rep.* **424**, 175–308 (2006).
4. D. Posada and K. Crandall, *Trends Ecol. Evol.* **16**, 37–45 (2001).
5. C. C. Cartozo, D. Garlaschelli, C. Ricotta, M. Barthelemy, and G. Caldarelli, Quantifying the taxonomic diversity in real species communities (2006), URL `http://arxiv.org/abs/q-bio.PE/0612023`.
6. A. O. Mooers and S. B. Heard, *Q. Rev. Biol.* **72**, 31–54 (1997).
7. T. J. Givnish and K. J. Sytsma, editors, *Molecular Evolution and Adaptive Radiation*, Cambridge University Press, Cambridge, 1997.
8. M. A. Hemminga and C. M. Duarte, *Seagrass Ecology*, Cambridge University Press, Cambridge, 2000.
9. S. Arnaud-Haond, F. Alberto, G. Procaccini, E. A. Serrão, and C. M. Duarte, *J. Heredity* **96**, 434–440 (2005).
10. A. F. Rozenfeld, S. Arnaud-Haond, E. Hernández-García, V. M. Eguíluz, M. A. Matías, E. Serrão, and C. M. Duarte, to appear in *J. Royal Soc. Interface* (2007), URL `http://arxiv.org/abs/q-bio.PE/0605050`.
11. E. Hernández-García, A. F. Rozenfeld, V. M. Eguíluz, S. Arnaud-Haond, and C. M. Duarte, *Physica D* **224**, 166–173 (2006).
12. J.-P. Onnela, A. Chakraborti, K. Kaski, J. Kertész, and A. Kanto, *Phys. Rev. E* **68**, 056110 (2003).
13. V. Eguíluz, D. Chialvo, G. Cecchi, M. Baliki, and A. Apkarian, *Phys. Rev. Lett.* **94**, 018102 (2005).
14. D. J. Watts and S. Strogatz, *Nature* **393**, 440–442 (1998).

Markovian Strategies for Spatial Dilemmas in Lattices and Networks

E. A. Sicardi* and H. Fort†

*Instituto de Física, Facultad de Ingeniería, Universidad de la República, Julio Herrera y Reissig 565, 11300 Montevideo, Uruguay.
†Instituto de Física, Facultad de Ciencias, Universidad de la República, Iguá 4225, 11400 Montevideo, Uruguay

Abstract. In a first part we analyze evolutionary spatial 2×2 games between heterogeneous agents using cellular automata (CA). Cells play repeatedly against their nearest neighbors pairwise games specified by a rescaled payoff matrix with two parameters. Each cell is governed by a deterministic binary Markovian strategy (BMS). The system evolves within strategy space according to the simple standard rule: each cell copies the strategy of the neighbor who got the highest score. In a second part we study the networks that form between populations of strategists, spatially randomly distributed, provided the cost of an edge is a function of the distance between agents.

Keywords: Complex adaptive systems, Agent-based models, Evolutionary Game Theory
PACS: 89.75.-k, 89.20.-a, 89.65.Gh, 02.50.Le, 87.23.Ge

INTRODUCTION

Many situations in nature and in social sciences are amenable to formulation in terms of 2×2 *non cooperative* and *non zero sum* games. That is, two players each confronting two choices: cooperate (C) or defect (D), each making its choice without knowing what the other will do and where it is possible that both players win (and, of course, that both lose).

The paradigmatic example is the *Prisoner's Dilemma* (PD) game. In this game the payoff of an agent in this kind of contest depend on what it plays as well as what its opponent plays: If both play C both get the "reward" R for mutual cooperation. If both play D both get the "punishment" P for mutual defection. If one player cooperates and its opponent defects, the agents get respectively the "sucker's payoff" S and the "temptation to defect" T. The four payoffs obey the relations: $T > R > P > S$. Hence defection D is the best strategy for a given agent: if its opponent plays D, playing D (C) it gets P (S) while if its opponent plays C, it gets T (R) playing D (C). The dilemma is that if both defect, both do worse than if both had cooperated: both players get P which is smaller than R.

A possible way out for this dilemma is to play the game repeatedly. In this iterated Prisoner's Dilemma (IPD), players meet several times and provided they remember the result of previous encounters, more complicated strategies than just the unconditional C or D are possible. Some of this *conditional* strategies outperform the dominant one-shot strategy, "always D".

The problem of how strategies that allow cooperation evolved in the IPD can be approached from a Darwinian perspective: diverse strategies are let to compete and the most successful propagate displacing the others. This *evolutionary game theory* [1], [2], inspired Axelrod [3], [4],[5], to organize tournaments between different strategies using genetic algorithms that mimic biological evolution. That is, the starting point is a population of strategies, with one representative of each 'species', or competitor. If a strategy performed well, in the next generation it would be represented more than once, and if a strategy did poorly, it would die off. The champion strategy was *Tit for Tat* (TFT): cooperate on the first move, and then cooperate or defect exactly as your opponent did in the preceding round. Afterwards another computer tournament was carried out between strategies described by four conditional probabilities: $[p_R, p_S, p_T, p_P]$ that determine, respectively, the probability that the strategy play C after receiving the payoff R, S, T, or P [6]. In this simulation, a different strategy was found to be the most stable in the evolutionary sense. This strategy was approximately [1,0,0,1]. It had previously been named *Pavlov* by D. and V. Kraines [7], because if its action results in a high payoff (T or R) it stays, but otherwise it changes its action.

The spatial structure by itself has also shown to be a sufficient element to produce cooperation. That is, unconditional players (who always play C or D no matter what their opponents play) without memory and no strategical elaboration can attain sustainable cooperation when placed in a two dimensional array and playing only against their nearest

CP913, *Nonequilibrium Statistical Mechanics and Nonlinear Physics, XV Conference*
edited by O. Descalzi, O. A. Rosso, and H. A. Larrondo

neighbors [8].

The combination of iterated competition between conditional strategies and spatial structure, was first studied in [9] using m-step memory strategies. Later on, Brauchli *et al* [10] studied the strategy space of all stochastic strategies with one-step memory. Here, in the first part, our approach is in a similar vein: we consider cellular automata (CA) with conditional strategies on their cells, specified by $[p_R, p_S, p_T, p_P]$. However, to provide a greater "microscopic" insight than just the four average values of the conditional probabilities (as is the case when continuous real conditional probabilities are considered), we resort to *binary Markovian strategies* (BMS). That is, the probability p_X of playing C after getting the payoff X is either 0 or 1, instead of real. There are thus only 2^4=16 possible BMS whose frequencies can be measured. Another simplification we introduced in is that at each time step a given agent plays the same action (C or D) against all its neighbors and takes account only its total score, collected by playing against them, instead of keeping track of each individual payoff. Therefore, we have *outer totalistic* CA [11] which, besides their simplicity and known properties of symmetry [12], show greater robustness against variations of the initial configuration. Besides deterministic automata, we explore a stochastic variant by introducing a (small) noise parameter ε and replacing the values of the conditional probabilities p_X, 0 or 1, by ε, 1-ε respectively. In a recent work [13] we studied the time evolution of the strategy frequencies in periodic lattices for several 2×2 games, including PD. In the first part of next section we resume some of those results for comparison purposes. In a second part, we study a similar system but agents have now a dynamic connectivity, i.e.: agents can create and break links with other agents in an attempt to improve their performance by changing the neighbors it plays with.

RESULTS

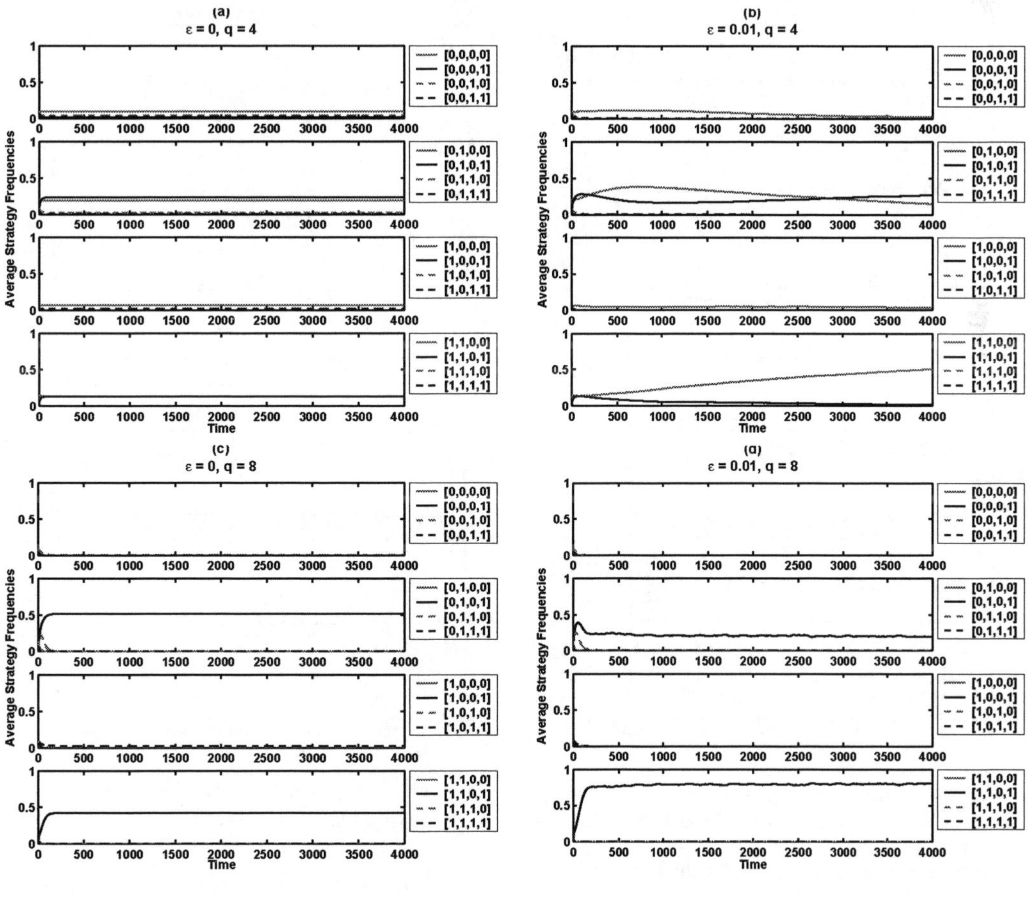

FIGURE 1. Average frequencies for the 16 competing Binary Markovian Strategies (BMS) vs the number of time steps for the Prisoner's Dilemma (PD) in a periodic lattice with $T = 1.333$, and $P = 0.5$. The four cases considered are (a) $q = 4$, $\varepsilon = 0$; (b) $q = 4$, $\varepsilon = 0.01$; (c) $q = 8$, $\varepsilon = 0$; (d) $q = 8$, $\varepsilon = 0.01$.

Periodic Lattice

We start with a periodic lattice of agents with a strategy random distribution of the 16 BMS. At each time each agent plays with all its nearest neighbors (we consider von Neumann neighborhood, $q = 4$ or Moore neighborhood, $q = 8$.), being its utility the sum of all the payoffs it obtains. Each agent compares its utility with the other neighbors utility, and it copies the strategy of the agent with the best performance. We study the time evolution of the strategy frequencies. We average the results for several different random seeds, to obtain results independent of the initial conditions. Figure 1 corresponds to the results for a periodic lattice of 100×100 and 1000 different random seeds [13].

For $q = 4$ and $\varepsilon = 0$ the system reaches quickly a dynamic equilibrium state in which several of the 16 strategies are present. As long as the noise increases ($\varepsilon = 0.01$), the number of strategies decreases, and the diversity without noise transforms into a few surviving strategies: TFT ($[1,0,1,0]$), $[0,0,1,1]$ and $[0,0,1,0]$. For $q = 8$, both with or without noise ($\varepsilon = 0$ or $\varepsilon = 0.01$), we have only two surviving strategies [0,0,1,1] and [1,0,1,1]. As we see, noise can change the performance of strategies, in the former [0,0,1,1] frequency is above [1,0,1,1], meanwhile in the latter the opposite happens. The behavior is robust with the change of the lattice size.

Random Evolving Network

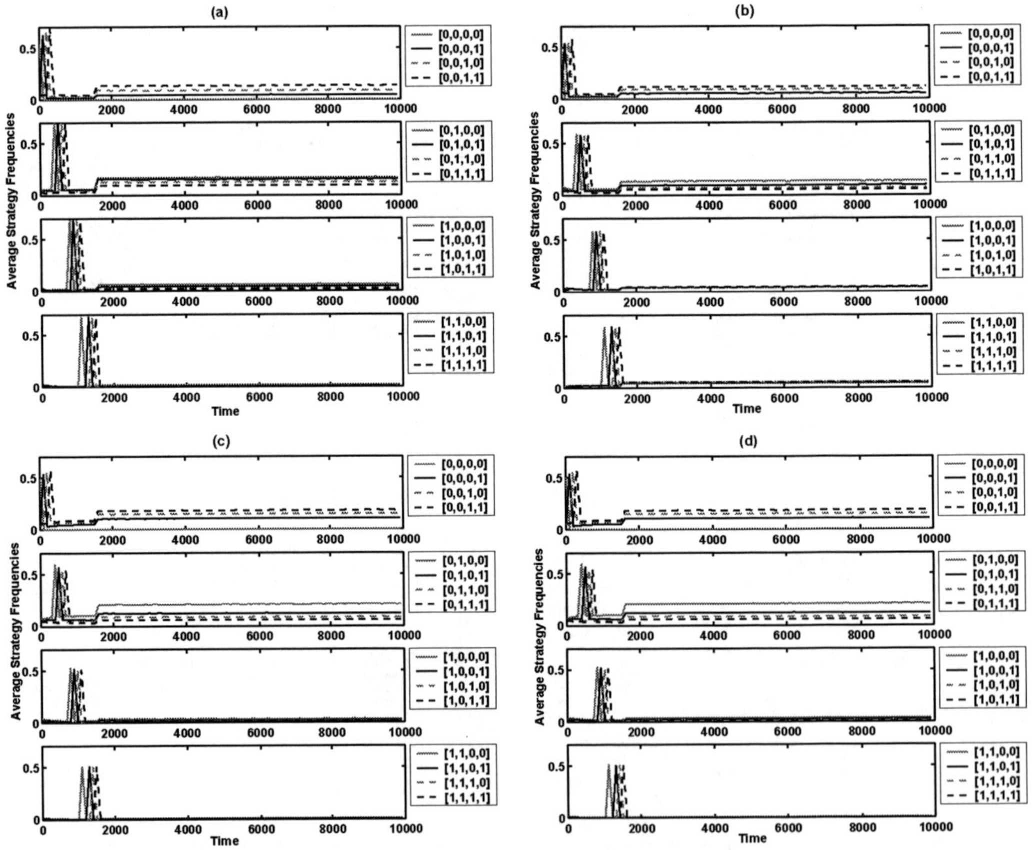

FIGURE 2. Average frequencies for the 16 competing BMS vs the number of time steps for the PD in a random evolving network with $R = 1, S = -2, T = 2$, and $P = -1$. The strategy references used in this figure (Fig. 2) are the same than in Fig. 1. The four cases considered are (a) $\varepsilon = 0$, the first move for all agents is playing C ; (b) $\varepsilon = 0$, the first move is random ; (c) $\varepsilon = 0.01$,the first move for all agents is playing C; (d) $\varepsilon = 0.01$.the first move is random.

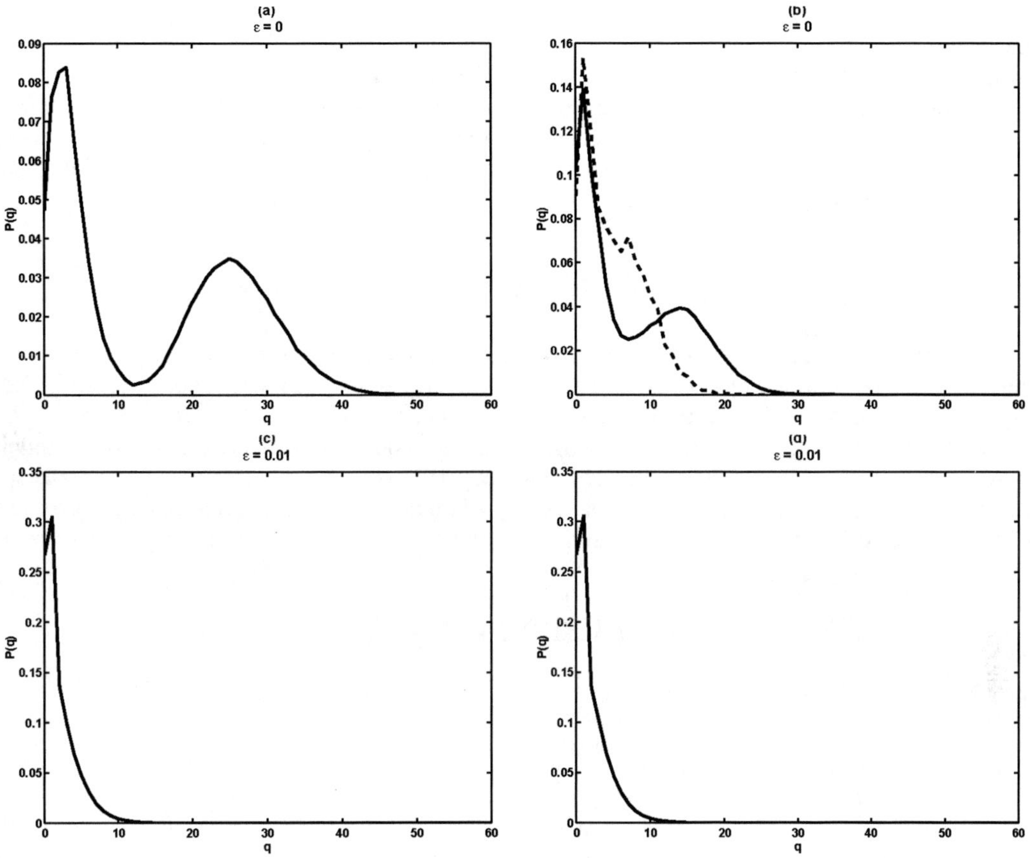

FIGURE 3. Connectivity distribution at the stationary state for all agents, regardless of their strategy,for the PD in a random evolving network with $R = 1, S = -2, T = 2$,and $P = -1$. The four cases considered are (a) $\varepsilon = 0$, the first move for all agents is playing C ; (b) $\varepsilon = 0$, the first move is random ; (c) $\varepsilon = 0.01$,the first move for all agents is playing C; (d) $\varepsilon = 0.01$.the first move is random. The dashed curve in (b) is the connectivity distribution for agents with strategy $[0,0,0,1]$. All the other distributions for each strategy considered alone are approximately the same than the general distribution.

We start with a completely disconnected network of agents and the strategies distributed at random among them as before. Unlike above, agents can play differently (C or D) to each neighbor. We used $R = 1, S = -2, T = 2$, and $P = -1$ instead of the normalized payoffs in this case. At each time step each agent tries to create a link if he is isolated or if its utility was satisfactory (above zero) in the previous time step. It selects at random one agent which was not in its neighborhood at the previous time step and only if the other agent was either satisfied with its own utilities or it was isolated, the link is created. If the agent's utility is not satisfactory, or it is connected to all the network, the agent attempts to break a link. It selects one of its neighbors at random and breaks the link if the individual payoff it got from that agent the previous time step was not satisfactory (above the distance factor, as we explain below). After the network is configured, like before each agent plays against all its neighbors, and copies the strategy of the neighbor with best performance. The utility is now the sum of all the payoffs obtained minus the a factor proportional to the sum of the distances between the agent and each of its neighbors. The distance is measured as the distance between two nodes of a periodic lattice. We introduce this distance factor to avoid a completely connected network as a result. Also we consider two cases: either all agents start cooperating or the first move is selected at random. As before we study the strategy frequency time evolution. The results for a network of 1024 agents are in Fig. 2.

After a short transitory the frequencies reach a stationary state. As in the case of a periodic lattice with $q = 4$ and $\varepsilon = 0$, in all the cases we have several coexisting strategies. We notice that as before $[0,0,1,0]$ and $[0,0,1,1]$ are among the most successful strategies. As expected, the cases for $\varepsilon = 0.01$ have an identical stationary state, which dependence with initial condition is erased by the introduction of noise. The connectivity distribution for the stationary state is given in Fig. 3.

For $\varepsilon = 0$ (see the two graphs above), there is a bimodal distribution of neighbors, meanwhile for the cases with $\varepsilon = 0.01$ (see the two lower graphs, which are identical for the reasons explained above) the distribution is fitted by a Laplace distribution $P(q) \approx 45.6 \exp(-0.486 \times q)$. In all the cases the neighbor distribution of each coexisting strategy is the same for all the strategies, being the only exception the distribution corresponding to $[0,0,0,1]$ for $\varepsilon = 0$ and the first move at random (dashed curve in Fig. 3 (b)).

CONCLUSIONS

Comparing the results for lattices [13] and evolving networks, we conclude the geometry is crucial for the selection of the surviving strategies. For the case of lattices the presence of noise can reduce the number of surviving strategies in a significant way, but this is not true for evolving networks. As expected, for the case of evolving networks, noise and random first moves reduce the connectivity, because agents are less cooperative and links are only maintained when both agents are satisfied. If we start with a completely disconnected network, for $\varepsilon = 0$ there is a bimodal distribution of neighbors, meanwhile for $\varepsilon = 0.01$ the distribution is fitted by a Laplace distribution. As a future work we are studying the case of evolving networks with different initial spatial configurations, for example starting from a periodic lattice.

REFERENCES

1. J. Maynard-Smith and G. Price, "The Logic of Animal Conflict", *Nature* (London) **146**,15,(1973).
2. J. Maynard-Smith,"Evolution and the Theory of Games", Cambridge Univ. Press, 1982.
3. R. Axelrod, in "The Evolution of Cooperation", Basic Books, New York, 1984
4. R. Axelrod, in "The Complexity of Cooperation", Princeton University Press, 1997.
5. R. Axelrod and W. D. Hamilton, "The evolution of cooperation", *Science*, **211**, (1981) 1390-1396.
6. M. A. Nowak and K. Sigmund, "A strategy of win-stay, lose-shift that outperforms tit for tat in Prisoner's Dilemma", *Nature*,**364**, 56-59, (1993).
7. D. Kraines and V. Kraines, *Theory Decision*, **26**, (1988) 47-79.
8. M.A. Nowak and R. May, "Evolutionary Games and Spatial Chaos", *Nature*, **359**, (1992) 826-829.
9. K. Lindgren and M. G. Nordahl, *Physica D*,**75**75 (1994) 292-309.
10. K. Brauchli, T. Killingback and M. Doebeli, *J. Theor. Biol.*, **200**, (1999) 405-417.
11. S. Wolfram, "Universality and Complexity in Cellular Automata", *Physica D*, **10**, (1984) 15-57.
12. S. Wolfram, "A New Kind of Science", Wolfram Media 2002.
13. H. Fort and E. Sicardi, "Evolutionary Markovian strategies in 2×2 spatial games", *Physica A*,**375**, (2007), 323-335

Bistability in Some 'Aggregates' of Logistic Maps

Ricardo López-Ruiz* and Danièle Fournier-Prunaret†

*DIIS and BIFI, Facultad de Ciencias, Universidad de Zaragoza, E-50009 Zaragoza, Spain.
†SYD and LESIA, Institut National des Sciences Appliquées, 31077 Toulouse Cedex, France.

Abstract. In general, the behavior of large and complex aggregates of elementary components can not be understood nor extrapolated from the properties of a few components. Following this insight, we have simulated different 'aggregates' of logistic maps according to a particular coupling scheme. All these aggregates show some common dynamical properties, concretely a bistable behavior that is sketched and collected here. A qualitative relationship with brain-like systems is suggested.

Keywords: Bistalibity, coupled logistic oscillators, neural networks
PACS: 07.05.Mh, 05.45.Ra, 05.45.Xt

INTRODUCTION

One of the most challenging scientific problems today is to understand how the millions of neurons of our brain give rise to the emergent property of thinking [1]. Different aspects of neurocomputation take contact on this problem: how brain stores information and how brain processes it to take decisions or to create new information. Other universal properties of this system are more evident. One of them is the existence of a regular daily behavior: the sleep-wake cycle [2, 3]. The internal circadian rhythm is closely synchronized with the cycle of sun light. Roughly speaking and depending on the particular species, the brain is awake during the day and it is slept during the night, or vice versa. All mammals and birds sleep. There is no a well established law relating the size of the animal with the daily time it spends sleeping, but, in general, large animals tend to sleep less than small animals. Hence, at first sight, the bistable sleep-wake behavior seems not depend on the precise architecture of the brain nor on its size. If we represent the brain as a complex network this property would mean that this possible bistability should not depend on the topology (structure) nor on the number of nodes (size) of the network.

So, on one side, it has been recently argued in [4] that the distribution of functional connections $p(k)$ in the human brain, where $p(k)$ represents the probability of finding an element with k connections to other elements of the network, follows the same distribution of a scale-free network [5]. Thus, in that work [4], the human brain is divided in 34*x*64*x*64 sites (called *voxels*) and the magnetic resonance activity of all the voxels is recorded. Let us observe at this point that if the human brain has around 10^{11} neurons, then a voxel has around 10^5 neurons. The calculation of the correlation matrix among the set of the voxels activities shows a power law behavior, $p(k) \sim k^{-\gamma}$, with γ around 2. This finding means that there are regions in the brain that participate in a large number of tasks while most of the regions are only involved in a tiny fraction of the brain's activities.

On the other side, it has been shown by Kuhn et al. [6] the nonlinear processing of synaptic inputs in cortical neurons. They studied the response of a model neuron with a simultaneous increase of excitation and inhibition. They found that the firing rate of the model neuron first increases, reaches a maximum, and then decreases at higher input rates. Functionally, this means that the firing rate, commonly assumed to be the carrier of information in the brain, is a non-monotonic function of balanced input. These findings do not depend on details of the model and, hence, are relevant to cells of other cortical areas as well.

Putting together all these facts, we arrive to the central question that we want to bring to the reader: are we able to reproduce the bistability in a complex network independently of the topology and of the number of nodes?. The answer is 'yes' (Fig. 1). What kind of local dynamics and coupling among nodes must be implemented in order to get this behavior?. In the next section, we give a possible strategy for the coupling and the local dynamics which should be implemented in a few or many units network in order to find bistable behavior. An example of this implementation is given in [7]. Here we center our attention in the bistability present in the case of a few coupled functional units. Four different discrete models in two and three dimensions are collected. In view of the results, we want to suggest with all these examples the possibility of 'constructivism' in the world of complex systems.

CP913, *Nonequilibrium Statistical Mechanics and Nonlinear Physics, XV Conference*
edited by O. Descalzi, O. A. Rosso, and H. A. Larrondo

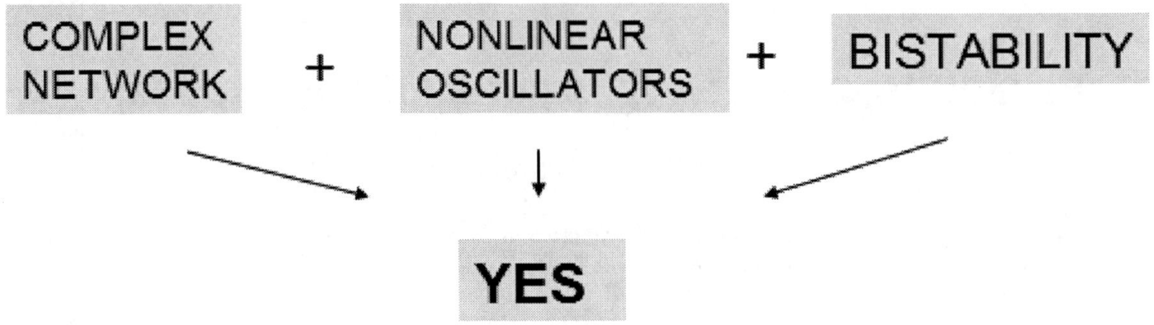

FIGURE 1. QUESTION: Is it possible to implement some kind of coupling and nonlinear dynamics in each node of a complex network in order to get bistability?. ANSWER: Yes.

MODELS OF A FEW COUPLED FUNCTIONAL UNITS

General model

Our approach consider the so called *functional unit*, i.e. a neuron or group of neurons (voxels), as a discrete nonlinear oscillator with two possible states: active (meaning one type of activity) or not (meaning other type of activity). Hence, in this naive vision of the brain as a networked system, if x_n^i, with $0 < x_n^i < 1$, represents a measurement of the *ith* functional unit activity at time n, it can be reasonable to take the most elemental local nonlinearity, for instance, a logistic evolution [8], which presents a quadratic term, as a first toy-model for the local neuronal activity:

$$x_{n+1}^i = \bar{p}_i x_n^i (1 - x_n^i). \tag{1}$$

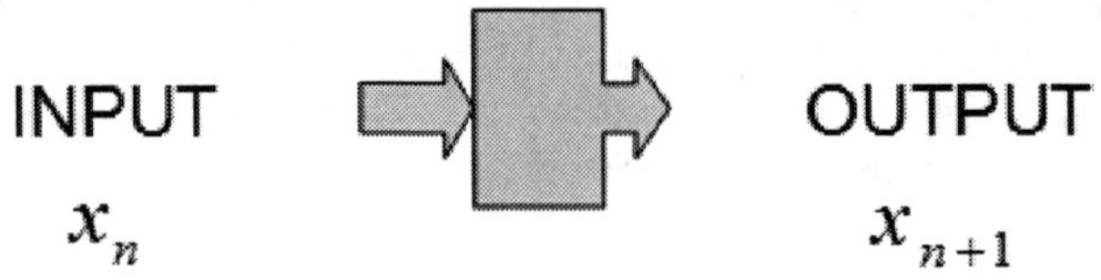

FIGURE 2. Discrete nonlinear model for the local evolution of a functional unit.

It presents only one stable state for each $\bar{p}_i$. Then, there is no bistability in the basic component of our models. For $\bar{p}_i < 1$, the dynamics dissipates to zero, $x_n^i = 0$, then it can represent the functional unit with no activity. For $1 < \bar{p}_i < 4$, the dynamics is non null and it would represent an active functional unit. This local transition is controlled by the parameter $\bar{p}_i$. The functional dependence of this local coupling on the neighbor states is essential in order to get a good brain-like behavior (i.e., as far as the bistability of the sleep-wake cycle is concerned) of the network. As a first approach, we can take $\bar{p}_i$ as a linear function depending on the actual mean value, X_n^i, of the neighboring signal activity and expanding the interval $(1,4)$ in the form:

$$\bar{p}_i = p_i\,(3X_n^i + 1), \quad (excitation\ coupling) \tag{2}$$

$$or$$

$$\bar{p}_i = p_i\,(-3X_n^i + 4), \quad (inhibition\ coupling) \tag{3}$$

with

$$X_n^i = \frac{1}{N_i} \sum_{j=1}^{N_i} x_n^j. \tag{4}$$

N_i is the number of neighbors of the *ith* functional unit, and p_i, which gives us an idea of the interaction of the functional unit with its first-neighbor functional units, is the control parameter. This parameter runs in the range $0 < p_i < p_{max}$, where $p_{max} \succeq 1$. When $p_i = p$ for all i, the dynamical behavior of these networks with the excitation type coupling [7] presents an attractive global null configuration that has been identified as the *turned off* state of the network. Also they show a completely synchronized non-null stable configuration that represents the *turned on* state of the network. Moreover, a robust bistability between these two perfect synchronized states is found in that particular model (see [7] for more details). For different models with a few coupled functional units we sketch in the next subsections the regions where they present a bistable behavior. The details of the complete unfolding of these dynamical systems can be found in the references [9, 10, 11].

Models of two functional units

Let us start with the simplest case of two interconnected (x_n, y_n) functional units. Three different combinations of couplings are possible: $(excitation, excitation)$, $(excitation, inhibition)$ and $(inhibition, inhibition)$.

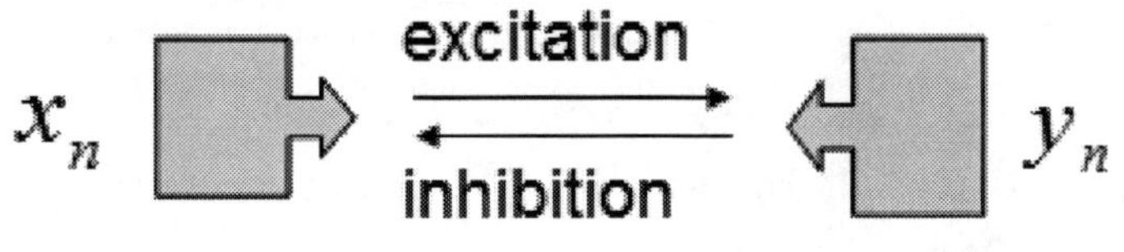

FIGURE 3. Two functional coupled units.

The first two cases of coupling present bistability in different regions of the parameter space. The third case also shows bistability in a very narrow interval of parameter space but it requires a very fine inspection [12] that we shelve for a further work.

Model with mutual excitation

The dynamics of the $(excitation, excitation)$ case [9] is given by the coupled equations:

$$x_{n+1} = p\,(3y_n + 1)x_n(1 - x_n), \quad (5)$$
$$y_{n+1} = p\,(3x_n + 1)y_n(1 - y_n). \quad (6)$$

The regions of the parameter space (Fig. 4) where we can find bistability are:

- For $0.75 < p < 0.86$, the synchronized state, $x_+ = (\bar{x}, \bar{x}) = P_4$, with $\bar{x} = \frac{1}{3}\{1 + (4 - \frac{3}{p})^{\frac{1}{2}}\}$, which arises from a saddle-node bifurcation for the critical value $p = 0.75$, is a stable *turned on* state. This state coexists with the *turned off* state $x_\theta = 0$. The system presents now bistability and depending on the initial conditions, the final state can be x_θ or x_+. Switching on the system from x_θ requires a level of noise in both functional units sufficient to render the activity on the basin of attraction of x_+. On the contrary, switching off the two functional units network can be done, for instance, by making zero the activity of one functional unit, or by doing the coupling p lower than 0.75.
- For $0.86 < p < 0.95$, the active state of the network is now a period-2 oscillation, namely the period-2 cycle (P_5, P_6) in Fig. 4. This new dynamical state bifurcates from x_+ for $p = 0.86$. A smaller noise is necessary to activate the system from x_θ. Making zero the activity of one functional unit continues to be a good strategy to turn off the network.
- For $0.95 < p < 1$, the active state acquires a new frequency and presents quasiperiodicity (the invariant closed curves of Fig. 4). It is still possible to switch off the network by putting to zero one of the functional units.

Model with excitation + inhibition

The dynamics of the $(excitation, inhibition)$ case [10] is given by the coupled equations:

$$x_{n+1} = p\,(3y_n + 1)x_n(1 - x_n), \quad (7)$$
$$y_{n+1} = p\,(-3x_n + 4)y_n(1 - y_n). \quad (8)$$

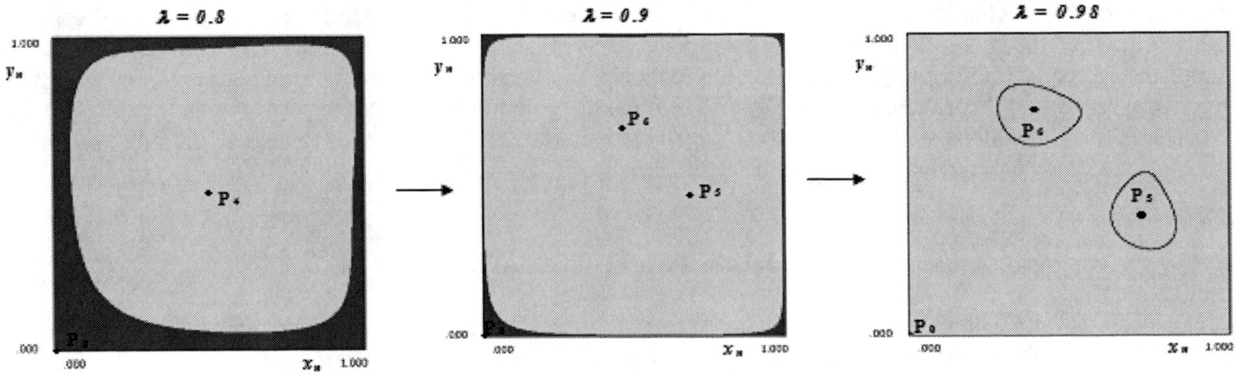

FIGURE 4. Bistability in 2 functional units with excitation type coupling.

The regions of the parameter space (Fig. 5) where we can find bistability are:

- For $1.051 < p < 1.0851$, a stable period three cycle (Q_1, Q_2, Q_3) appears in the system. It coexists with the fixed point P_4. When p is increased, a period-doubling cascade takes place and generates successive cycles of higher periods $3x2^n$. The system presents bistability. Depending on the initial conditions, both populations (x_n, y_n) oscillate in a periodic orbit or, alternatively, settle down in the fixed point. The borders between the two basins are complex.
- For $1.0851 < p < 1.0997$, an aperiodic dynamics is possible. The period-doubling cascade has finally given birth to an order three cyclic chaotic band(s) (A_{31}, A_{32}, A_{33}). The system can now present an irregular oscillation besides the stable equilibrium with final fixed populations. The two basins are now fractal.

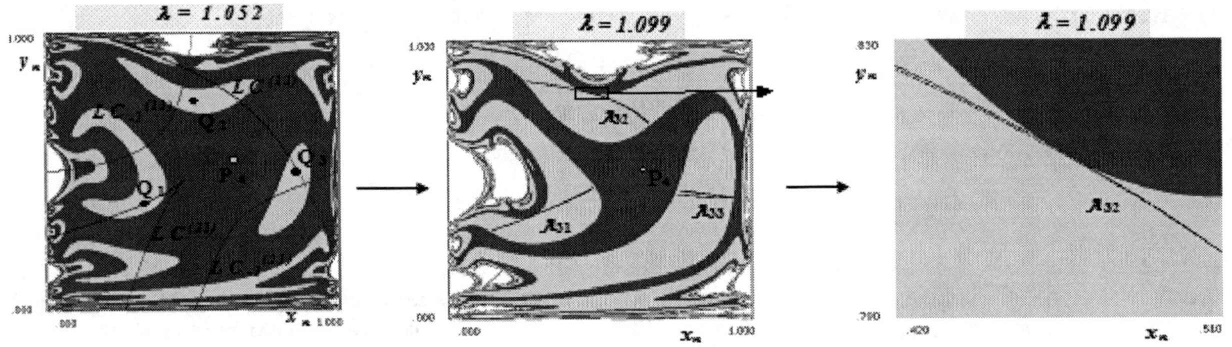

FIGURE 5. Bistability in 2 functional units with excitation+inhibition type coupling.

Models of three functional units

Following the strategy given by relation (2-3) several models with three functional units can be established. We have studied in detail two of them [11] and their bistable behavior is reported here.

Model with local mutual excitation

Let us start with the case of three alternatively interconnected (x_n, y_n, z_n) functional units under a mutual excitation scheme.

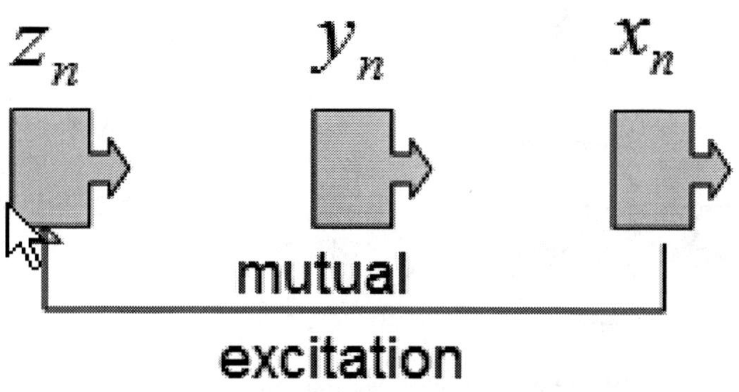

FIGURE 6. Three alternatively coupled functional units under the excitation scheme.

Then the dynamics of the system is given by the coupled equations:

$$x_{n+1} = p\,(3y_n+1)x_n(1-x_n), \tag{9}$$
$$y_{n+1} = p\,(3z_n+1)y_n(1-y_n), \tag{10}$$
$$z_{n+1} = p\,(3x_n+1)z_n(1-z_n). \tag{11}$$

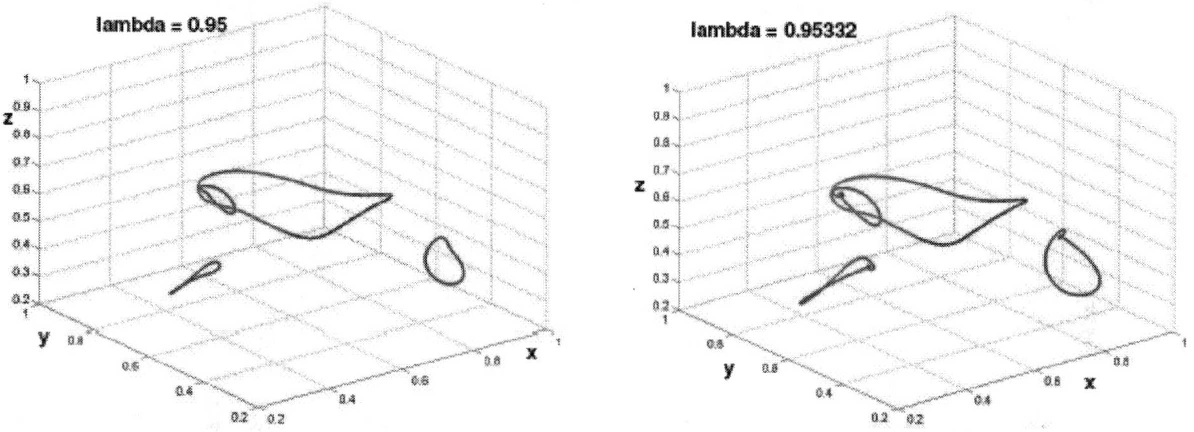

FIGURE 7. Bistability in 3 functional units with local excitation type coupling.

The regions of the parameter space where we have found bistability are:

- For $0.93310 < p < 0.95334$, a big invariant closed curve (ICC) $C1$ coexists with a period-3 orbit that bifurcates, first to an order 3-cyclic ICC (Fig. 7), and finally to an order-3 weakly chaotic ring (WCR) before disappearing.
- For $0.98418 < p < 0.98763$, the ICC $C1$ coexists with another ICC $C2$ (see Ref. [11]) that becomes chaotic, by following a period doubling cascade of tori, before disappearing.
- For $1.00360 < p < 1.00402$, the ICC $C1$ coexists with a high period orbit that gives rise to an ICC $C3$. This ICC also becomes a chaotic band (see Ref. [11]) by following a period doubling cascade of tori before disappearing.

Model with global mutual excitation

We expose now the case of three globally interconnected (x_n, y_n, z_n) functional units under a mutual excitation scheme.

Then the dynamics of the system is given by the coupled equations:

$$x_{n+1} = p\,(x_n+y_n+z_n+1)x_n(1-x_n), \tag{12}$$

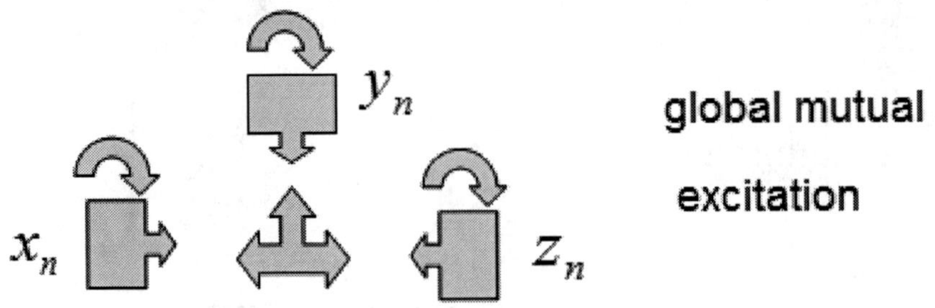

FIGURE 8. Three globally coupled functional units under the excitation scheme.

$$y_{n+1} = p\,(x_n + y_n + z_n + 1) y_n (1 - y_n), \tag{13}$$
$$z_{n+1} = p\,(x_n + y_n + z_n + 1) z_n (1 - z_n). \tag{14}$$

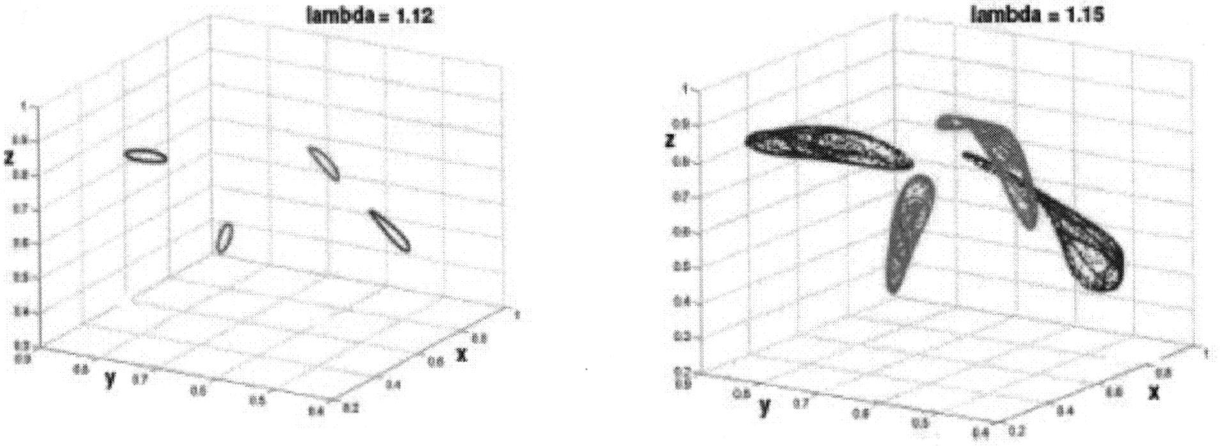

FIGURE 9. Bistability in 3 functional units with global excitation type coupling.

For the whole range of the parameter, $0 < p < 1.17$, bistability is present in this system:

- Firstly, two order-2 cyclic ICC coexist before becoming two order-2 cyclic chaotic attractors (Fig. 9) by contact bifurcations of heteroclinic type. Finally the two chaotic attractors become a single one before disappearing.

CONCLUSIONS

One of the more challenging problems in nonlinear science is the goal of understanding the properties of neuronal circuits [13]. Synchrony and multistability are two important dynamical behaviors found in those circuits. In this work, different coupling schemes for networks with local logistic dynamics are proposed. It is observed that these types of couplings generate a global bistability between two different dynamical states. This property seems to be topology and size independent. This is a direct consequence of the local mean-field multiplicative coupling among the first-neighbors. If a formal and naive relationship is established between these two states and the sleep-wake states of a brain, respectively, one would be tempted to assert that these types of couplings in a network, regardless of its simplicity, give us a good qualitative model for explaining that specific bistability. Following this insight, different low-dimensional systems with logistic components coupled under these schemes have been presented. The regions where the dynamics shows bistability have been identified. Other low and high dimensional models merit a similar detailed inspection in the future.

ACKNOWLEDGMENTS

R. L.-R. acknowledges some financial support of the Spanish DGICYT Projects FIS2005-06237 and FIS2006-12781-C02-01.

REFERENCES

1. J. Carey (Editor), *Brain Facts*, fifth revised edition, Publications of the Society for Neuroscience, Washington, 2006.
2. A.T. Winfree, *The timing of biological clocks*, Scientific American Library, New York, 1986.
3. Y. Bar-Yam, *Dynamics of Complex Systems*, Westview Press, New York, 1997.
4. V.M. Eguiluz, D.R. Chialvo, G. Cecchi, M. Baliki, and A.V. Apkarian, *Phys. Rev. Lett.* **94**, 018102(4) (2005).
5. A.L. Barabási and R. Albert, *Science* **286**, 509–512 (1999).
6. A. Kuhn, Ad. Aertsen, and S. Rotter, *J. Neurosci.* **24**, 2345–2356 (2004).
7. R. Lopez-Ruiz, Y. Moreno, A. F. Pacheco, S. Boccaletti, and D.-U. Hwang, *Neural Networks* **20**, 102–108 (2007).
8. R.M. May, *Nature* **261**, 459–467 (1976).
9. R. Lopez-Ruiz and D. Fournier-Prunaret, *Math. Biosci. Eng.* **1**, 307–324 (2004).
10. R. Lopez-Ruiz and D. Fournier-Prunaret, *Chaos, Solitons and Fractals* **24**, 85–101 (2005).
11. D. Fournier-Prunaret, R. Lopez-Ruiz, and A.K. Taha, *Grazer Mathematische Berichte*, **350**, 82–95 (2006).
12. C. Mira, L. Gardini, A. Barugola, and J.-C. Cathala, *Chaotic Dynamics in Two-Dimensional Noninvertible Maps*, series A, vol. 20, World Scientific Publishing, Singapore, 1996.
13. M.I. Rabinovich, P. Varona, A.I. Selverston, and H.D.I. Abarbanel, *Rev. Mod. Phys.*, **78**, 1213–1265 (2006).

Topology for Dominance for Network of Multi-Agent System

K. Y. Szeto

Department of Physics, Hong Kong University of Science and Technology.
Clear Water Bay, Hong Kong SAR, China

Abstract. The resource allocation problem in evolving two-dimensional point patterns is investigated for the existence of good strategies for the construction of initial configuration that leads to fast dominance of the pattern by one single species, which can be interpreted as market dominance by a company in the context of multi-agent systems in econophysics. For hexagonal lattice, certain special topological arrangements of the resource in two-dimensions, such as rings, lines and clusters have higher probability of dominance, compared to random pattern. For more complex networks, a systematic way to search for a stable and dominant strategy of resource allocation in the changing environment is found by means of genetic algorithm. Five typical features can be summarized by means of the distribution function for the local neighborhood of friends and enemies as well as the local clustering coefficients: (1)The winner has more triangles than the loser has. (2) The winner likes to form clusters as the winner tends to connect with other winner rather than with losers; while the loser tends to connect with winners rather than losers. (3) The distribution function of friends as well as enemies for the winner is broader than the corresponding distribution function for the loser. (4) The connectivity at which the peak of the distribution of friends for the winner occurs is larger than that of the loser ; while the peak values for friends for winners is lower. (5) The connectivity at which the peak of the distribution of enemies for the winner occurs is smaller than that of the loser ; while the peak values for enemies for winners is lower. These five features appear to be general, at least in the context of two-dimensional hexagonal lattices of various sizes, hierarchical lattice, Voronoi diagrams, as well as high-dimensional random networks. These general local topological properties of networks are relevant to strategists aiming at dominance in evolving patterns when the interaction between the agents is local.

Keywords: Topology, Econophysics, Resource Allocation, Evolutionary Computation, Multi-Agent Systems.
PACS: 89.20.-a; 89.65.; 05.70.Fh

INTRODUCTION

The analysis of empirical point patterns is an active branch of science and engineering research with applications in many areas of artificial intelligence [1-3]. One important application in operational research on urban studies and location theory in economics [4] is resource allocation on evolving point patterns. We introduce our approach to this problem via a simple example. Let's consider a game of chess defined with fair rule of play on a network. Assume that there are two chess players with equal number of pieces: one with white and the other with black pieces. The player can control only the initial assignment of chess pieces as the game will automate its evolution according to the local rule, which determines the color change probabilistically according to the local environment of the piece under scrutiny. The question is whether there is a way to assign the pieces in the finite lattice so that it will dominate the lattice after some time of evolution? For example, consider a hexagonal lattice as shown in Fig.1. We have four different ways of initial assignment of the chess pieces for the two players occupying 50% of the lattice sites. We like to ask which of the four topological arrangements will give high probability of the white pieces dominating, defined as occupying about 90% of the lattice sites. A more difficult question is to ask which topology will give the higher speed to dominance, measured as the first passage time of the white pieces to achieve dominance over many runs of the evolution. Here, the evolution of color is defined by a local Hamiltonian, which in a typical Multi-agent system depends on the model of interaction. The discussion of the Hamiltonian in the context of econophysics has been made in the case of supermarket chain in a city map [5]. However, the result of the present work on the topological assignment problem is expected to be valid for pattern evolution as long as rules of the chess game are local ones. This example will illustrate the importance of topology in pattern evolution.

From this introduction, we identify the key problem as one of optimal strategy of resources allocation. We have raised two issues in competing multi-agent systems in econophysics. The first issue concerns the dynamics. Given a static point pattern in a given network, we can design a model that includes competition, cooperation, and alliance to describe the dynamics of multi-agents. In this paper, we use the spin model with local interaction, and employ statistical mechanics to describe the patterns of various states in this model. The second issue is the resource allocation

CP913, *Nonequilibrium Statistical Mechanics and Nonlinear Physics, XV Conference*
edited by O. Descalzi, O. A. Rosso, and H. A. Larrondo

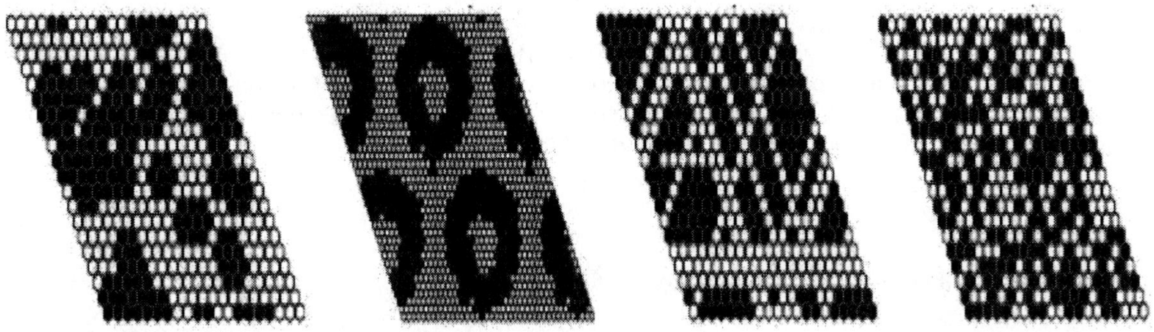

FIGURE 1. Four examples of different topologies of initial spin configurations (Cluster, Ring, Line, and Random Graph). All figures have equal number of white and black hexagons. By convention, the player with white hexagons is more likely to be the winner, (i.e. the dominant color after some time of evolution).

problem based on the model of interacting multi-agent on the given network. The objective is to search for a stable and dominant strategy of allocation in the network so as to achieve dominance. In two-dimensional physics, many works have addressed the relevance of topology [6-12]. It is natural for us to expect that local arrangement of the resource also plays a crucial role in the fast dominance of a particular group of agents in this problem of evolving patterns in networks.

The dynamics of the chess pieces on the network can be described by a Hamiltonian with short-ranged interaction, such as the Ising model for the case of two competing companies of agents. Leaving the justification in econophysics aside, as it has been discussed elsewhere [5], we can consider the interaction energy J_{ij} between agents in two neigbouring sites (i,j) with the total energy of the system defined on a network given by

$$E = -\sum_{\langle i,j \rangle} J_{ij}\sigma_i\sigma_j \quad \text{with } J_{ij} = \begin{cases} 0 & if \quad i = j \\ 1 & if \quad i \neq j \end{cases} \tag{1}$$

Here σ_i represents occupancy of the branch of the two competing company, which takes the value 1 or -1. This is the Ising spin variable describing the two colour chess problem. It means that a node i in the network has to be occupied by one and only one agent of either company, but not both or none. Note that the interaction between two agents of the same company lowers the energy and vice versa. The evolution of the pattern is based on the metropolis algorithm to switch the affiliation of agents between the two companies. In statistical physics, the probability of switching from one state to another in a canonical ensemble depends on the Boltzmann factor,$\exp(-\beta\Delta E)$, where $\beta = 1/T$ is related to thermal noise and ΔE is the energy difference between two states: energy after switching minus energy before switching. Here a large β corresponds to low noise or high loyalty for the agent. In our simulation we've tried a temperature below the critical temperature of ordering, (for the case the hexagonal lattice in two dimension and in general higher dimensional networks, such ordered phase exists for this Ising Hamiltonian), thereby the existence of one spin dominance is guaranteed. Our Monte Carlo simulation is conducted by repeating the metropolis steps, until one spin dominates 87.5% of the lattice. The choice of 87.5% as the definition of dominance is by design and the result is not very sensitive to this choice. It only serves as a stopping criterion for the Monte Carlo simulation.

MODEL IN TWO-DIMENSION

We evolve a given initial configuration with the Metropolois algorithm for the simple example of Ising Hamiltonian defined on the hexagonal lattice. We design a measure, called fitness, for the quality of the given configuration. The first element of this fitness measure is to rate the speed in achieving dominance, defined by a fixed fraction $b > 0.5$ of the white pieces on the lattice. Dominance of the white (black) pieces is declared once there are more than b fraction of the lattice turn white (black). The stopping criterion for the evolution of the pattern is defined by the maximum time step $t < 50000$, or when one colour has reached b percent of the sample in a particular trial, whichever is smaller.

For each of the four patterns in Fig.1 as initial configurations, we generate 200 copies and for each copy we repeat the Monte Carlo simulation with the given Hamiltonian for N' trials. Some of these N' trials may enter a deadlock, in which neither colour can dominate the other within 50000 MC steps. In this situation we reject these trials. Assuming that there remain N $(< N')$ simulations with clear winners, with n_1 times having white dominance and n_{-1} having black dominance, $(N = n_1 + n_{-1})$,we define the measure of fitness by

$$F(x,b) = \frac{n_x}{N}\left\langle \frac{1}{t} \right\rangle_x \quad where \quad x \in \{1,-1\} \tag{2}$$

Here x can be 1 or -1 and t is the number of MCS for configuration i to evolve to a state with a winner. Here $<1/t>_x$ is the average of $1/t$ over n_x trials where spin x dominates the lattice. After evaluating 200 copies of the same topological configuration, we observe that one colour (which we choose to be white) has a higher probability to dominate, except for the case of Random Graph. In our simulations we set $b = 0.75$ and $N' = 2000$, which is chosen so that the error of $<1/t>x$ is less than 10%. The probability for white dominance is close to certainty for the ring, and 0.94 for the straight line, 0.60 for the cluster, and as expected 0.50 for the random graph. Although our numerical works are performed on the hexagonal lattice, we expect our findings to be generally applicable to the Voronoi diagrams constructed from random point patterns in two-dimension. This expectation is based on our recent work on damage spreading in various kinds of two-dimensional cellular structures [13-15].

GENETIC ALGORITHM TO SEARCH FOR WINNIG CONFIGURATION

In order to find the topological characteristics of winning strategies for resource allocation in more complex networks, we must design an efficient algorithm to search for initial configuration with high fitness. Note that the search space is in general too big for exhaustive enumeration. For example, for an 8×8 hexagonal lattice, the number of possible our spin configuration is C^{64}_{32}. In this regards, evolutionary computational techniques [16,17] have been quite successful in many problems of this kind, such as the travelling salesman problem. From our work on genetic algorithm using the formalism of mutation matrix [18-20], we find that the knapsack problem can be solved quite efficiently. We therefore use this technique to search for good initial configurations that lead to dominance. Before we begin, we first give a more useful measure of fitness of configuration.

Consider the example of the 8×8hexagonal lattice. We search for good configuration of initial spin assignment: good in the sense that the winning probability of spin up is high. (Here the winner by convention is associated with spin up). Using genetic algorithm, we first map each spin configuration onto a chromosome with 64 loci, with half of the loci assigned 1 and the rest 0. Secondly, we provide an operational definition of fitness value for each chromosome. This definition should reflect the quality of the configuration such as the speed to dominance and probability of domination of the winner. Specifically, given a chromosome c representing a particular initial spin configuration, we perform N' trials of Monte Carlo simulation, among which there are N trials with clear winners. Furthermore, let's assume that spin up achieves (b=87.5%) dominance in $u$out of the N *successful* trials, while $N - u$ times spin down is dominant. For each successful trial, we denote by s as the metropolis steps required to evolve the initial configuration to one that has 87.5% dominance by one spin type, say the spin up type. Thus, among N trials, the average speed to dominance by spin up is given by $<1/s>_u$. For each initial spin configuration represented by chromosome c, we define its fitness f_c by the average speed to dominance by spin up times the probability that spin up dominates among Ntrials,

$$f_c = \left\langle \frac{1}{s} \right\rangle_u \times \frac{u}{N} \tag{3}$$

Note that this definition of fitness of spin assignment reflects the two desirable qualities of the configuration: speed to dominance and probability of dominance. The choice of 87.5% is for convenience only, as it corresponds to 56 out of 64 sites to be dominated by one spin type. To determine a suitable N' for a typical chromosome, we first use a few thousands trials to obtain a distribution of speed $(1/s)$. For an 8×8 hexagonal lattice at a temperature $T = 1$, we observed from our numerical experiment that a precision of two significant figures can be achieved with N' around 3000. We therefore set N'=6000 and truncate all the digits after the second significant number. For higher temperature, such as T=3 or 4, which is above the critical temperature for the ordered state, we observed that the required $N'$is too large to carry out simulation in a reasonable time. However, we have tested the good configurations found at $T = 1$ and test them at *T=3, or 4*, and found that these configurations are still "good", meaning that their fitness is still significantly higher than those random configurations by one standard deviation.

The type of the genetic algorithm we use is mutation only genetic algorithm (MOGA) [19,20] which features an adaptive parameter control that provides an efficient method for search in large configuration space. Since we want to compare the strategies of two companies with initially equal market share, we must make sure that in the construction of initial spin configurations there are equal number of spin up and spin down. However, the mutation operation in genetic algorithm will not conserve the number of spin up and down (represented by 1 and 0). We thus need to modify MOGA [21]. In our evolutionary computation, we set the population size of chromosome to be 50 and the number of generation for the chromosomes to evolve to be 200. In order to draw conclusion on the topological characteristics of the winners, we collect a handful of configurations with high fitness by repeating the simulation many times. The number of generation and the corresponding population size are : (number of generation for evolution, population size)={(200,50), (200,30), (50,30),(10,50)} and the best results are found for the 50 chromosomes evolving for 200 generations. The simulation is repeated until we have 10 chromosomes of fitness which are within 10% of the highest fitness observed.

GENERAL MEASURES OF TOPOLOGICAL FEATURES

Even though our evolutionary computation method produces a set of good configurations, we still need to find the reason behind their success. Fig.1 provides some hint to measure these good configurations in terms of their topological features. Here we make use of the concept of connectivity. For a site $i$with degree k_i, the connectivity with friends is denoted by C_i , which is the number of nearest neighbors of the site i that have the same spin as$i$. We define the connectivity with enemies for a site i as $k_i - C_i$. We first introduce some definitions for the local configuration of spins at site i: C_i^S is defined as the connectivity with the same spin type, which is the number of edges connected to sites with the same spin type as the spin at site i;and C_i^O is defined as the connectivity with the opposite spin type, which is the number of edges connected to sites with the opposite spin type as the spin at site i. Next we define the set $R(\sigma,K,x) = \{i|\text{spin type of site } i = \sigma \text{ and } C_i^K = x;\ K = S,O\}$, which contains the sites of spin type σ with connectivity $C_i^K = x$.The number of elements of $R(\sigma,K,x)$ is denoted by$N(\sigma,K,x)$. We then define the distribution functions,

$$D(\sigma,S,x) \equiv \frac{N(\sigma,S,x)}{\sum_y N(\sigma,S,y)}, \qquad D(\sigma,O,x) \equiv \frac{N(\sigma,O,x)}{\sum_y N(\sigma,O,y)}. \tag{4}$$

For spin type $\sigma, D(\sigma,S,x)$ and $D(\sigma,O,x)$are the distribution function of connectivity x with same (S) spin and opposite spin (O) respectively.The local neighborhood described by the distribution function of connectivity of the same spin type (friend) $D(\sigma,S,x)$or opposite spin type (enemy) $D(\sigma,O,x)$for enemies are informative about the local arrangement of spin types, as shown in Fig.2. We see that the winner tends to connect to the sites with same spin, but the loser tends to connect to sites with opposite spin.

Another useful local measure is the number of triangles T^σ for monitoring the clustering of friends. A triangle is defined as three interconnected sites with the same spin type σ. The total number of triangleT_i at a site i is defined as the number of triangles that has site i as one of its vertices. The total number of triangle of winner and loser, $T^W$and T^L, are defined as:

$$T^W = \frac{1}{3}\sum_{i\in W} T_i \quad and \quad T^L = \frac{1}{3}\sum_{i\in L} T_i \quad . \tag{5}$$

The difference in triangles $\tau \equiv T^W - T^L$is also a useful measure to distinguish winners from losers. Our numerical experiments show consistently that $\tau > 0$ with an averaged value of 3.33 for the 8×8 hexagonal lattice.

Based on results of the 8×8 hexagonal lattice with periodic boundary condition, we summarize our results into five observations.

1. The winner has more triangles than the loser has.
2. The winner likes to form clusters as the winner tends to connect with other winner rather than with losers; while the loser tends to connect with winners rather than losers.
3. The distribution function of friends as well as enemies for the winner $D(+1,K,x)$ is broader than the corresponding distribution function for the loser$D(-1,K,x)$.
4. The connectivity at which the peak of the distribution of friends for the winner occurs is larger than that of the loser $(x^*(+1,S) = 4 > x^*(-1,S) = 3)$; while the peak values for friends for winners is lower $(D(+1,S,4) < D(-1,S,3))$. Here $x^*(\sigma,S)$is the location of the peak value of $D(\sigma,S,x)$ versus x.

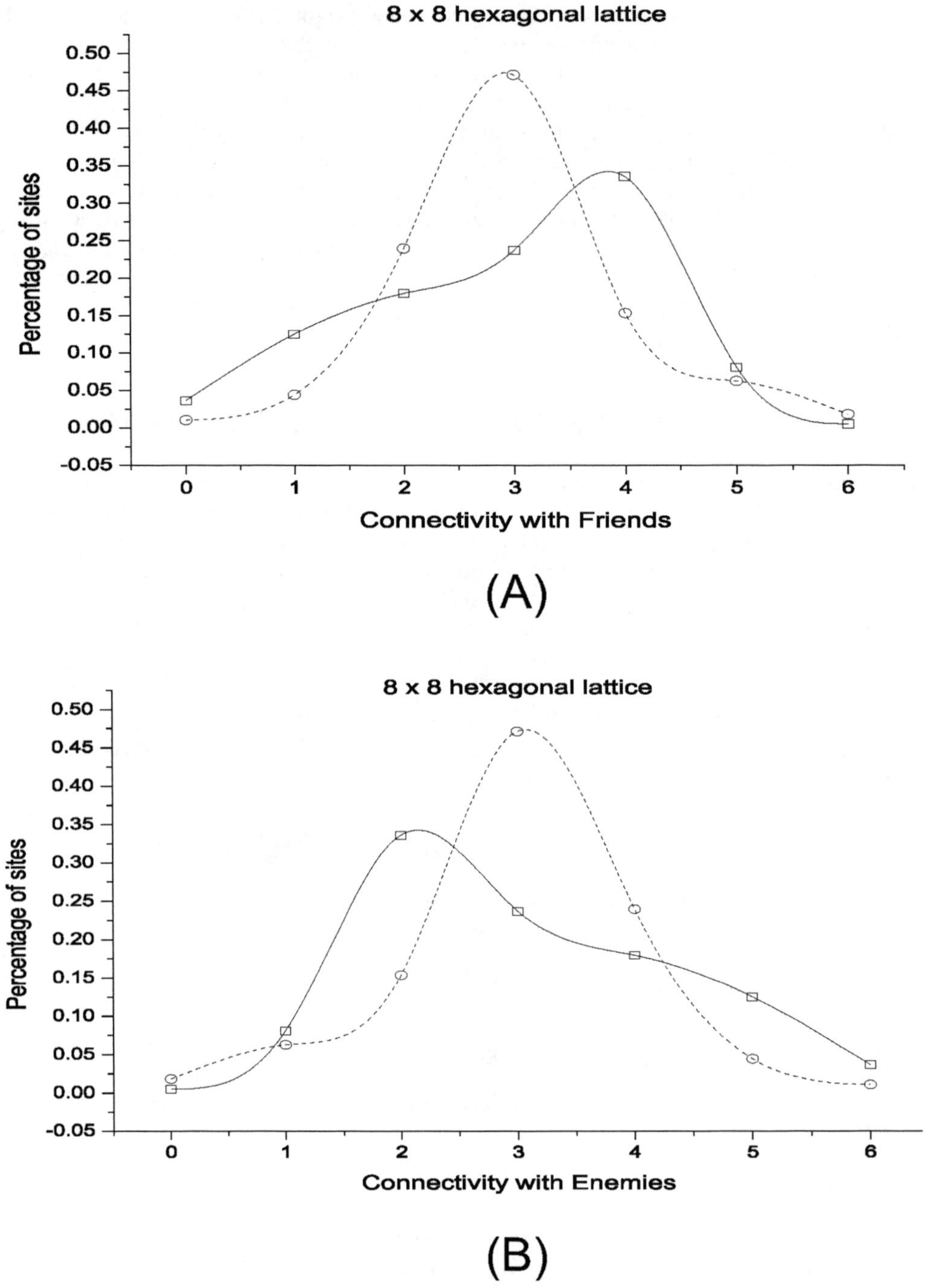

FIGURE 2. The distribution of *(a) friendly, (b) unfriendly* connectivity averaged over twelve best configurations for 8×8 hexagonal lattice with periodic boundary condition. The winner is represented by squares and solid lines while the loser is represented by circles and dash line.

5. The connectivity at which the peak of the distribution of enemies for the winner occurs is smaller than that of the loser ($x^*(+1,O) = 2 < x^*(-1,O) = 3$); while the peak values for enemies for winners is lower ($D(+1,O,2) < D(-1,O,3)$). Here $x^*(\sigma,O)$ is the location of the peak value of $D(\sigma,O,x)$ versus x.

We also repeat our simulation on the hexagonal lattice of different sizes (4×4, 6×6, 8×8 and 10×10) with periodic boundary condition. The results are similar to the ones shown in section 2. Other than hexagonal lattice, the simulation was also performed on 3–12 lattice[15], 70 sites two-dimensional Voronoi pattern and 8×8 square lattice. The above five features for the 8×8 hexagonal lattice are observed in these lattice.

CONCLUSION

We have used the statistical physics of spin model to describe the competition and cooperation between several companies aiming at market dominance in the framework of Multi-Agent System. Our focus is on the search for strategies that lead to optimal resource allocation in a noisy environment for a variety of lattice structures and networks. Through special design of numerical experiments, we demonstrate the existence of a good strategy of initial spin assignment that leads to market dominance. A systematic approach with evolutionary computation (MOGA [19,20]) has been used to search for a stable and dominant strategy of resource allocation in the changing environment. Several general topological features of the initial configuration leading to market dominance with high probability and speed are identified in section 4. These features have positive correlation with the clustering coefficients of the network. Our results appear to be general, within the confine of the two-dimensional hexagonal lattices of various sizes, hierarchical lattice[15], Voronoi diagrams, as well as some high-dimensional random networks[21]. Extension to the problem of more than two competing companies has been addressed in a model Hamiltonian similar to the Potts model [22]. This interdisciplinary research on the resource allocation problem in evolving two-dimensional point patterns hopefully inspires potential applications in econophysics.

ACKNOWLEDGMENTS

K.Y. Szeto acknowledges the help from Pang Tin Yau and financial support by CERG Grant HKUST 6071-02P and DAG04/05.SC23.

REFERENCES

1. A. Okabe, B. N. Boots and K. Sugihara, *Spatial tessellations: concepts and applications of Voronoi diagrams*, Wiley, 1992
2. B. N. Boots and A. Getis, *Point Pattern Analysis*, Sage Scientific Geography Series Vol 8, Newbury Park, California, Sage Publications, 1988
3. K. Inoue et al. Editors, *Parallel image analysis and processing*, Singapore, World Scientific, 1994
4. J. F. Thisse, K. J. Button, and P. Nijkamp Ed., *Location theory*, Cheltenham, UK, E. Elgar Pub., 1996
5. Kwok Yip Szeto and Chiwah Kong, *Computational Economics* **22** 163–172 (2003)
6. K. Y. Szeto, *Current Topics in Physics*, edited by Y.M. Cho, J.B. Hong and C.N. Yang, World Scientific Vol.1, 1998, pp. 361–375
7. B. Dubertret, K. Y. Szeto and W. Y. Tam, *Europhysics Letter* **45**, 143–148 (1999)
8. K. Y. Szeto, T. Aste and W. Y. Tam, *Phys. Rev. E* **58**, 2656–2659 (1998)
9. T. Aste, K. Y. Szeto and W. Y. Tam, *Phys. Rev. E* **54**, 5482–5492 (1996)
10. K. Y. Szeto, Xiujun Fu and Wing Yim Tam, *Phys. Rev. Lett.* **88**, 138302 (2002)
11. G. LeCaer and R. Delannay, *J. Phys. A* **26**, 3931–3940 (1993)
12. G. LeCaer and R. Delannay, *Phys. Rev. Lett.* **73**, 1553–1556 (1994)
13. Z. Z. Guo and Kwok Yip Szeto, *Physica A* **374**, 471–477 (2007)
14. Z. Z. Guo and Kwok Yip Szeto, *Phys. Rev. E* **71**, 066115 (2005)
15. Z. Z. Guo, K. Y. Szeto, and Xiujun Fu, *Phys. Rev. E* **70**, 016105 (2004)
16. J. H. Holland, *Adaptation in Natural and Artificial Systems*. Ann Arbor, Michigan, University of Michigan Press, 1975
17. D. E. Goldberg, *Genetic Algorithms in Search, Optimization, and Machine Learning*. Addison-Wesley, Reading, MA, 1989
18. Chun Wai Ma and Kwok Yip Szeto, *Proceeding of The 5th International Conference on Recent Advances in Soft Computing*, RASC2004 Nottingham, UK, 2004, pp. 410–415
19. K. Y. Szeto and J. Zhang, "Adaptive Genetic Algorithm and Quasi-Parallel Genetic Algorithm: Application to low-dimensional Physics", in *LSSC2005*, edited by I. Lirkov, S. Margenov, and J. Wasniewski, LNCS3743, Springer-Verlag, 2006, pp. 189–196
20. N. L. Law and K. Y. Szeto; *Proceeding of the 12th International Joint Conference on Artificial Intelligence (IJCAI-07) 2007 (Volume II)*, Hyderabad, India, 2007, pp. 2330–2333
21. Pang Tin Yau and K.Y. Szeto, *Proceeding of the 7th International Conference on Intelligent Data Engineering and Automated Learning*, edited by E.Corchado et al., IDEAL2006, LNCS4224, 2006, pp. 1191–1198.
22. J. Zhang and K.Y. Szeto, *International Journal of Information Technology and Intelligent Computing* **1**, 355–364 (2006)

Modelling Neural Systems with Short-Term Depression and Facilitation

Jorge F. Mejías, Joaquín J. Torres, and J. Marro

Institute "Carlos I" for Theoretical and Computational Physics, and Department of Electromagnetism and Matter Physics, University of Granada, 18071 Granada, Spain

Abstract. We review some computational properties of neural model systems with activity-dependent synaptic connections. In particular, we study the interplay between short-term depression and facilitation concerning signal detection, stochastic resonance and global behaviour of attractor neural networks. We discuss on the computational and biological relevance of the competition between depression and facilitation.

Keywords: short-term depression and facilitation, coincidence detection, stochastic resonance, unstable memories
PACS: 87.18.Sn, 87.18.Bb, 87.19.La

INTRODUCTION

It has recently been reported that postsynaptic potentials recorded in cortical neurons present dynamical properties which depend on the presynaptic activity [1, 3, 2]. This behaviour can be understood by the action of different mechanisms at the synaptic level, as for instance, short-term depression and/or facilitation. Depression seems to the due to a limited amount of available neurotransmitters in the synaptic button. Thus, the neuron needs some time to recover before transmiting the next incoming spike. The synapse dynamics is thus an activity-dependent mechanism producing a non-trivial effect in the postsynaptic response. This picture differs from the classical synaptic description which considers the synaptic strengths as static identities with only possible time modifications due to the (relatively) very slow learning processes, as described for instance by Hebb [24]. Synaptic facilitation takes into account that the influx of calcium ions through voltage-sensitive channels favours the neurotransmitter vesicle depletion. Moreover, it is well known that short-term depression plays an important role in several emerging phenomena in the brain [22, 21], such as selective attention [5, 6], cortical gain control [2], and complex switching behaviour between activity patterns in neural network models [7, 9]. Facilitation yields to relevant behaviour in synchrony and selective attention [5], in detection of bursts of action potentials (AP) [10, 11] and in the efficient transmission of temporal correlations between spikes trains arriving from different synapses [12]. However, a full description of the consecuences of competition between the mechanisms present during synaptic transmission on pyramidal neurons is still lacking.

We shall describe here some of the computational properties found in realistic models of dynamic synapses and their implications in the global behaviour of neural networks. In particular, we analyse the interplay between short-term depression and facilitation among different computational tasks. We start focusing at detection of weak signals, detection of changes in the input properties (e.g. input frequency) and synchronization of the postsynaptic neuron with the input for different frequencies. We then review the dynamical properties found in attractor neural networks for the case in which the synapses present short-term depression and facilitation. We show in examples that the effect caused by the competition between the two mechanisms could be of computational and biological relevance in a wide variety of systems.

DETECTING AND PROCCESING SIGNALS WITH DYNAMIC SYNAPSES

In a real tissue, neurons are usually exposed to hundreds of stimuli. Each neuron must be able to distinguish the relevant information, and also to separate it from the noisy environment caused by the presence of other neurons. It is also known that the presence of noise can enhance the signal level and allow the neuron to detect weak signals. This phenomena, known as stochastic resonance (SR), has been measured in experiments in neurons in different areas of the brain [13, 14]. In this context, the dynamic properties of the synapses seems to be relevant. Since the behaviour of real synapses depends on presynaptic activity, this characteristic could regulate the level of the noisy environment

CP913, *Nonequilibrium Statistical Mechanics and Nonlinear Physics, XV Conference*
edited by O. Descalzi, O. A. Rosso, and H. A. Larrondo

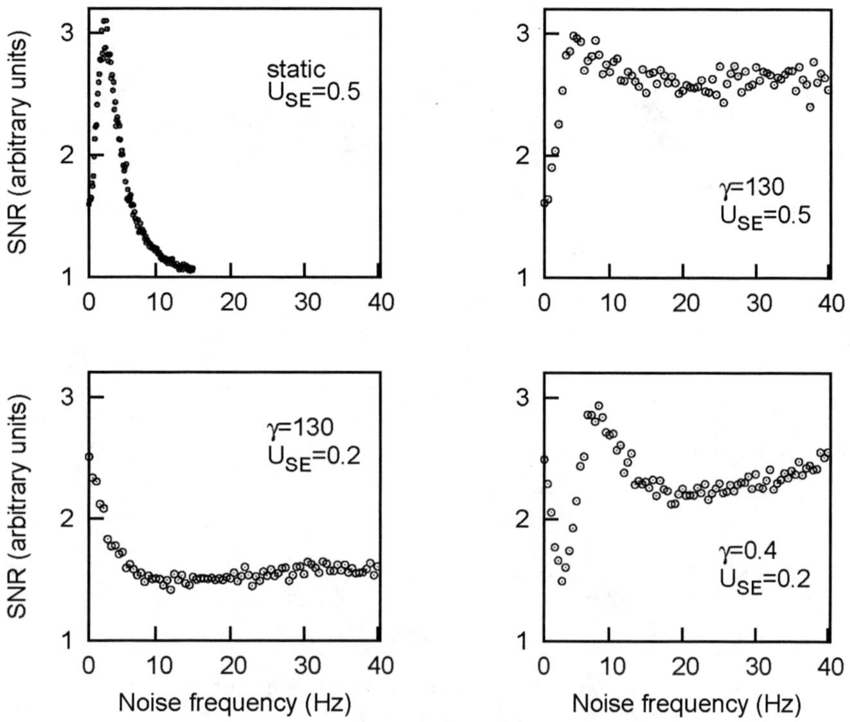

FIGURE 1. SNR as a fuction of the mean frequency of the neurons that constitutes the noisy background. One can see that the effect of including depression is to enhance the response of the neuron to the weak signal, via stochastic resonance. This effect is maintained when facilitation is added, even for low values of U_{SE}.

and allows the neuron to detect a weak signal. This is particularly interesting under high levels of input noise, that is, when the mean firing rate of the neurons which contribute to the noisy background is high. To analyse the influence of dynamic synapses on the detection of weak signals, we employ the model and parameters of dynamic synapses presented in [1] and a standart FitzHugh-Nagumo model, in order to qualitatively reproduce the basic features of neurons as excitable elements. The synapse model is

$$\begin{aligned}
\frac{dx_i}{dt} &= \frac{z_i}{\tau_{rec}} - u_i x_i \delta(t-t_{sp}) \\
\frac{dy_i}{dt} &= -\frac{y_i}{\tau_{in}} + u_i x_i \delta(t-t_{sp}) \\
\frac{dz_i}{dt} &= \frac{y_i}{\tau_{in}} - \frac{z_i}{\tau_{rec}} \\
\frac{du_i}{dt} &= \frac{U_{SE}-u_i}{\tau_{fac}} + U_{SE}(1-u_i)\delta(t-t_{sp}),
\end{aligned} \tag{1}$$

where x_i,y_i,z_i are the fraction of neurotransmitters in a recovered, active and inactive state, respectively. Here, τ_{in}, τ_{rec} and τ_{fac} are the inactivation, recovery and facilitating time constants, respectively. Depressing synapses are obtained for $u(t) = U_{SE}$ constant, which represents the maximum amount of neurotransmitters which can be released (activated) after the arrival of each presynaptic spike. The facilitation mechanism is introduced with the dynamics of u_i, which is a dynamical variable which takes into account the influx of calcium ions into the neuron near the synapse through voltage-sensitive ion channels [15]. The postsynaptic current can be written as $I_{noise} = \sum_i^M A_{SE}\, y_i$, where $A_{SE} \approx 42\ pA$ is the maximum postsynaptic current and M is the number of presynaptic afferents [8]. It is useful to define the balance

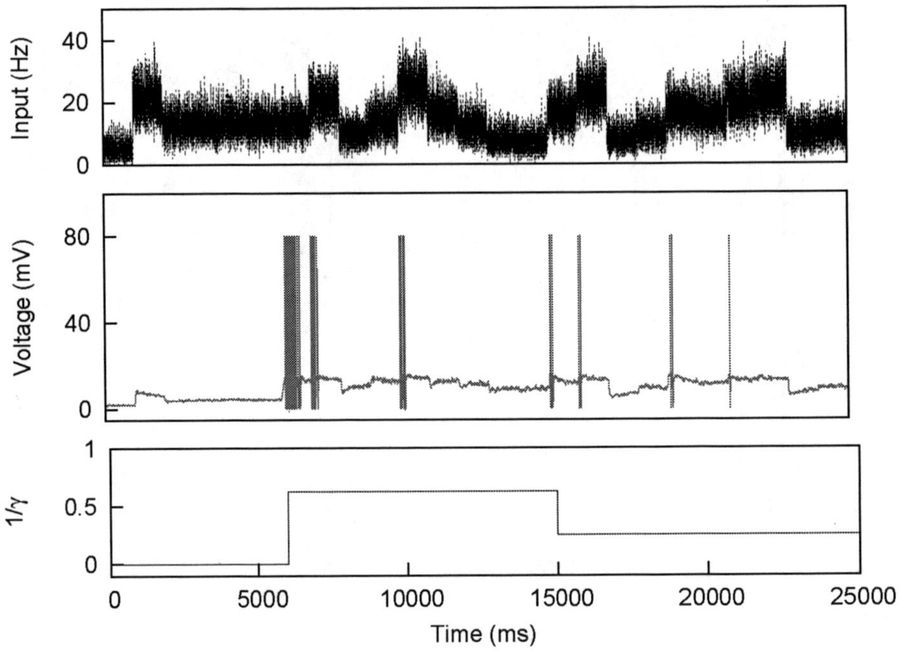

FIGURE 2. Detection of fi ring rate changes with depressing and facilitating synapses. The top panel shows the mean fi ring rate of the $M = 1000$ presynaptic neurons as a function of time. Middle panel shows the response of the postsynaptic membrane potential for different levels of balance (bottom panel),with $U_{SE} = 0.1$. Detection of variations onto lower frequencies are not possible with these synaptic mechanisms.

factor $\gamma \equiv \tau_{rec}/\tau_{fac}$, since our goal is to study the interplay between depression and facilitation. The associated neuron model is

$$\begin{aligned} \varepsilon \frac{dv}{dt} &= v(v-a)(1-v) - w + S(t) + R_{in} I_{noise}(t) \\ \frac{dw}{dt} &= bv - w, \end{aligned} \tag{2}$$

where v is the postsynaptic membrane potential and w is a slow recovery variable. In these equations, $S(t)$ is the signal term, $R_{in} I_{noise}$ is the noise term and $a = 0.001$, $b = 3.5$ and $\varepsilon = 0.0015$ are other neuron parameters to put the model in the excitable regime.

In order to study the detection of a weak signal of low frequency ($f_s = 1\ Hz$) we compute the signal-to-noise ratio (SNR) of the output [16], for different values of the synaptic parameters. The result is shown in Fig. 1. For static synapses (top left panel) the system has a well known SR peak around $\sim 3Hz$ for the noise frequency (f_N). If we increase the depression characteristic time (top right) we can see that the resonance is maintained for much higher values of f_N ($\sim 40\ Hz$ in the plot), allowing the neuron to detect a weak signal over a wide range of noise intensity. However, this is only valid for high values of U_{SE}. If it takes low values, around 0.1, the neuron becomes unable to detect the input signal for any value of noise intensity (bottom left). By including the facilitation mechanism (bottom right) one can recover the maintained resonance. Therefore, depending on the value of U_{SE} a balance between depression and facilitation should be useful to obtain optimal SNR values over a wide range of noise frequencies [19]. Since these frequencies are within the physiological range, dynamic synapses could allow neurons to work even with high frequency values without annoying to other neurons in the network.

Dynamic synapses are also relevant in order to respond to changes in the input properties of the signals. Related to this, a realistic assumption is to consider the presynaptic firing rate as a dynamic variable as it happens in real neuronal tissue. The rate changes during normal functioning of neural systems in the presynaptic current leads to a transient behaviour in the excitatory postsynaptic potential (EPSP) which could cause a burst or an AP in the postsynaptic neuron [1, 12, 2]. The question that arises is if the postsynaptic neuron is able to detect synchronous

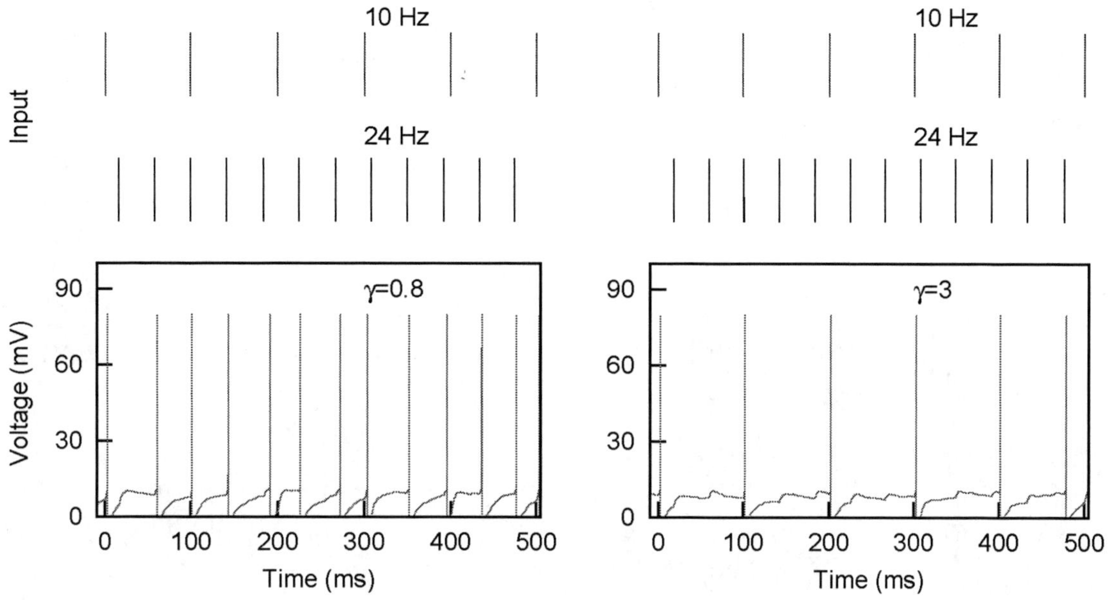

FIGURE 3. Selectivity among different frequencies embedded in the same input signal. Depending on the level of facilitation, the neuron synchronizes its activity with the higher frequency (right panel, for high γ values) or the lower frequency (left panel, low γ values).

changes (increases) in the afferent firing rates, as a function of the facilitation parameters. Here, we try to answer this question by studying the effect of increasing facilitation in spite of depression in the optimal detection of rate changes in the presynaptic current.

We assume a population of M afferents firing uncorrelated Poisson spike trains with a certain frequency f (which changes its value every 1000 ms) into a postsynaptic neuron. We use an integrate-and-fire model $dv/dt = -v + R_{in}I(t)$ for the postsynaptic membrane potential. The Fig. 2 shows the behaviour of the postsynaptic neuron for different levels of facilitation. Simulations show that facilitating synapses ($\tau_{fac} = 500\,ms$) allow for a better detection of rate changes, and over a large range of frequencies, than depressing synapses. In general, the regions in which depressing and facilitating synapses perform well can vary. For instance, there are particular situations where facilitation is needed to detect presynaptic rate changes.

Another situation that should be taken into account is that presynaptic spike trains could contain much information in it, and in some cases the neuron must be able to distinguish between different patterns of activity encoded in a single signal. Since short-term depression works as a gain modulator [2], one could expect that the inclusion of facilitation has also an effect in the signal detection depending on the embedded frequency. To test this hypothesis, we have considered two different population of correlated neurons firing spikes periodically to a postsynaptic one, including also a noisy background with fixed frequency f_N. The neurons of a given population j fire spikes simultaneously with a certain frequency f_j. Therefore, the postsynaptic neuron receives signals from both populations, resulting in a nontrivial input spike train. The results of this study are shown in Fig. 3. We take a fixed value for $\tau_{rec} = 800$ ms and $U_{SE} = 0.05$. From the figure, one can see that the effect of modifying the balance between depression and facilitation is the synchronization of the postsynaptic neuron with the frequency of population 1 or 2. That is, depending on the level of facilitation, the neuron will detect the spikes of one of the populations and ignore the spikes from the other. Therefore, facilitation works as a selectivity mechanism between signals with different frequencies encoded in the same input spike train.

Finally, we explore the effects of including dynamic synapses in recurrent networks, as a first approach to model the behaviour of real systems constituted by a large number of neurons. The most simple way to do this is to work with an attractor neural network, which is able to capture some basic features of real systems, such as associative memory, with an extremely simple neuron dynamics. This simplicity allows to obtain approximate theoretical results to compare with the simulations [17]. The inclusion of depression and facilitation in attractor neural networks leads to a phenomenon of switching between patterns of activity, previously stored in the network via the Hebb rule [17, 7].

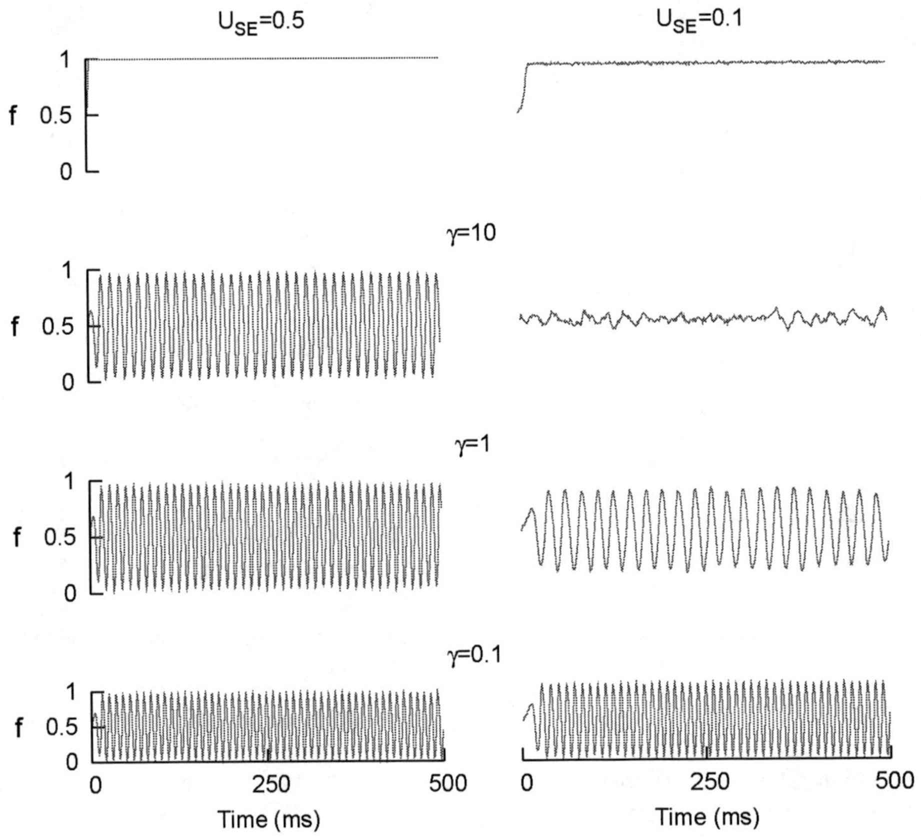

FIGURE 4. Time evolution of the mean firing rate of the network for U_{SE} high (left panels) and low (right) and different levels of facilitation, in several phases. One can see that low values of U_{SE} lead to a richer phenomenology, including oscillations with different mean frequencies (the two lowest panels on the right).

Here, we focus on the role of facilitation in this switching phenomena. An example of this behaviour is shown in Fig. 4. The state of a network of N neurons at instant t is characterized by his mean firing rate $f \equiv (1/N)\sum_i s_i$, where $s_i = 0, 1$ denotes if the neuron i is silent or firing, respectively. One can see that for high values of U_{SE} (when only depression is relevant) and increasing the level of facilitation, the system can only be in two phases, the retrieval phase (top left panel) and the oscillatory phase (the other panels on the left). However, if U_{SE} takes lower values, the facilitation mechanism becomes relevant, and phases of retrieval, non-retrieval and oscillatory behaviour appear (right panels, from top to bottom). One may think, therefore, that the inclusion of facilitation could help to explain the rich phenomenology of oscillations measured in different brain areas [18]. Detailed observations [17] reveal that facilitation also contributes to enhance the sensibility to external stimuli and to obtain a better recovery of the patterns (in a short time scale) than depression.

DISCUSSION

In the last years, there has been an increasing interest in the study of the computational functionality of synaptic activity dependent processes, as synaptic facilitation and depression, in real systems and neural networks models [20, 11, 17, 12]. In this work we have reviewed some of the computational properties found in models of neural systems which exhibit activity-dependent synaptic mechanisms. In particular, we studied the interplay between short-term depression and facilitation for different computational tasks. It has been shown that dynamic synapses enhance the SNR allowing the neurons to have a good performance in signal detection over a wide range of noisy levels, via stochastic resonance [19]. This fact could be of interest in order to understand how networks self-organize to allow neurons work freely within an apropiate noise level, that is, without interfere between them. Dynamic synapses also appear useful to detect changes in the presynaptic firing rate, which is a realistic assumption and could help to understand the behaviour of real systems in non-stationary conditions. In addition, the role of facilitation as an active selectivity mechanism between signals of different frequencies should be of much importance in actual problems in neural computation, to clarify how the neuron encode and filter the information embedded in spike trains. Looking at the level of networks, dynamic synapses reveal a rich phenomenology observed in some cortical areas [18], and a clear role for facilitation mechanism appears. In particular, the transition between the different possible dynamical phases and their nature depends strongly on the interplay between depression and facilitation. This could be of computational and biological relevance for a wide variety of neural systems.

ACKNOWLEDGMENTS

This work was supported by the MEyC-FEDER project FIS2005-00791 and the Junta de Andalucía project FQM-165.

REFERENCES

1. M. V. Tsodyks, and H. Markram, *Proc. Natl. Acad. Sci. USA* **94**, 719–723 (1997).
2. L. F. Abbott, J. A. Valera, K. Sen, and S. B. Nelson, *Science* **275**, 220–224 (1997).
3. A. Thomson, *J. Physiol. (London)* **502**, 131–147 (2006).
4. J. J. Hopfi eld, *Proc. Natl. Acad. Sci. USA* **79**, 2554–2558 (1982).
5. C. I. Buia, and P. H. E. Tiesinga, *Neurocomputing* **65-66**, 809–815 (2005).
6. C. McAdams, and J. Maunsell, *Journal of Neuroscience* **19**, 431–441 (1999).
7. L. Pantic, J. J. Torres, H. J. Kappen, and S. C. A. M. Gielen, *Neural Comp.* **14**, 2903–2923 (2002).
8. L. Pantic, J. J. Torres, and H. J. Kappen, *Network: Comput. Neural Syst.* **14**, 17–33 (2003).
9. J. M. Cortes, J. J. Torres, J. Marro, P. L. Garrido, and H. J. Kappen, *Neural Comp.* **18**, 614–633 (2006).
10. V. Matveev, and X. J. Wang, *Cerebral Cortex* **10**, 1143–1153 (2000).
11. A. Destexhe, and E. Marder, *Nature* **431**, 789–795 (2004).
12. J. F. Mejias, and J. J. Torres, *Neurocomputing*, (2007), in press.
13. A. Longtin, A. Bulsara, and F. Moss, *Phys. Rev. Letters* **67**, 656–659 (1991).
14. J. K. Douglas, L. Wilkens, E. Pantazelou, and F. Moss, *Nature (London)* **365**, 337–340 (1993).
15. R. Bertram, A. Sherman, and E. F. Stanley, *J. Neurophysiol.* **75**, 1919–1931 (1996).
16. B. McNamara, and K. Wiesenfeld, *Phys. Rev. A* **39**, 4854–4869 (1989).
17. J. J. Torres, J. Cortes, J. Marro, and H. Kappen, *Neural Comp.* (2007), in press, q-bio.NC/0604019.
18. D. Holcman, and M. Tsodyks, *PLoS Comput. Biol. 2(3)* **2**, 174–181 (2006).
19. J. F. Mejias, et al., *preprint* (2007).
20. L. F. Abbott, and W. G. Regehr, *Nature* **431**, 796–803 (2004).
21. D. Bibitchkov, J. M. Herrman, and T. Geisel, *Network* **13**, 115–129 (2002).
22. S. Romani, D. J. Amit, and G. Mongillo, *J. Comp. Neurosci.* **20**, 201–217 (2006).
23. Y. Wang, H. Markram, P. H. Goodman, T. K. Berger,J. Ma, et al., *Nature Neurosci.* **9**, 534–542 (2006).
24. D. O. Hebb, *The organization of Behavior: A Neuropschycological Theory*, Wiley (1949).

Consequences of Quadratic Frictional Force on the One Dimensional Bouncing Ball Model

Edson D. Leonel[1] and Danila F. Tavares[2]

[1]Departamento de Estatística, Matemática Aplicada e Computação, Universidade Estadual Paulista – UNESP Av. 24a, 1515 – Bela Vista – CEP:13.506-700 – Rio Claro – SP – Brazil
[2]Departamento de Física, Universidade Estadual Paulista – UNESP – Rua 10, 2527 – Santana – CEP:13.500-230 – Rio Claro – SP – Brazil

Abstract. Some dynamical properties of the one dimensional Fermi accelerator model, under the presence of frictional force are studied. The frictional force is assumed as being proportional to the square particle's velocity. The problem is described by use of a two dimensional non linear mapping, therefore obtained via the solution of differential equations. We confirm that the model experiences contraction of the phase space area and in special, we characterized the behavior of the particle approaching an attracting fixed point.

Keywords: Fermi accelerator model; classical billiard; nonlinear mapping
PACS: 05.54.-a, 05.45.Ac, 05.45.Pq

1. INTRODUCTION

The one dimensional bouncing ball model, sometimes also referred as to the Fermi accelerator model, consists of a classical particle confined between two rigid walls and suffering elastic collisions with them. One wall is assumed to be fixed while the other one is periodically time varying. The fundament of this model back to earlies 1949 when Enrico Fermi [1] proposed a simple model as an attempt to describe a mechanism in which cosmic rays would be accelerated by intense time varying magnetic fields. After that, many other different versions of the model including e.g. external fields, quantum effects, damping forces and many other modifications were considered [2, 3, 4, 5, 6, 7, 8, 9] (see also refs. [10, 11, 12] for recent results of the Fermi accelerator model). Moreover, the model may also be considered as a toy model and many tools developed to study and characterize such system can be extended to more complex models including classical static and time dependent billiard problems.

For the non dissipative case, i.e. the case where dissipative forces are absent in the model, the phase space is of mixed kind in the sense that there are a set of Kolmogorov-Arnold-Moser (KAM) islands surrounded by a chaotic sea (characterized by a positive Lyapunov exponent) which is also confined by a set of invariant spanning curves. The introduction of damping forces however yield profounds consequences on the dynamics of the problem. It is worth stressing that there are different ways of introducing dissipation in the model. One of them is to consider inelastic collisions with the walls [13, 14, 15]. Thus, after the colision, the particle experiences a fractional loss of energy. Some consequences of such a dissipation include: events of crisis [16, 17] yielding the particle to experience the phenomenon of transient [18]; annihilation of fixed points [19]; depending on the strength of the damping coefficient, the particle can also shows the phenomenon of locking [20, 21]. Other kind of dissipation which is eventually present in real experiments is the effect of drag force [22]. Such force is generated by a relative motion of a particle inside a fluid, like a gas. Thus the particle looses energy proportionally to a power law of its velocity. The phase space of the problem no longer present a mixed form with chaos and regularity. Besides, it is possible to observe attracting sinks and a chaotic attractors. Recently, Leonel and McClintock [23, 24] had shown that, for a damping force proportional the velocity and considering a simplified version,[1] the system has a determinant of the Jacobian matrix equals to the unity, thus confirming that parts of the phase space are area preserving. This apparent paradox was explained by using the well known Poincaré recurrence theorem (see [25] for specific details).

---

[1] The simplified version assumes that both walls are fixed but that, after the collisions with one wall, the particle suffers an exchange of energy and momentum as if the wall were moving. Such a simplification was often used to speed up numerical simulations when computers were far slower. It also makes easier some analytical treatment.

CP913, *Nonequilibrium Statistical Mechanics and Nonlinear Physics, XV Conference*
edited by O. Descalzi, O. A. Rosso, and H. A. Larrondo

In this paper, we revisit the one dimensional Fermi accelerator model seeking to understand and describe some of its dynamical properties in the presence of quadratic frictional force. In our approach, we are considering that the drag force is proportional to the square particle's velocity. As it is so usual, the model is described by a two dimensional non linear mapping for the variables velocity of the particle and corresponding time. The dynamical equations that describes all the possible motion of the model were obtained by solution of the Newton's law. The paper is organized as follows, in section 2 we describe the procedures used in the construction of the mapping. Section 3 is devoted to discuss some numerical results while final remarks and conclusions are drawn in section 4.

2. THE MODEL AND THE MAPPING

Let us now describe the model and all the details needed to construct the equations of the mapping. The model consists of a classical particle of mass m confined between two rigid walls and suffering elastic collisions with them. The particle also suffers the effect of a frictional force whose magnitude is assumed to be of the type $F = -\eta v^2$, where η is the strength of the viscosity and v is the particle's velocity. In our approach on this paper, we will consider only the simplified version (see footnote 1). Thus we assume that either walls are fixed. One of them is at $x = l$ while the other is at the origin $x = 0$. However, when the particle hits the wall located at the origin, it suffers a change of energy and momentum as if the wall were moving according to $x_w(t) = \varepsilon' \cos(\omega t)$, where ε' is the amplitude of oscillation and ω is the frequency of oscillation. This approximation retain the nonlinearity of the problem and avoid the inconveniency of finding numerical solutions from transcendental equations.[2]

The problem is them described in terms of a two dimensional mapping $T(v_n, t_n) = (v_{n+1}, t_{n+1})$. In order to construct the map, we will suppose that in the instant $t = t_n$ the particle is at the position $x = 0$ with velocity $v = v_n > 0$. After solving the Newton's law $-\eta v^2 = m dv/dt$, consider the effects of elastic collisions with both wall and define the dimensionless variables $V_n = v_n/(\omega l)$, $\delta = \eta l$, $\varepsilon = \varepsilon'/l$ and $\phi_n = \omega t_n$ we find that the mapping is given by

$$\mathrm{T}: \begin{cases} V_{n+1} = |V_n^* - 2\varepsilon \sin(\phi_{n+1})| \\ \phi_{n+1} = \phi_n + 2\left[\frac{e^{\delta}-1}{V_n \delta}\right] \mod 2\pi \end{cases}, \tag{1}$$

where the auxiliary term is $V_n^* = V_n/(2e^{\delta} - 1)$. It is important to stress that in the simplified model, non-positive velocities are forbidden because they are equivalent to the particle traveling beyond the wall. In order to avoid such problems, if after the collision the particle has a negative velocity, we inject it back with the same modulus of velocity. This procedure is effected by use of a module function. Note that the velocity of the particle is reversed by the module function only if, after the collision, the particle remains traveling in the negative direction. The module function has no effect on the motion of the particle if it moves in the positive direction after the collision. We stress that this approximation is valid only for small values of ε.

We now describe some details of the Jacobian matrix for the mapping (1). It is defined as

$$J = \begin{pmatrix} \frac{\partial \phi_{n+1}}{\partial \phi_n} & \frac{\partial \phi_{n+1}}{\partial V_n} \\ \frac{\partial V_{n+1}}{\partial \phi_n} & \frac{\partial V_{n+1}}{\partial V_n} \end{pmatrix}, \tag{2}$$

with coefficients given by

$$\frac{\partial \phi_{n+1}}{\partial \phi_n} = 1 \ , \quad \frac{\partial \phi_{n+1}}{\partial V_n} = -\frac{2e^{\delta} - 2}{V_n^2 \delta} \ ,$$

$$\frac{\partial V_{n+1}}{\partial \phi_n} = \mathrm{sign}\left[V_n^* - 2\varepsilon \sin(\phi_{n+1})\right] \times \left[-2\varepsilon \cos(\phi_{n+1}) \frac{\partial \phi_{n+1}}{\partial \phi_n}\right] \ ,$$

$$\frac{\partial V_{n+1}}{\partial V_n} = \mathrm{sign}\left[V_n^* - 2\varepsilon \sin(\phi_{n+1})\right] \times \left[\frac{1}{2e^{\delta} - 1} - 2\varepsilon \cos(\phi_{n+1}) \frac{\partial \phi_{n+1}}{\partial V_n}\right] \ ,$$

where the function $\mathrm{sign}(u) = 1$ if $u > 0$ and $\mathrm{sign}(u) = -1$ if $u < 0$.

---

[2] In the complete model, the solutions of the transcendental equations furnish the time that the particle hits the periodically varying wall. Such time is obtained by matching the condition that, at the instant of the colision, the particle's position is the same as that of the time varying wall.

After a careful investigation, we find that the determinant of the Jacobian matrix yields that

$$det\ J = sign[V_n^* - 2\varepsilon \sin(\phi_{n+1})] \times \frac{1}{(2e^{\delta}-1)} ,$$

thus confirming that the system now shrinks area in the phase space.

3. NUMERICAL RESULTS

Let us now present and discuss some of our numerical results. We begin discussing the behavior for large values of the particle's velocity. Considering the first equation of the mapping (1), we find that

$$V_1 = \frac{V_0}{2e^{\delta}-1} - 2\varepsilon \sin(\phi_1) ,$$

$$V_2 = \frac{V_0}{(2e^{\delta}-1)^2} - \frac{2\varepsilon \sin(\phi_1)}{(2e^{\delta}-1)} - 2\varepsilon \sin(\phi_2) ,$$

$$V_3 = \frac{V_0}{(2e^{\delta}-1)^3} - \frac{2\varepsilon \sin(\phi_1)}{(2e^{\delta}-1)^2} - \frac{2\varepsilon \sin(\phi_2)}{(2e^{\delta}-1)} - 2\varepsilon \sin(\phi_3) .$$

Finally, the general expression can be written as

$$V_n = \frac{V_0}{(2e^{\delta}-1)^n} - 2\varepsilon \sum_{i=1}^{n} \frac{\sin(\phi_i)}{(2e^{\delta}-1)^{n-i}} . \quad (3)$$

Before present the behavior of the velocity V as function of the iteration number, let us discuss some approximations of the eq. (3). Firstly, it is important to say that the second term in eq. (3) contributes with an oscillation on the velocity with maximum amplitude of 2ε (recall that ε is small). However, it is easy to see that on the average, the second term can be disregarded[3]. If we then expand the first term of eq. (3) in powers of δ, we obtain that

$$V_n \cong V_0 \times \left[1 - 2n\delta + (n+2n^2)\delta^2 + \left(-n - 2n^2 - \frac{4}{3}n^3\right)\delta^3 + \left(\frac{13}{12}n + \frac{5}{2}n^2 + 2n^3 + \frac{2}{3}n^4\right)\delta^4 + O(\delta^5)\right] . \quad (4)$$

Moreover, if we consider that n is relatively large[4], one might conclude that only the larger terms inside each brackets dominate. Thus, disregarding the lower powers of n inside the brackets, we can rewrite eq. (4) as

$$V_n \cong V_0 \times \left[1 - 2n\delta + 2n^2\delta^2 - \left(\frac{4}{3}n^3\right)\delta^3 + \left(\frac{2}{3}n^4\right)\delta^4 + O(\delta^5)\right] . \quad (5)$$

The power series shown in eq. (5) is the own definition of the exponential given by

$$V_n = V_0 e^{-2\delta n} . \quad (6)$$

In order to confirm our approach, it is shown in figure 1 the behavior of the velocity V as function of iteration number n. The control parameters and initial conditions used in the construction of the figure were $\varepsilon = 10^{-2}$, $\delta = 10^{-4}$ and $V_0 = 3$. After doing an exponential fit of the type $Y = Ae^{Bx}$, we found that $A = 3.02(1)$ and $B = -1.997(1) \times 10^{-4}$. Such a result thus confirms that the particle approaches the fixed point exponentially, as it was expected in eq. (6).

Suppose now that the particle is evolving in time and then suddenly it is captured by an attracting region, like if the particle was moving in a basin of attraction of a sink. The particle will approach it and we will describe such approaching as function of the iteration number. As an attempt to investigate the asymptotic behavior to the attracting sink, we first define a set of different initial conditions and then allow the system to evolve in time. Moreover, we have to stablish a convergency criterion in order to check whether the particle is near enough to the attracting fixed

[3] Such property can be considered since that ϕ_i might be assumed as being uniformally distributed in $\phi_i \in [0, 2\pi)$.

[4] As for example, consider that $n > 4$

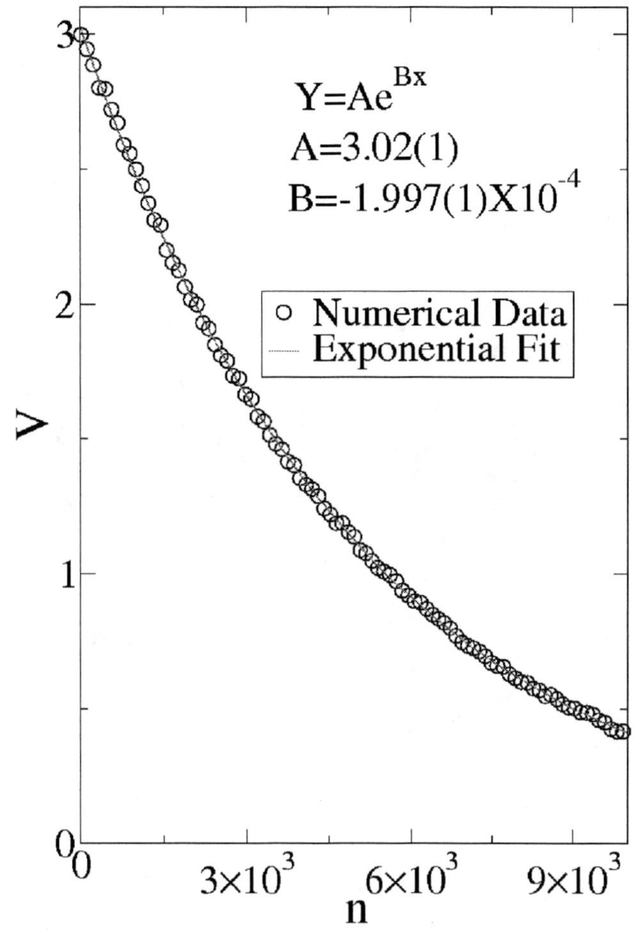

FIGURE 1. Behavior of the velocity V plotted against n. The control parameters used were $\varepsilon = 10^{-2}$ and $\delta = 10^{-4}$.

point. The criterion lies basically in checking the distance of the particle to the fixed point. We thus define a critical radius $r_c = 10^{-6}$ and evolve each set of initial condition. If the particle is sufficiently close to the attracting sink, say less then r_c, we thus save, in an array, the corresponding iteration number spent until match such condition and start a new initial condition. The evolution of an ensemble of different initial conditions allow us to define the average iteration number $\bar{n_x} = 1/M\sum_{i=1}^{M} n_i$. It is shown in figure 2(a) the behavior of the average iteration number $\bar{n_x}$ plotted against the proximity of the attracting sink for a set of different trajectories approaching it. The horizontal axis denotes the distance $r(n) = \sqrt{(V_n - V^*)^2 + (\phi_n - \phi^*)^2}$, where the coordinates of the fixed point are given by V^* and ϕ^*. The average was made using the 500 different initial conditions in the range $(V_0, \phi_0) = ([0.325, 0.33], \pi)$.

We have fitted each curve of figure 2(a) by the function $\bar{n_x} = A + B\ln(r)$. In figure 2, we assume as fixed the control parameter $\varepsilon = 10^{-2}$ and we obtain that, for $\delta = 10^{-4}$, $A = -4.680(4) \times 10^4$ and $B = -9.9991(4) \times 10^4$. On the other hand, using $\delta = 5 \times 10^{-5}$, we found that $A = -9.379(8) \times 10^4$ and $B = -1.99984(4) \times 10^4$. Finally, for $\delta = 10^{-5}$, we obtain $A = -4.702(3) \times 10^5$ and $B = -9.996(3) \times 10^5$. We can conclude that trajectories approach the attracting sink exponentially as the iteration number evolves.

Let us now describe how the trajectory evolves towards the fixed point as function of the drag coefficient. Thus, it is shown in figure 2(b) the behavior of $\bar{n_x} \times \delta$. In that figure, we have evolved our simulations up to $r < 10^{-6}$. Such a behavior can be described as

$$\bar{n_x} \propto \delta^{\mu} \ . \tag{7}$$

After fitting a power law, as shown in figure 2(b), we obtain that $\mu = -0.9993(7)$. It is worth stressing that, in the

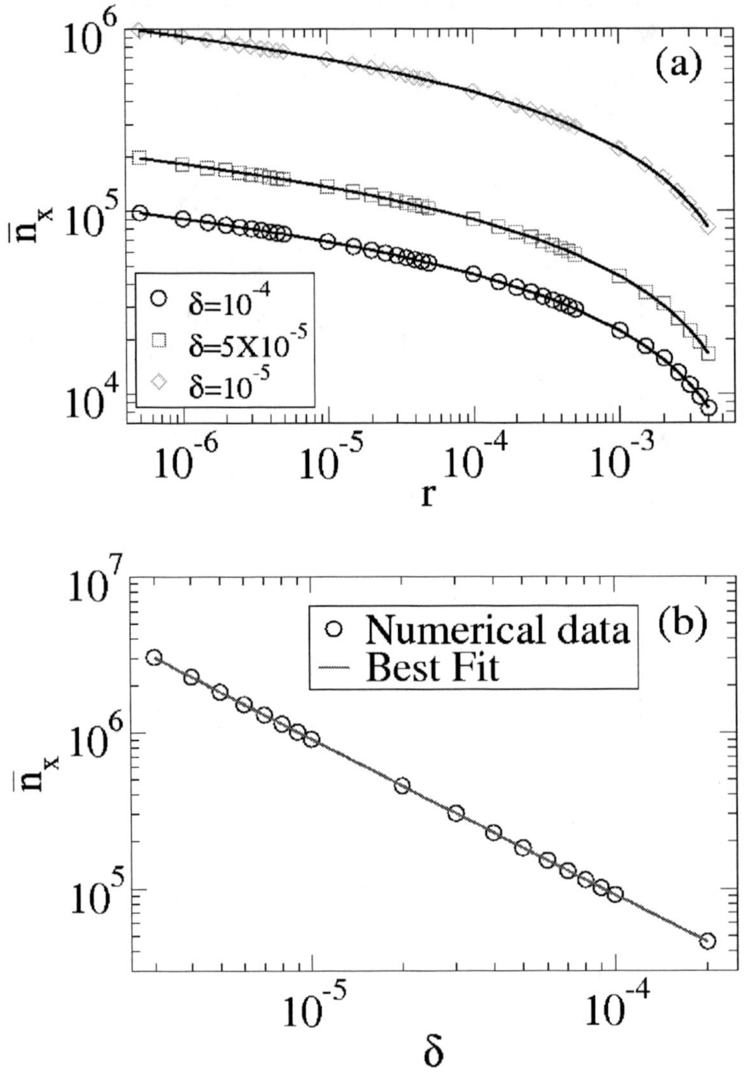

FIGURE 2. (a) It is shown the behavior of the average $\bar{n_x} \times r$ while (b) plots the transient $\bar{n_x} \times \delta$. The control parameter used was $\varepsilon = 10^{-2}$.

limit $\delta \to 0$, equation (7) gives us that $\bar{n_x} \to \infty$, thus confirming that no convergency to the fixed point is observed. Note however that the preservation of the phase space measure is recovered for that limit of δ. We can also conclude that, beyond the particle approach exponentially the attracting sink, the speed of the approach depends on the strength of the damping coefficient.

4. CONCLUSIONS

As a final remark, we have studied the one dimensional Fermi accelerator model in the presence of frictional force proportional to the square particle's velocity. Our results confirm that the model experiences contraction of the phase space area. We have also characterized the behavior of the particle approaching an attracting sink. It was shown that the particle approaches a fixed point exponentially as the iteration number evolves and with a speed of approach that depends on the strength of the drag force.

ACKNOWLEDGMENTS

E. D. L. thanks Prof. T. Yokoyama and Dr. N. Callegari Jr for fruitfull discussions. D. F. T. acknowledges Fundação do Amparo a Pesquisa do Estado de São Paulo – FAPESP; E. D. L. gratefully acknowledges support from FAPESP, CNPq and FUNDUNESP, Brazilian agencies.

REFERENCES

1. E. Fermi, *Phys. Rev.* **75**, 1169-1174 (1949).
2. A. J. Lichtenberg and M. A. Lieberman, *Regular and Chaotic Dynamics*, Appl. Math. Sci. **38**, Springer Verlag, New York, 1992.
3. M. A. Lieberman and A. J. Lichtenberg, *Phys. Rev. A* **5**, 1852-1866 (1972).
4. K. Wiesenfeld and N. B. Tufillaro, *Physica D*, **26**, 321-335 (1987).
5. E. D. Leonel, P. V. E. McClintock and J. K. L. da Silva, *Phys. Rev. Lett.*, **93**, 014101(1-4) (2004).
6. M. A. Lieberman and K. Y. Tsang, *Phys. Rev. Lett.*, **55**, 908-911 (1985).
7. K. Y. Tsang and K. L. Ngai, *Phys. Rev. E*, **56**, R17-R20 (1997).
8. E. D. Leonel, J. K. L. da Silva and S. O. Kamphorst, *Physica A*, **331**, 435-447 (2004).
9. A. J. Lichtenberg, M. A. Lieberman and R. H. Cohen, *Physica D*, **1**, 291-305 (1980).
10. D. G. Ladeira and J. K. L. da Silva, *Phys. Rev. E*, **73**, 026201(1-6) (2006).
11. J. K. L. da Silva, D. G. Ladeira, E. D. Leonel, P. V. E. McClintock and S. O. Kamphorst, *Braz. J. Phys.*, **36**, 700-707 (2006).
12. A. K. Karlis, P. K. Papachristou, F. K. Diakonos, V. Constantoudis and P. Schmelcher, *Phys. Rev. Lett.*, **97**, 194102(1-4) (2006).
13. E. D. Leonel and P. V. E. McClintock, *J. Phys. A: Math. and Gen.*, **38**, L425-L430 (2005).
14. P. J. Holmes, *J. Soun. Vib.*, **84**, 173-189 (1982).
15. R. M. Everson, *Physica D*, **19**, 355-383 (1986).
16. C. Grebogi, E. Ott, J. A Yorke, *Phys. Rev. Lett.*, **48**, 1507-1510 (1982); *Physica D*, **7**, 181-200 (1983).
17. C. Grebogi, E. Ott, J. A Yorke, *Physica D*, **7**, 181-200 (1983).
18. E. D. Leonel and R. E. de Carvalho, *Phys. Lett. A*, To be published (2007) http://dx.doi.org/10.1016/j.physleta.2006.11.097
19. D. G. Ladeira and E. D. Leonel, *Chaos*, **17**, 013119(1-7) (2007).
20. J. M. Luck and A. Mehta, *Phys. Rev. E*, **48**, 3988-3997 (1993).
21. A. Mehta and J. M. Luck, *Phys. Rev. Lett.*, *65*, 393-396 (1990).
22. G. A. Luna-Acosta, *Phys. Rev. A*, **42**, 7155-7162 (1990).
23. E. D. Leonel and P. V. E. McClintock, *Phys. Rev. E*, **73**, 066223(1-5) (2006); *J. Phys. A*, **39**, 11399-11415 (2006).
24. E. D. Leonel and P. V. E. McClintock, *J. Phys. A*, **39**, 11399-11415 (2006).
25. G. J. Sussman, J. Wisdom and M. E. Mayer, *Structure and Interpretation of Classical Mechanics*, MIT Press, Cambridge - MA, 2001.

Dynamics of a Single Tethered Polymer under Shear Flow

Rafael Delgado-Buscalioni

Departamento de Física Teórica de la Materia Condensada, Universidad Autónoma de Madrid, Ciudad Universitaria de Cantoblanco, 28049 Madrid.

Abstract. This work considers the dynamics of a flexible chain attached to a surface and interacting with a shear flow. Results from molecular dynamics with explicit (Lennard-Jones) solvent and Brownian dynamics simulations in the free draining limit are compared. We present results for the polymer extension, mean squared displacement (MSD) and chain relaxation times. We show that the MSD in gradient direction is closely related to the chain relaxation time in flow direction. We also discuss the behaviour of the polymer at long-times, and the so called "cyclic motion" of the chain, resulting from its interaction with the (thermal) noise and the flow drag. A well defined characteristic frequency is revealed in this cyclic motion, whose period is about 14 times larger than the chain relaxation time in flow.

Keywords: Polymer dynamics, Brownian dynamics, molecular dynamics
PACS: 36.20.-r, 83.50.Ax, 83.10.Rs

INTRODUCTION

Many different technological applications and biological processes involve grafted polymers. To cite a few, tethered chains are used for stabilisation of colloidal suspensions, lubrication, chromatography, adhesion [1] or drag reduction. Also, tethers are relevant in several biological processes, such as the ligand-receptor binding [2] whereby membrane proteins grafted to the cell membrane need to extend well beyond their equilibrium conformation in order to promote the adhesion with adjacent cells. In this particular process, like in many others, knowledge on the chain dynamics is essential.

In an experimental study on the dynamics of individual DNA grafted chains under shear flow, Doyle *et al.* [1], demonstrated that tethered chains perform a recursive cyclyc motion arising from the interaction between the flow drag, the thermal noise and the chain's entropic spring. Recent works [3, 4, 5] have shown that these "cyclic dynamics", initially considered to be aperiodic, do in fact present a characteristic frequency in their associated power spectra. Free chains under shear perform also perform a cyclic motion (tumbling [1]) which has been observed in experiments [4]. In the case of free chains, recent experiments have studied the statistics of the chain tumbling time [6]. Also, a theorical derivation of the effect of shear on the relaxation time and other structural features of a free chain under flow has been recently presented [7]. The relaxation time is an important quantity as it turns out that the preferential cycling period decreases with shear rate in a similar way [5]. However the characteristic tumbling period is much longer than the longest chain relaxation time, as it shall be shown below.

It well known that polymer dynamics are closely related to polymer structure. While many previous works have focused on the chain structure [8, 9, 10], this work intends to highlight one of these connections between dynamics and structure in this particular problem of a tether polymer under shear flow. We show that the structure in the gradient direction (normal-to-wall) is closely related with the observed characteristic times of the cycle. To that end, after presenting the methods and models used; we start by presenting some results on the polymer extension along gradient direction. This information is then used in the explanation of the chain dynamics. We show that a central quantity is the mean squared displacement of the center of mass of the chain along gradient direction, because chain dynamics are ruled by the balance of its convection along flow direction and its diffusion in gradient direction. Finally, some conclusions and final remarks are made.

---

[1] An important difference is that free chains in shear can rotate 360°.

CP913, *Nonequilibrium Statistical Mechanics and Nonlinear Physics, XV Conference*
edited by O. Descalzi, O. A. Rosso, and H. A. Larrondo

METHODS AND MODELS

The physical problem we want to model is relatively simple: a polymer chain is tethered to a surface and in contact with a liquid solvent which is subject to a constant shear [2]. The interaction with the flow induces a complex motion of the polymer which is the subject of this research. In order to discern to what extent hydrodynamic interactions (HI) between monomers are important in these dynamics, we considered two different kind of simulation models: molecular dynamics with explicit solvent (MD) and Brownian dynamics (BD) in the free draining limit, where only excluded volume interactions were allowed. In both cases (MD and BD) the polymer considered is flexible and solved using a Kremer and Grest bead-spring model [11]. The chain of lenght $L = bN$ ($b = 0.96\sigma$) consist on N monomers (we considered $N = \{20, 30, 60\}$) bonded via the well known Finite Extension Non-linear Elastic (FENE) potential. We refer to [12, 5] for details. The atomistic fluid solvent is a Lennard-Jones fluid at density $\rho = 0.8\sigma^{-3}$ and temperature $T = 1.0$ in reduced LJ units (mass, length, energy and time scales are given in standard LJ units: m monomer mass, σ, ε and $\tau = \sigma\sqrt{m/\varepsilon}$, respectively). Molecular dynamics were solved by two different means: either via fully fledged MD or by the hybrid particule-continuum model [12, 13], which solves the flow outside the polymer via continuum fluid dynamics (CFD) and dynamically couples this CFD region with the MD domain, where the chain and surrounding solvent molecules are solved via standard atomistic techniques. Hybrid and full MD simulations were found to be statistically indistinguishable so in the foregoing discussion no distinction shall be made between them, and both will be named as MD results. Brownian dynamics (BD) simulations were performed using a diffusion constant equal to 0.424, time step ranged between 0.0005 and 0.0002. We note that to gather enough statistics of long-time chain dynamics, simulation times were required to be quite long, more than at least 10^3 chain relaxation times.

The effect of hydrodynamic interactions (HI) is readily observed in the decay of the chain normal coordinates. In the case of a tether a good (linear) approximation to the normal modes is given by [14] $\hat{\mathbf{R}}_p(t) = (1/N)\sum_{n=0}^{N} \mathbf{R}_n(t)\sin[(p+1/2)\pi n/N]$, where $\mathbf{R_n}$ denotes the positions of the chain beads. In the presence of HI the normal coordinates relax with a smaller exponent. For a free chain, the Zimm model predicts, $\tau_p \sim p^{-1.5}$, while in the free draining (FD) case (i.e., without HI) the Rouse picture gives $\tau_p \sim p^{-2}$. In the case of a tether these exponents varies (one expects $p^{-1.8}$ for HI and $p^{-2.2}$ for FD, see [15]), but indeed, HI still affect the polymer relaxation spectra. We measured the relaxation time spectra $\{\tau_p\}$ for $p \leq 10$ by fitting the autocorrelation of the p-th normal coordinate in x direction $\hat{X}_p \equiv \hat{\mathbf{R}}_p \cdot \mathbf{i}$ to a single exponential function [3]. The longest relaxation times were found to be $\tau_0(0) = (4000 \pm 100)\tau$ and $(2000 \pm 500)\tau$ in BD and MD simulations respectively. Consistently with the theorical prediction, our BD simulations yield $\tau_p(0) \sim (p+1/2)^{-2.1\pm0.1}$ while MD results provide $\tau_p(0) \sim (p+1/2)^{-1.6\pm0.1}$, indicating the presence of HI mediated by the explicit solvent molecules. Interestingly enough, we shall see that the main features of the long-time dynamics exposed here are not related to HI.

The main parameter which determines the dynamics of a single polymer under flow is the Weissenberg number defined as $\mathrm{Wi} \equiv \dot{\gamma}\tau_0(0)$. This nondimensional number gives the ratio between the longest polymer relaxation time at equilibrium $\tau_0(\dot{\gamma} = 0)$ and the flow characteristic time $1/\dot{\gamma}$. At $\mathrm{Wi} > 1$ the flow starts to significantly affect the polymer structure, as it cannot relax as fast as the flow deforms its surounding space.

STRUCTURE

In studying how the polymer extension varies with the flow strengh our main objective is to connect the polymer structure with its dynamics. In this sense, under shear flow, one of the most relevant magnitude is the chain extension in the gradient (normal to wall) direction Y. For instance, the chain contribution to the fluid viscosity η_p is ruled by the breath of the chain in gradient direction because $\eta_p \sim \tau_{xy}/\mathrm{Wi}$ and the stress tensor scales like $\tau_{xy} \sim \mathrm{Wi}\langle Y \rangle^2$ [16]. Similarly, we shall see that the typical time scales of the chain are ruled by Y and in this section we focus on the relation between $\langle Y \rangle$ and the Weissemberg number, Wi. Several theoretical approaches have been proposed in the literature to describe this relation (see e.g., [9, 17]). One of the most elegant theories is the Graetz-Leveque argument (see e.g. [17]), which uses a diffusion-convection balance of the monomers in the flow field. This sort of balance is

---

[2] An illustration of the set-up is shown in Fig. 3

[3] In equilibrium, i.e. $\dot{\gamma} = 0$, all directions yield similar relaxation times, but the chain relaxation dynamics become anisotropic at increasing shear rates

borrowed from the thermal or concentration boundary layer theory in fluid dynamics. It reads,

$$D_{ef}\frac{\partial^2 Y}{\partial y^2} \sim v_f(Y)\frac{\partial Y}{\partial x} \tag{1}$$

where D_{ef} is an "effective" diffussion constant of monomers, which needs to be determined and $v_f(Y) = \dot{\gamma}Y$ is the fluid velocity at $y = Y$. Order of magnitude analysis yields, $\partial^2 Y/\partial y^2 \sim 1/\langle Y\rangle$ and, taking into account that the polymer is stretched by the flow, $\langle X\rangle \sim L$, and $\partial Y/\partial x \sim \langle Y\rangle/L$. One thus gets,

$$\langle Y\rangle \sim (D_{ef}/\dot{\gamma})^{1/3} \tag{2}$$

In order to close Eq. (2) one need to propose a relation for the effective diffusion of the type $D_{ef} \sim \langle Y\rangle^n$. The literature has not reached a full agreement in this issue (see [5]). Theoretical scalings propose either $n = 0$ [17], $n = -2/3$ [18] or $n = -1$ [19]. Also, experiments and simulations yield somewhat different scalings for the chain extension $\langle Y\rangle \sim \mathrm{Wi}^m$. Discrepancies may come out from differences in the chain flexibility or in the solvent behaviour, although a deep study on this issue is lacking in the literature. By introducing $D_{ef} \sim \langle Y\rangle^n$ with $n = 0, -2/3$ and -1, into Eq. (2) one gets $\langle Y\rangle \sim \mathrm{Wi}^m$ with $m = -1/3$ $m = -3/11$ and $m = -1/4$ respectively. Simulations in this work yield $\langle Y\rangle \sim \mathrm{Wi}^{-0.23}$ and $\langle Y\rangle \sim \mathrm{Wi}^{-0.32}$ for BD and MD respectively (see Fig. 2 left)). Note that both exponents are close to the limits of the range of values predicted by theoretical guesses.

In an attempt to understand the origin of this discrepancy between BD and MD we measured the mean squared displacement (MSD) of the position of the chain center of mass (CoM) in the gradient direction, i.e,

$$\langle \Delta\bar{Y}(t)^2\rangle \equiv \langle(\bar{Y}(t) - \bar{Y}(0))^2\rangle \tag{3}$$

where $\bar{Y} = (1/N)\sum_n^N Y_n$. Figure 1 illustrates the behaviour of $\langle \Delta\bar{Y}(t)^2\rangle$ versus time. Three regions can be distinguished. At short times τ_s monomers diffuse almost independently to each other and $\langle \Delta\bar{Y}(t)^2\rangle \sim t$, while at times larger than the longest relaxation time in flow direction $t > \tau_0^{(x)}$ the chain displacement $\langle \Delta\bar{Y}(t)^2\rangle$ collapses, simply because a tethered polymer cannot diffusse away above a certain distance. We are more interested in the intermediate regime $\tau_s < t < \tau_0^{(x)}$, where collective motions and constraints impede a normal diffusion and a subdiffusive regime is expected, $\langle \Delta\bar{Y}(t)^2\rangle \sim t^\alpha$ with $\alpha < 1$. Following a similar argument by Doyle *et al.* [18] [4], this exponent α provides the relation between the effective diffusion constant D_{ef} and $\bar{Y}$. First, by definition $\langle \Delta\bar{Y}(t)^2\rangle \simeq D_{ef}t$; secondly, as illustrated in Fig. 1, at intermediate times $\tau_s < t_m < \tau_0^{(x)}$, the mean squared deviation of the chain CoM becomes similar to its variance $\langle \Delta\bar{Y}(t_m)^2\rangle \sim \langle \bar{Y}^2\rangle$, which, for a tether scales like the squared mean $\langle \bar{Y}^2\rangle \sim \langle \bar{Y}\rangle^2$. Thus $\langle \bar{Y}\rangle^2 \sim t_m^\alpha$ and, using the previous relations one gets $D_{ef} \sim \langle \bar{Y}\rangle^{(2-2/\alpha)}$. Finally, using Eq. 2, one obtains the desired relation, $\langle \bar{Y}\rangle \sim \dot{\gamma}^{-\alpha/(\alpha+2)}$. In other words, the diffusion-advection balance of Eq. (2) with a general subdiffusive regime $\langle \Delta\bar{Y}(t)^2\rangle \sim t^\alpha$, predicts $\langle \bar{Y}\rangle \sim \mathrm{Wi}^n$ with $n = -\alpha/(2+\alpha)$. Using the values of α obtained from the MSD (see Fig. 1) we get $n = -0.27 \pm 0.1$ and $n = -0.324 \pm 0.05$ respectively for BD and MD [5]. As shown in Fig. 2(left) these scalings are in good agreement with those directly obtained from simulations. Interestingly enough, MD simulations with explicit solvent provide a quasi-normal diffusion of the chain CoM ($\alpha \simeq 0.95$) , while it is clearly subdiffusive in free draining BD simulation with excluded volume interactions ($\alpha \simeq 0.75$). The reason to this fact could be due to the effect of the explicit solvent (which may favour diffusion across density layers near the wall, see [12]). However, this issue requires further studies on the rôle of the chain flexibility, hydrodynamic interactions and solvent quality in the chain MSD.

DYNAMICS

Relaxation times. This section discusses how the chain relaxation times depends on the shear rate and how the relaxation dynamics are connected with the chain structure discussed above. In equilibrium (no flow) the relaxation

---

[4] Doyle *et al.* [18] deals with free chains and measure the MSD of the monomer position with respect of the center of mass. Here the polymer is fixed by its end, so the relevant quantity is given by Eq. 3

[5] It is noted that the value of α remains almost independent of the shear rate (a slight decrease with $\dot{\gamma}$ is observed, however)

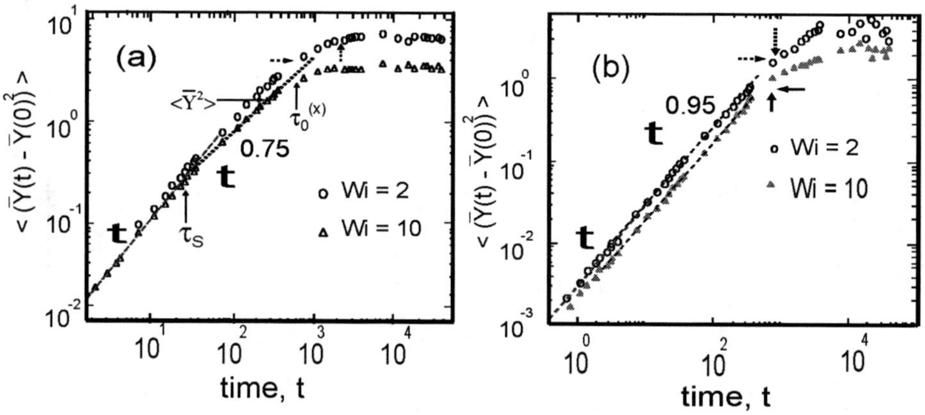

FIGURE 1. Mean squared displacement (MSD) of the chain center of mass alogn gradient direction $\langle \Delta \bar{Y}(t)^2 \rangle = \langle (\bar{Y}(t) - \bar{Y}(0))^2 \rangle$, versus time. (a) Results for free draining BD and (b) for MD with explicit solvent. In both cases, the chain is flexible with $N = 60$ monomers. The value of the Weissemberg number is indicated. Vertical arrows indicate values of $\tau_0^{(x)}$ and τ_s (explained in the text). Horizontal arrows indicate the variance of the center of mass position $\langle \bar{Y}^2 \rangle$ (dashed arrows corresponds to Wi $= 2$).

times of the chain arises from the equilibrium between the entropic forces (which tend to drive the chain to its most probably, coiled, stated) and the Brownian forces, which tend to diffuse each monomer away. This balance is altered by the presence of an extra driving force; the flow drag acting along flow direction. As stated above, one of the consequences is that the relaxation spectra of the chain is no longer isotropic, but depends on the direction considered. Briefly, the relaxation along gradient and vorticity directions is still ruled by the balance of the entropic spring and Brownian forces. An order of magnitude analysis provides, $\tau_0^{\perp} \simeq \mathrm{Wi}^{-2/3}$, where $\perp$ indicates perpendicular to the flow. Here, the dependence with Wi comes out from the non-linear nature of the FENE chain [6]. This theoretical scaling is in good agrement with our BD results and shall be discussed elsewhere. In this work we focus on the longest relaxation time along flow direction, $\tau_0^{(x)}$ because, as shown in the next section, this time is closely related to the characteristic frequency of the cyclic dynamics, at a larger time scale. Thus from now on τ_0 indicates $\tau_0^{(x)}$.

Relaxation times in flow direction; cyclic dynamics. Under large enough shear, the flow drag is the dominant force along flow direction, and in average, it is balanced by the spring force. As shown by Ladoux et al. this average balance provides the relation between the elongation of the chain and the shear rate [9]. However, in order to discern the relaxation rate one needs to include dynamical information on the order of magnitud analysis. To that end let us introduce the mechanism of the so called chain "cyclic dynamics", first discovered by Doyle *et al.* [1] in experiments with tethered DNA. This continuous (non-periodic) recirculating motion develops at time scales larger than the longest relaxation time. Figure 3 illustrates one cycle of a chain at Wi $= 2$ obtained from MD simulations, covering a time interval of about 9000τ [for Wi $= 2$ $\tau_0 = (1250 \pm 100)\tau$]. In Fig. 3a the polymer starts in a coiled configuration and eventually experiences a thermal fluctuation along y-direction. This leads to a slightly extended configuration, like that in Fig. 3b. The flow drag increases linearly with the gradient direction (y-coordinate) so any expansion in y direction induces a rapid elongation in flow direction (Fig. 3c,d). While being extended, the flow drag and the constraint to the wall creates a pair of forces which induce the subsequent rotation of the polymer towards the surface (Fig. 3e). There, the flow drag vanishes and the polymer coils back to the configuration of Fig. 3f, statistically similar to that of Fig.3a.

There are two important steps in the cycle: the elongation of the chain by the flow and the diffusion of the chain in positive gradient direction above a limiting distance, where convection overpowers diffusion $Y \sim \langle Y \rangle$, where $\langle Y \rangle$ is given in Eq. (2). Let us now discuss the time scales for each process. Assuming that the leading force during the elongation step is the flow drag (proportional to $dX/dt - Y\dot{\gamma}$), and taking into account that inertia is negligible,

[6] As shown in Ref. [20], above a certain elongation the effective spring constant of the chain starts to depend on its elongation.

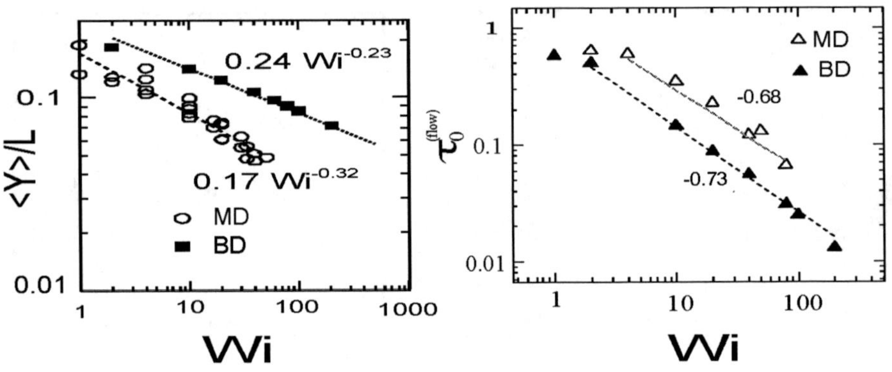

FIGURE 2. (left) Mean extension of the chain in gradient direction as a function of the Weissemberg number. Results corresponds to molecular dynamics (MD) and free draining Brownian dynamics with excluded volume interactions (BD). (right) Relaxation times of the fundamental Rouse mode along flow direction, $\tau_0^{(x)}$, against the Weissemberg number. Lines corresponds to the theoretical predictions (see text).

then while elongating, the polymer moves affinely with the fluid. The typical convective time τ_c thus follows from $\langle X \rangle / \tau_c \sim \langle Y \rangle \dot{\gamma}$. We note that above a certain $\dot{\gamma}$ the mean elongation $\langle X \rangle$ remains close to L and thus, $\tau_c \sim 1/(\langle Y \rangle \mathrm{Wi})$. On the other hand, the time needed for the chain to diffuse in y direction above $\langle Y \rangle$ is $\tau_d \sim \langle Y \rangle^2 / D_{ef}$. It is easy to check that convection and difussion times are similar; in fact, this follows from the diffusion-convection balance assumed in Eq. (1). The chain relaxation time $\tau_0^{(x)}$ is thus governed by the balance between convection along flow direction and diffusion along gradient direction, which provides $\tau_c \sim \tau_d \sim \mathrm{Wi}^{\beta}$. Using the scaling derived above $\langle Y \rangle \sim \mathrm{Wi}^{-\alpha/(2+\alpha)}$ one gets $\beta = 2/(2+\alpha)$. In the case of BD simulations (see Fig. 3a) $\alpha = 0.75$ and $\beta = -0.73$, while for MD (Fig. 3b) $\alpha = 0.95$ and $\beta = -0.68$. As shown in Fig. 2(right), these values of β are in good agreement with those obtained directly from the decorrelation time of X_0 in simulations, for which best fits are $\beta = -0.77$ (BD) and $\beta = -0.69$ (MD).

Characteristic period of cyclic dynamics. It is noticed that this recursive coil-stretching transition does not produce any periodic behaviour in the time series associated to the polymer extension along any direction. In fact, this was pointed out by Doyle *et al.* [1]. However, some years after Doyle's *et al.* paper, several independent works [4, 5, 3] discovered the presence of a characteristic time in the polymer "cyclic dynamics". Gratton and Slater [3] found that the cross correlation of the polymer extension in flow and gradient direction reaches a secondary peak at a certain lagging time, which was associated to the typical period of the cyclic dynamics. The experiments with DNA strands by Shaqfeh group clearly showed a main peak in the the power spectra of the autocorrelation of the chain extension, for both free and tethered polymers [4]. Present simulations concerning flexible tethers, also provide a clear main peak in the cross-power spectra associated to extension in the flow and gradient directions, as reported in a recent publication [5]. Figure 4 show the cross-power spectra obtained from the lowest Rouse modes in flow and gradient directions. Frequencies have been scaled with the relaxation times in flow τ_0 (whose dependence with Wi is shown in Fig. 2(left)). Clearly, the frequency of the main peak ω_{peak}, related to long time recirculation, scales like the inverse of the relaxation time in flow τ_0. Actually, the main cycle is about 14 times the relaxtion time, and interestingly enough the ratio $\omega_{peak} \tau_0$ remains similar in BD and MD simulations, and even with different chain lengths. We also note that, upon frequency scaling, the shape of spectra from BD and MD simulations at different Wi are quite similar (even at large frequencies). These fact indicate that these recursive dynamics are quite general and mainly determined by the lowest mode of the chain. In fact, our results indicate that hydrodynamic interactions do not afect the ratio between the main period of the cycle and the chain relaxation time.

CONCLUDING REMARKS

In conclusion, a close connection exists between the structure of the chain along gradient direction and its characteristic times in flow direction. The chain configuration is determined by the diffusion-convection balance and a central

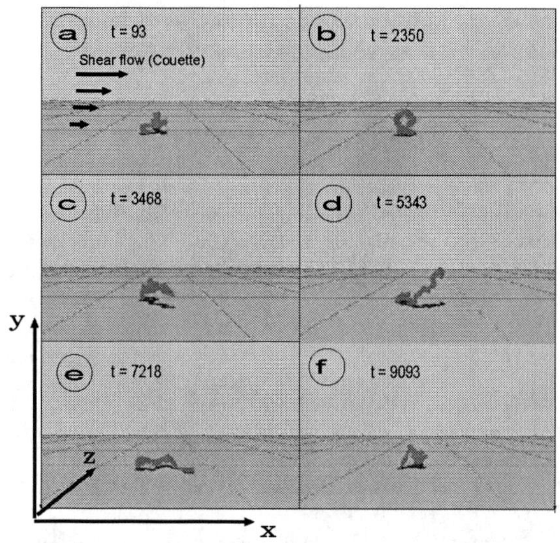

FIGURE 3. A sketch of the polymer cycle taken from MD simulations of a flexible chain with $N = 60$ monomers at $\mathrm{Wi} = 2$.

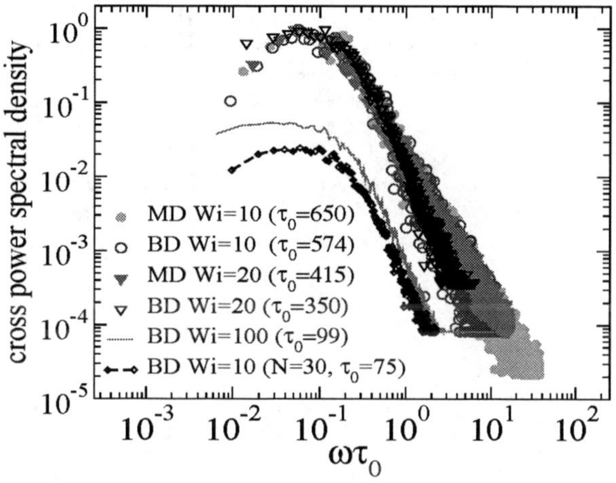

FIGURE 4. The cross power spectra of the lowest Rouse modes X_0 and Y_0 for MD and BD simulations at different Wi. Frequency ω is scaled with the fundamental relaxation time in flow $\tau_0^{(x)}(\mathrm{Wi})$.

quantity is the mean squared displacement (MSD) of the chain in gradient direction $\langle \Delta \bar{Y}(t)^2 \rangle$. We showed that the scaling of the relaxation time in flow direction $\tau_0^x \sim \mathrm{Wi}^{\beta}$, can be predicted from the time-scaling of the MSD $\langle \Delta \bar{Y}(t)^2 \rangle \sim t^{\alpha}$: the order of magnitude analysis prediction $\beta = 2/(2+\alpha)$ agrees with simulations. It is important to stress that α clearly depends on the chain and solvent details: free draining BD simulations with excluded volume interaction provide a subdiffusive regime $\langle \Delta \bar{Y}(t)^2 \rangle \sim t^{0.75}$, while MD with explicit LJ solvent yields a quasi-normal diffussion $\langle \Delta \bar{Y}(t)^2 \rangle \sim t^{0.95}$. At longer time scales, the chain performs cyclic dynamics whose main period is about 14 times larger than the chain relaxation time in flow τ_0. This ratio is fixed as Wi increased and interestingly enough it takes a similar value in BD and MD simulations, and does not seems to depend so much on the chain lenght [5].

The relevance of the relation $\langle \Delta \bar{Y}(t)^2 \rangle \sim t^{\alpha}$ at intermediate times is thus highlighted: further work to understand the dependence of the exponent α on the chain and solvent properties is required.

ACKNOWLEDGMENTS

I would like to acknowledge Osvaldo Rosso for inviting me to this interesting congress, Juan J. Freire for many useful comments and J. L. Cabrera for discussion about stochastic dynamics. This work is supported by Spanish grants CTQ2004-05706/BQU and FIS2004-01934 and also had support from the Marie-Curie European Reintegration Grant MERG-CT-2004-006316.

REFERENCES

1. P. Doyle, B. Ladoux, and J.-L. Viovy, *Phys. Rev. Lett.* **84**, 4769 (2000).
2. C. Jeppesen, J. Wong, T. L. Kuhl, J. N. Israelachvili, N. Mullah, S. Zalipsky, and C. M. Marques, *Science* **293**, 465 (2001).
3. Y. Gratton, and G. Slater, *Eur. Phys. J. E* **17**, 455 (2005).
4. C. M. Schroeder, R. E. Teixeira, E. S. G. Shaqfeh, and S. Chu, *Phys. Rev. Lett.* **95**, 018301 (2005).
5. R. Delgado-Buscalioni, *Phys. Rev. Lett.* **96**, 088303 (2006).
6. S. Gerashchenko, and V. Steinberg, *Phys. Rev. Lett.* **96**, 038204 (2006).
7. R. G. Winkler, *Phys. Rev. Lett.* **97**, 128301 (2006).
8. R. Lehner, J. Koota, G. Maret, and T. Gisler, *Physical Review Letters* **96**, 107801 (2006).
9. B. Ladoux, and P. Doyle, *EuroPhys. Lett.* **52**, 511 (2000).
10. A. S. Lemak, N. K. Balabaev, Y. N. Krnet, and Y. G. Yanovsky, *J. Chem. Phys.* **108**, 797 (1998).
11. K. Kremer, and G. Grest, *J. Chem. Phys.* **92**, 5057 (1990).
12. S. Barsky, R. Delgado-Buscalioni, and P. V. Coveney, *J. Chem. Phys.* **121**, 2403 (2004).
13. G. D. Fabritii, R. Delgado-Buscalioni, and P. Coveney, *Phys. Rev. Lett* **97**, 134501 (2006).
14. Y. Marciano, and F. Brochard-Wyart, *Macromolecules* **28**, 985 (1995).
15. R. Rzehak, and W. Zimmermann, *Phys. Rev. E* **68**, 021804 (2003).
16. R. Teixeira, H. P. Babcock, E. S. G. Shaqfeh, and S.Chu, *Macromolecules* **38**, 581 (2005).
17. R. Jendrejack, J. de Pablo, and M. Graham, *J. Chem. Phys.* **116**, 7752 (2002).
18. P. S. Doyle, E. S. G. Shaqfeh, and A. P. Gast, *J. Fluid Mech.* **334**, 251–291 (1997).
19. J. Hur, E. S. G. Shaqfeh, and R. G. Larson, *J. Rheol.* **44**, 713 (2000).
20. J. W. Hatfield, and S. R. Quake, *Phys. Rev. Lett.* **82**, 3548–3551 (1999).

Influence of the Substrate on the Pattern Formation of a Surface Reaction

Stefan Wehner*, Patrick Hoffmann†, Dieter Schmeisser†, Helmut R. Brand,** and Jürgen Küppers*,‡

*Experimentalphysik III, Universität Bayreuth, 95440 Bayreuth, Germany
†Angewandte Physik II, Brandenburgische Technische Universität Cottbus, 03013 Cottbus, Germany
**Theoretische Physik III, Universität Bayreuth, 95440 Bayreuth, Germany
‡Max-Planck-Institut für Plasmaphysik (EURATOM Association), 85748 Garching, Germany

Abstract. The catalytic CO oxidation on Iridium(111) surfaces was investigated experimentally under ultrahigh vacuum (UHV) conditions using mass spectroscopy to detect gaseous products and photoelectron emission microscopy (PEEM) to visualize surface species. The underlying reaction-diffusion system based on the Langmuir-Hinshelwood mechanism was analyzed numerically.
The effect of noise on this bistable surface reaction was examined. In a surface science experiment the effects on product formation and the development of spatio-temporal patterns on the surface were explored. The influence of noise on the reaction rates and the formation of spatio-temporal patterns on the surface was analyzed by superposing noise of Gaussian white type on the feed gas composition, characterized by the molar fraction of CO Y (deviation ΔY), which represents multiplicative and additive noise.
PEEM was used to visualize spatio-temporal adsorbate patterns on the surface as expected from the observations in the CO_2 rate measurements. CO- and oxygen-covered regions on the Ir(111) surface are visible in PEEM images as gray and black areas as a consequence of their work function contrast. Islands of the adsorbate, corresponding to the globally stable branch, are formed in a background of the other one. The long transient times are the result of the extremely slow domain wall motion of these islands (around 0.05 μms^{-1}). For small noise only a few islands nucleate and grow until they merge and finally fill the whole surface. With increasing noise the number of islands increases and their maximum size decreases. At constant ΔY the island wall velocity and the number of islands increases when Y approaches the hysteresis boundary.
In the hysteresis region CO oxidation on Iridium(111) surfaces is dominated by domain formation and wall motion for small to moderate noise strength. The island density increases with noise, but the wall velocity is independent of applied ΔY. The recorded islands show an elliptic shape and not a circular one as expected from the symmetries of the Ir(111) surface. Experiments on CO oxidation on an intentionally stepped surface (Ir(977)) revealed that this observation is the result of an anisotropy in surface diffusion due to the limited terrace width on a real surface. The elliptic islands are aligned with the terraces and their aspect ratio is given by the corresponding wall velocities parallel and normal to the terraces.

Keywords: Iridium, pattern formation, surface reaction, reaction-diffusion-system
PACS: 82.40.Ck, 82,40.Np, 05.40.-a

INTRODUCTION

Platinum group metal surfaces in ultrahigh vacuum (UHV) are one possible system to study pattern formation in reaction-diffusion-systems in situ. They can be established with reproducible surface conditions using appropriate standard cleaning methods. Very precise control of reactants in a continuous-feed reactor and of all other parameters is possible using available UHV equipment.
In addition to this, surfaces are two-dimensional systems, which makes them easier to handle in modeling, since numerical techniques for solving partial differential equations in two dimensions are less complex than their three dimensional counterparts. Thus two-dimensional modeling can describe surfaces and a direct comparison with experiment is possible.
Recently it was shown that the CO oxidation reaction on Iridium(111) surfaces is a bistable surface reaction [1]. The rate hysteresis is formed by a branch with high CO_2 formation on a predominantly oxygen covered surface (upper rate, UR) and a branch of low CO_2 formation on a mainly CO covered surface (lower rate, LR). This catalytic reaction fulfills several conditions necessary for a long time analysis of pattern formation. For the (111) surface no reconstruction by adsorption of CO or oxygen and no dissociation of CO is known. Iridium surfaces do not oxidize strongly during this reaction. The steady state product formation rate is therefore constant and not oscillatory. The

CP913, *Nonequilibrium Statistical Mechanics and Nonlinear Physics, XV Conference*
edited by O. Descalzi, O. A. Rosso, and H. A. Larrondo

underlying reaction-diffusion system is based on the Langmuir-Hinshelwood mechanism [2, 3] and is in addition not excitable, see [4] for more details.
This makes the CO oxidation reaction on Ir(111) surfaces an ideal model system, which can be studied in surface science experiments as well as by numerically solving the underlying reaction-diffusion equations.
Because the system and its parameters are known with a high degree of accuracy and the vacuum experiments can be controlled with high precision, it is furthermore possible to study the influence of noise on the reaction rates and the formation of spatio-temporal adsorbate patterns.
This system was explored in different studies [1, 4, 5, 6, 7] using a quadrupol mass spectrometer (QMS) for data acquisition. The mass spectrometer measures the partial pressure of the reaction product CO_2 present in the UHV chamber. The CO_2 rate is averaged over the surface, since all molecules formed contribute to the partial pressure wherever the reaction took place.
From a numerical analysis of the underlaying reaction-diffusion system it is known that spatial degrees of freedom are necessary to explain the observations, especially long time transients. Long transition times are connected with the slow growth of domains of one species on a surface dominantly covered with the other [4].
Here the spatio-temporal adsorbate patterns are analyzed and the influence of the substrate is explored. This study is limited to small and moderate noise superposed on the feed gas. The results for large noise are reported elsewhere and those adsorbate patterns are not as dominated by the substrate as they are for small to moderate noise [8, 9].

EXPERIMENTAL SETUP

An experimental microscopy technique suitable for surfaces in UHV is photoelectron emission microscopy (PEEM). This technique utilizes the work function contrast of different adsorbates on a surface to obtain a two-dimensional picture of the spatial distribution of these adsorbates in a reasonably wide field of view. Photoelectron emission microscopes have already been used before as a suitable tool to visualize different adsorbate coverage on other platinum group surfaces. Some examples are CO oxidation on Pt(100) [10], Pt(110) [11] and Pt(111) [12] as well as NO reduction on Rh(111) [13].
It is known from the literature [14] that a clean Iridium(111) surface has a work function of 5.76 eV. Adsorbed oxygen will increase the work function to over 6.3 eV depending on coverage. CO will also increase the work function, but less. When the energy of the incoming photons exceeds the work function in an area electrons could be released. By irradiation with UV photons of an energy around 6.2 eV an oxygen-covered surface will appear dark gray to black, CO-covered areas will appear light gray and the free Iridium surface white.
Brightness of the PEEM image increases with the number of electrons collected. Less electrons are emitted from areas with a high work function, in this case oxygen-covered patches, and more from areas with a smaller work function, in this case mainly CO-covered areas. Therefore oxygen- and CO-covered patches appear on a PEEM image in different brightness - oxygen dark and CO bright. A connection between pattern brightness scale and adsorbate coverage is not known.
Photons of the energy required are provided by a Deuterium discharge lamp, for this study a lamp of the type Heraeus D200 was used. The emitted electrons are recorded spatially resolved by a commercial Omicron/Focus PEEM setup. The maximum accessible field of view was around 300 μm with a resolution of about 0.3 μm limited mainly by increasing distortion, decreasing intensity and pixel resolution (1404 $\times$ 1050 pixel) of the recording CCD camera. All PEEM images were taken and stored at 3 s intervals. The integration time of the CCD camera was 2.86 s for all images shown here. The actual resolution of the image is about 3 μm due to limiting finite-size effects of work-function variations and the extended integration time.
Conformal mapping of the microscope was checked by imaging a Cu mesh (7.5 $\times$ 7.5 μm^2 holes and 12.5 μm repetition length). Since the crystal has to be heated all the time to obtain a temperature around 500 K, the clamping wires conduct some ten Ampere all the time. It was checked that this moving current and resulting fields do not distort the picture, only a lateral shift is produced.
All PEEM experiments were done in a UHV system located at BTU Cottbus (Angewandte Physik II), which is suitable for long-term operation and possesses the high stability needed to follow the long transients observed earlier, see [8] for detailed description. This setup was as automated as the one used for the studies described in [1, 4, 5, 6, 7] and measurements were done nonstop up to one week, mainly interrupted by the need to provide enough data storage (up to 40 GByte per day).
Unlike the studies listed in the beginning of this section, in all experiments described here the incoming gas flux

$\Phi_{total} = \Phi_{CO} + \Phi_{O_2}$ was kept constant during the runs at 1.0 ML s^{-1}. One monolayer (ML) equals the number of atoms on the surface, for Ir(111) it is 1.56×10^{15} cm^{-2}. Only the fractions of CO and oxygen in the feed gas were varied. The composition of the feed gas can then be characterized solely by the molar fraction of CO; this is normally labeled as $Y_{CO} = \Phi_{CO}/\Phi_{total}$, but only Y is used here, since it is the only independent molar fraction used.
$Y = 0$ therefore represents pure oxygen as feed gas, and $Y = 1$ represents pure CO. Both gases were delivered through reducing valves from pressurized gas cylinders containing gases in the highest available purity (CO: 99.997%, O_2: 99.9999%) and were controlled by mass flow controllers (MFC MKS 200), with a precision of ± 0.001 of full scale and therefore ± 0.001 in Y.
In the experimental setup noisy Y of Gaussian shape was achieved by programming the MFC in a specific manner. Y(t) stands for the actual programmed CO fraction at time t, Y_0 for the preselected CO fraction average. $Y(t) = Y_0$ for all times when no noise is applied. When noise is applied a Y value is picked out randomly from an interval around Y_0 and a fixed width of $2 \cdot \Delta Y$ controlling the noise strength. The maximum ΔY and therefore the strongest noise possible is ΔY equals Y_0, since no negative gas composition is possible. $Y_{average}$ is the long time average of the CO fraction distribution. For infinite time $Y_{average}$ equals Y_0.
Y(t) is sent to the MFC controlling the CO channel and (1-Y(t)) to the oxygen-regulating MFC. One Y(t) was used for an interval of 3 s duration, before another CO fraction was generated and used for the next three second interval. This mixed gas with its specific composition is supplied for three seconds at the back of a capillary, delivering the mixed feed gas on the surface. Then it is replaced by the next mixture. The composition of the mixture outside the chamber is not only affected by MFC response times but also by the pumping speed present.
The relevant gas composition of the feed gas is analyzed inside the UHV system with the mass spectrometer. The shape of the realized Y_{exp} distribution is further influenced by the conductivity of the doser tube and the gas throughput of the vacuum system. After passing the inlet tube both the CO signal (amu 28) and the oxygen signal (amu 32) are described very well by a Gaussian distribution around Y_0 with deviation $\Delta Y_{exp} = \Delta Y/2$. Since $\Delta Y = Y_0$ is the largest ΔY which could be meaningfully applied, the strongest noise that could be used for experiments is given by $\Delta Y_{exp} = Y_0/2$.
Obviously a small error is made anyhow, since a Gaussian distribution has infinite wings, but Y is finite and could never be negative. A specific portion of the Gaussian distribution ($Y < 0$), is not meaningful in reality. Calculating the shift in the realized $Y_{average}$ from this missing piece reveals that the error is very small for the relevant region of ΔY. Even for the maximum value of $\Delta Y_{exp} = Y_0/2$ (set $\Delta Y = Y_0$) it is only about $0.03 \cdot Y_0$.
Altogether, noisy Y up to its maximum possible deviation $\Delta Y = Y_0$ was found to be usable for experiment. The intrinsic noise of the MFC is about 0.001 in Y. ΔY given below in text or figures does not include the intrinsic noise of MFCs, but only the externally applied noise. Therefore ΔY and Y_0 are always used to characterize noise strength and average, instead of ΔY_{exp} and $Y_{average}$.
All data shown were taken at a sample temperature $T = 500$ K and a total feed gas flux $\Phi_{total} = 1.0$ ML s^{-1}. The experimental study covered a wide range of Y and ΔY values, which is presented in detail in [8]. Here only the formation of oxygen islands near the low-Y boundary of the hysteresis is compared for two different substrates (Iridium single crystals) .

EXPERIMENTAL RESULTS

Experiments were done according to the following scheme: preparing the crystal by oxygen treatment, shifting to the branch of low CO_2 formation (LR) by application of a 60 s CO pulse, then Y was set to the selected value Y_0 for 300 s to establish a steady state, and noise of strength ΔY was superposed on Y_0 till the end of the respective experiment, for further details see [4, 1].
Figure 1 shows some photoelectron emission microscopy images of growing oxygen islands on a Iridium(111) surface for three representative experimental runs (Y_0 inside hysteresis near the low-Y boundary, see icon for orientation). These series of pictures span times between several hours and a few minutes.
The icon in the top-left corner of both figures shows schematically the CO_2 rate hysteresis, see [1] for further details. The CO_2 rate is therefore plotted as a function of the feed gas composition characterized by the CO fraction Y. The open circle represents the initial condition (LR), the filled circle the final state (UR) and the arrow the increase of CO_2 rate during each experimental run.
The top row shows pictures taken with $\Delta Y = 0.02$ for a surface initially prepared in LR (CO covered) at $Y_0 = 0.11$.

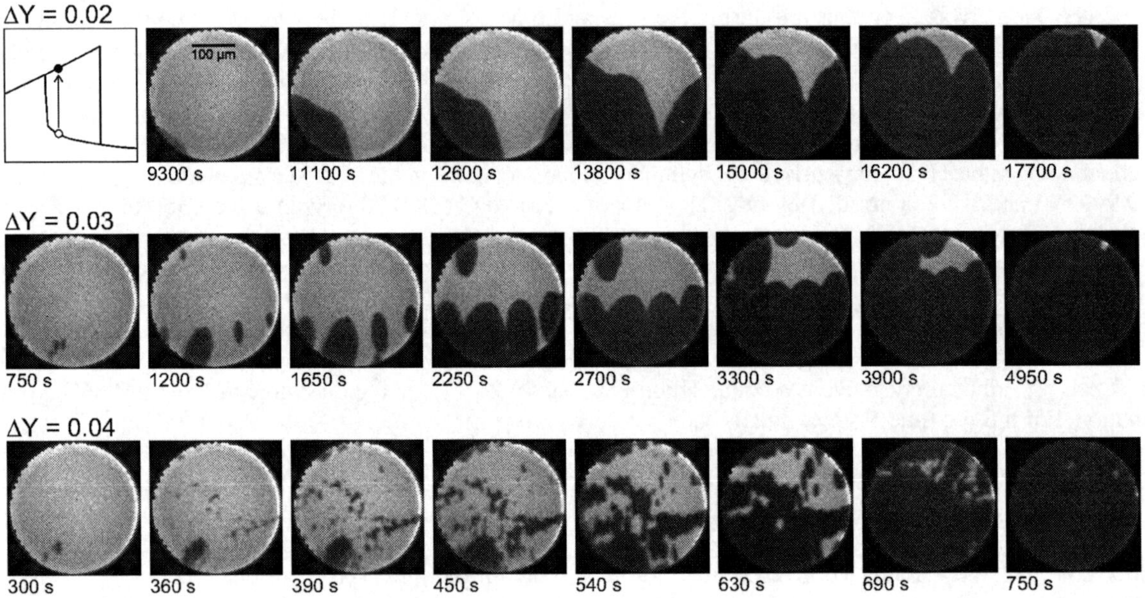

FIGURE 1. Three series of PEEM taken on a Ir(111) surface for different Y noise strength (top $\Delta Y = 0.02$, middle $\Delta Y = 0.03$, bottom $\Delta Y = 0.04$) superposed on $Y_0 = 0.11$ (see icon for orientation). The time given below each picture specifies the time that has passed since the experiment started in LR. Growth of oxygen islands (dark) on an initially CO-covered surface (gray) is recorded. The number of islands and therefore the transient time depends on Y noise strength. The field of view is 310 μm (black bar in left top picture is 100 μm). Icon in top-left corner shows schematically the CO_2 rate hysteresis (CO_2 rate vs. Y). [Modified version of figure 1 in [15]]

Over 9000 s after the application of noise started, the wall of a dark oxygen island enters the field of view from the lower left. Due to the expansion of this island the border moves slowly inward in the PEEM images. At 12600 s a second oxygen island has reached the field of view at the lower right. Around 13800 s the two islands have merged through steady wall motion and eventually fill the field of view about five hours after the experiment was started. The state of the surface outside of the field of view remains unknown. When the whole surface is finally filled with oxygen, the transient between LR and UR is completed. The Iridium surface stayed oxygen-covered and therefore in the high reactivity steady state branch (UR) for the next 12000 s until the run was stopped. Since the oxygen domains for $Y_0 = 0.11$ grow with less than 0.05 μms^{-1} sharp black-gray boundaries are recorded.
An experiment with slightly higher noise $\Delta Y = 0.03$ is shown in the middle row. After 750 s small dark spots have appeared on the CO-covered surface. These islands merge and further islands have nucleated. All these islands grow and others enter the field of view from outside. Around 4950 s the whole surface appears dark, the transient has finished much faster than for smaller noise.
The third row gives PEEM images for $\Delta Y = 0.04$. Many very small islands nucleate on the surface and merge before they have grown to a large size. The surface is completely filled with oxygen in about 450 s. For the remaining time of the experiment the field of view stayed black, the surface was oxygen-covered and exhibited a high CO_2 rate.

Figure 2 shows images taken on a Ir(977) surface during the transition from lower to upper rate branch near the low-Y boundary of the hysteresis region after preparing the CO-covered surface (LR) at the beginning as described above. Three runs for different Y noise strength are shown, ΔY increases from top to bottom.
For small $\Delta Y = 0.03$ (top row) only one large oxygen island enters the field of view around 7200 s, growing over time until more than 2 hours later it fills the whole area.
The second row shows a run with $\Delta Y = 0.04$. Some islands nucleate inside the field of view. The overall number of dark areas has increased and the time for the low-to-high-rate transition is reduced to less than one hour.
The bottom row includes images taken for $\Delta Y = 0.06$. Many quite small but elongated and aligned oxygen islands are observed and the transient is finished in less than 1000 s.

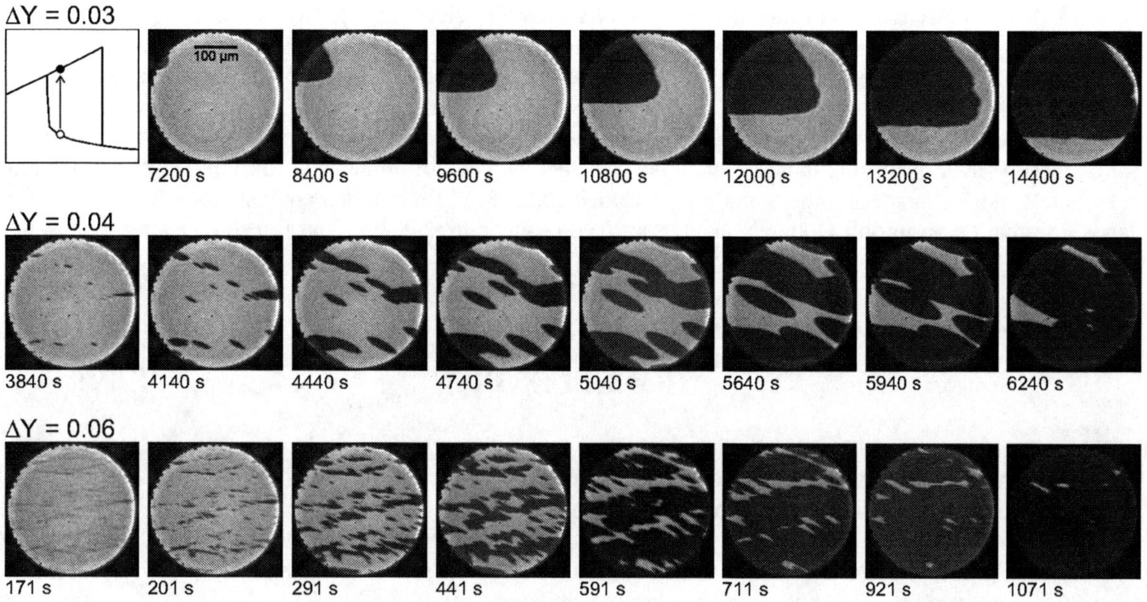

FIGURE 2. Three series of PEEM images taken on a Ir(977) surface for different Y noise strength (top $\Delta Y = 0.03$, middle $\Delta Y = 0.04$, bottom $\Delta Y = 0.06$) superposed on $Y_0 = 0.11$ (see icon for orientation). The time given below each picture specifies the time that has passed since the experiment started in LR. Growth of oxygen islands (dark) on an initially CO-covered surface (gray) is recorded. The number of islands and therefore the transient time depends on Y noise strength. The field of view is 300 μm. The black bar in the top left picture corresponds to 100 μm. Icon in top-left corner shows schematically the CO_2 rate hysteresis (CO_2 rate vs. Y). [Modified version of figure 2 in [9]]

DISCUSSION

Large growing dark oxygen islands with a sharp border are seen in experiments with small Y noise on Iridium(111) as shown in figure 1, and on Iridium(977) as shown in figure 2. The island density is increasing, island size and transient time are decreasing with increasing ΔY.

Due to the sixfold symmetry of the Iridium(111) surface, no directional diffusion anisotropy is present and the shape of growing islands in the background of the other adsorbate is expected to be circular. For very low Y noise amplitudes the islands are so large that the shape is hard to judge, compare the top row in figure 1. But in the second row of figure 1 for example, it is clearly seen that the oxygen islands exhibit an elliptic shape. The higher aspect ratio of the elliptic oxygen islands on Ir(977) is obvious compared to those seen on the Ir(111) surface.

It is known that stepped surfaces exhibit diffusional anisotropy. Diffusion along the terraces is faster than normal to the steps [16]. Other experiments on reactions on Platinum [17] and Rhodium [18] surfaces have also found anisotropy on those surfaces.

Steps on the flat Ir(111) surface are the consequence of unavoidable miscut during crystal disc production. The direction of the cutting is not perfectly aligned on an atomic scale. The Iridium(977) surface is intentionally stepped, it consists of (111) terraces of 8 atoms width followed by a one atom high (100) step [19]. Oxygen islands in PEEM images taken on an Iridium(977) surface, see figure 2, are more elongated than those observed on the Iridium(111) single crystal, since the terraces are smaller.

All ellipses are aligned preferentially in the 10 o'clock to 4 o'clock direction on Ir(977) and 7 o'clock to 1 o'clock direction on Ir(111). This alignment between terraces following low energy electron diffraction (LEED) measurements and the long axis of the ellipsis following PEEM imaging is better than 5°.

The wall velocity of oxygen islands was determined by comparing the size of a specific island in two images. On Iridium(111) the wall velocity parallel to the long axis of the island $v_{long} = 0.055\ \mu ms^{-1}$ and along the short axis $v_{short} = 0.032\ \mu ms^{-1}$; on Iridium(977) $v_{long} = 0.043\ \mu ms^{-1}$ and $v_{short} = 0.012\ \mu ms^{-1}$. Within the experimental errors the wall velocity parallel to the long axis is the same for both surfaces about 0.05 μms^{-1}, but the velocity along the short axis is significantly smaller on Ir(977) than on Ir(111). The measured wall velocities are independent of the

applied ΔY. A detailed discussion of these obviously very slow wall velocities is given in [9].
An analysis of the underlying reaction-diffusion (RD) equations in two spatial dimensions as presented in [4, 8] has shown that the aspect ratio of growing adsorbate islands could also in this case be explained by anisotropic diffusion [9].
As already observed in corresponding CO_2 rate measurements, increasing noise makes the transients shorter [4]. In the light of these PEEM experiments it is concluded that this happens by increasing the number of islands that are formed and grow, not by increasing the domain wall velocity, see [8, 9] for detailed analysis.
CO islands growing on an initially oxygen covered surface near the high-Y boundary of the hysteresis show similar behavior [8].

CONCLUSION

The PEEM experiments on Ir(111) and Ir(977) have shown that in the hysteretic region the dynamics of CO oxidation on Iridium(111) surfaces is dominated by domain formation and wall motion for small to moderate noise strength. The conclusions from former QMS measurements [4] and numerical calculations of the RD model [8, 9, 15] are confirmed. For intrinsic and/or small strength of the external Y-noise transitions from the locally stable to the globally stable branch take place via slow nucleation and growth of oxygen or CO islands of the globally stable state. The island density increases with noise, but the wall velocity is independent of applied ΔY.
Increasing the number of steps on a surface impedes diffusion normal to these steps and results in an elliptic shape of the growing islands. The more steps on the substrate, the more ellipticity of the adsorbate islands.

REFERENCES

1. S. Wehner, F. Baumann, and J. Küppers, *Chem. Phys. Lett.* **370**, 126-131 (2003).
2. T. Engel, and G. Ertl, *J. Chem. Phys.* **69**, 1267-1281 (1978).
3. T. Engel, and G. Ertl, *Adv. Catal.* **28**, 1-78 (1979).
4. Y. Hayase, S. Wehner, J. Küppers, and H.R. Brand, *Phys. Rev. E* **69**, 021609 (2004).
5. S. Wehner, F. Baumann, M. Ruckdeschel, and J. Küppers, *J. Chem. Phys.* **119**, 6823-6831 (2003).
6. Y. Hayase, S. Wehner, J. Küppers, and H.R. Brand, *Physica D* **205**, 15-24 (2005).
7. S. Wehner, Y. Hayase, H.R. Brand, and J. Küppers, *J. Phys. Chem. B* **108**, 14452-14461 (2004).
8. P. Hoffmann, S. Wehner, D. Schmeisser, H.R. Brand, and J. Küppers, *Phys. Rev. E* **73**, 056123 (2006).
9. S. Wehner, P. Hoffmann, D. Schmeisser, H.R. Brand, and J. Küppers, *Chem. Phys. Lett.* **423**, 39-44 (2006).
10. H. H. Rotermund, W. Engel, M. Kordesch, and G. Ertl, *Nature* **343**, 355-357 (1990).
11. H. H. Rotermund, M. Pollmann, and I. G. Kevrekidis, *Chaos* **12**, 157-163 (2002).
12. M. Berdau, G. G. Yelienin, A. Karpowicz, M. Ehsasi, K. Christmann, and J. H. Block, *J. Chem. Phys.* **110**, 11571-11573 (1999).
13. A. Schaak, and R. Imbihl, *J. Chem. Phys.* **116**, 9021-9027 (2002).
14. H.P. Bonzel, editor, *Landolt-Börnstein - Group III Condensed Matter Volume 42 A2 and references therein*, Springer, Heidelberg, 2002.
15. S. Wehner, P. Hoffmann, D. Schmeisser, H.R. Brand, and J. Küppers, *Phys. Rev. Lett.* **95**, 038301 (2005).
16. H.P. Bonzel, editor, *Landolt-Börnstein - Group III Condensed Matter Volume 42 A1 and references therein*, Springer, Heidelberg, 2001.
17. H. H. Rotermund, S. Nettesheim, A. von Oertzen, and G. Ertl, *Surf. Sci. Lett.* **275**, L645-L649 (1992).
18. F. Mertens, and R. Imbihl, *Chem. Phys. Lett.* **242**, 221-227 (1995).
19. G.A. Somorjai, *Introduction to surface chemistry and catalysis*, Wiley, New York, 1994.

Existence Range of Pulses in the Quintic Complex Ginzburg-Landau Equation

Pablo Gutiérrez* and Orazio Descalzi†

*Departamento de Física, FCFM, Universidad de Chile, Santiago, Chile
†Facultad de Ingeniería, Universidad de los Andes, Santiago, Chile

Abstract. Given a system in the vicinity of an oscillatory subcritical bifurcation (Hopf) that presents localized structures, we model its dynamics with the Quintic Complex Ginzburg-Landau Equation. For pulse-type structures, we study the bifurcation to fronts via a quasi-analytical approach.

Keywords: Ginzburg-Landau equation, Localized structures
PACS: 47.20.Ky, 82.40.Bj, 05.70.Ln

INTRODUCTION

In the last decades the importance of studying pulse-type localized structures has been shown, because they are observed in different physical contexts: hydrodynamics, chemical reactions, population dynamics, plasma physics, granular media, etc. The presence of pulses has been discovered particularly in systems whose dynamics is oscillatory, like binary-fluids [1] and chemical surface reactions [2].

In 1988, Thual and Fauve started a line of research on pulses showing the existence of pulses by using as a model the Quintic Complex Ginzburg-Landau Equation (QCGLE) [3]. The QCGLE arises generically as an envelope equation for a weakly inverted bifurcation associated with traveling waves. This equation has a non-variational nature and coexistence of stable solutions.

From that pioneer work a wide field of investigation was opened, finding new solutions for QCGLE [4–8] and in the creation of analytical methods that allow us understand the properties of those solutions. From an analytical point of view we can remark the limits of QCGLE, as the variational limit [9] or the perturbative analysis of solitons in the non-linear Schrödinger equation (conservative limit) [10]. Recently Descalzi et al. proposed a quasi-analytical approach that approximate pulses on the whole range between the above referred limits. In addition, they show that the appearance of pulses is related to a saddle-node bifurcation [11, 12].

In this paper, we focus on the pulse-front bifurcation and, in this way, it is possible characterize the existence range of pulses.

QUASI-ANALYTICAL STUDY OF PULSES

Our starting point is the QCGLE:

$$\partial_t A = \mu A + \beta |A|^2 A + \gamma |A|^4 A + D\partial_{xx} A, \tag{1}$$

where the subscripts x and t denote partial derivatives with respect to space and time respectively. $A(x,t) = r(x,t)e^{i\phi(x,t)}$ is a complex field. The control parameter μ is considered as real without losing generality. The parameters $\beta = \beta_r + i\beta_i$, $\gamma = \gamma_r + i\gamma_i$ and $D = D_r + iD_i$ are in general complex and contain the physical information of the problem under study. In order to assure that the bifurcation is subcritical and saturates the quintic order it is necessary to demand $\beta_r > 0$ and $\gamma_r < 0$. For the homogeneous case, the system shows coexistence of stable solutions in the range $\beta_r^2/4\gamma_r \leq \mu \leq 0$.

If the localized structure is stationary which means r is a function depending only on the space, then it is possible to make the following Ansatz [3]:

$$A(x,t) = R_0(x)\exp\{i(\Omega t + \theta_0(x))\}, \tag{2}$$

CP913, *Nonequilibrium Statistical Mechanics and Nonlinear Physics, XV Conference*
edited by O. Descalzi, O. A. Rosso, and H. A. Larrondo

where Ω is a unknown parameter. Replacing the Ansatz in (1) one obtain two real equations, and after a simple algebra results:

$$0=\mu_+R_0+\beta_+R_0^3+\gamma_+R_0^5+R_{0xx}-R_0\theta_{0x}^2, \tag{3}$$

$$\mu_-R_0=\beta_-R_0^3+\gamma_-R_0^5+2R_{0x}\theta_{0x}+R_0\theta_{0xx}, \tag{4}$$

with

$$\begin{aligned}\mu_+ &= \frac{D_r\mu-D_i\Omega}{|D|^2}, \quad \beta_+=\frac{D_r\beta_r+D_i\beta_i}{|D|^2}, \quad \gamma_+=\frac{D_r\gamma_r+D_i\gamma_i}{|D|^2},\\ \mu_- &= \frac{D_i\mu+D_r\Omega}{|D|^2}, \quad \beta_-=\frac{D_r\beta_i-D_i\beta_r}{|D|^2}, \quad \gamma_-=\frac{D_r\gamma_i-D_i\gamma_r}{|D|^2},\end{aligned} \tag{5}$$

and $|D|^2=D_r^2+D_i^2$. It is important to notice that if θ_{0x} is constant, one can integrate explicitly the equation (3). If one observe the associated numerical simulations [7, 8], becomes clear the validity of this condition on θ_{0x} in almost the whole space, but around the centre of the localized structure. This numerical fact allowed Descalzi et. al to propose the approximation scheme introduced in [11, 12]: the space is divided in one region near the center of the pulse (*core*) and another region out of the center of the pulse (*outside the core*).

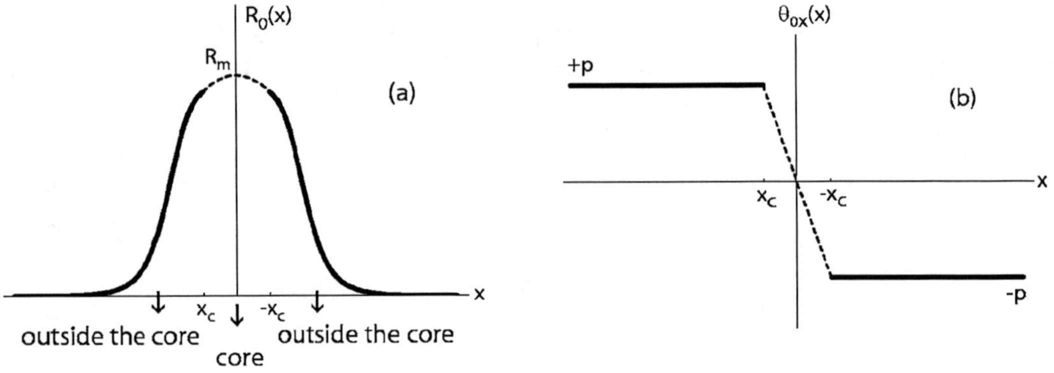

FIGURE 1. Regions of approximation for a pulse. The thick line is the *outside the core* region. The dotted line is the *core* region. (a) Modulus of the pulse R_0. (b) Phase gradient θ_{0x}.

Inside the *core* one approximate the functions $R_0(x)$ and $\theta_{0x}(x)$ by their first terms in Taylor expansion writing:

$$R_0(x)=R_m-\varepsilon x^2, \quad \theta_{0x}=-\alpha x, \tag{6}$$

where (Rm,ε,α) are unknown parameters. R_m is the maximum height of the pulse at $x=0$.

When the equations (6) are included in (3) and (4) it is obtained: $\varepsilon=(\mu_+R_m+\beta_+R_m^3+\gamma_+R_m^5)/2$ and $\alpha=\beta_-R_m^2+\gamma_-R_m^4-\mu_-$. ε and α are now expressed in terms of (R_m,Ω) and of the original parameters of the equation (1).

Outside the core, where θ_{0x} is constant ($+p$ for $x<x_c$ and $-p$ for $x>x_c$), one can integrate explicitly equation (3). The solution is:

$$R_0(x)=\frac{2b^{\frac{1}{4}}\exp\{\sqrt{-\mu_++p^2}(|x|+x_0)\}}{\sqrt{(\exp\{2\sqrt{-\mu_++p^2}(|x|+x_0)\}+\frac{a}{\sqrt{b}})^2-4}}, \tag{7}$$

where $a=-3\beta_+/2\gamma_+$, $b=-3(-\mu_++p^2)/\gamma_+$ and x_0 is a constant that emerges from the translation symmetry of the QCGLE.

Ω can be asymptotically evaluated and we obtain: $\Omega=(-D_i(\mu D_r-2p^2|D|^2)+2p|D|^2\sqrt{-\mu D_r+p^2|D|^2})/D_r^2$, that only depends on p (and on the parameters of (1)). Now ε and α depends on (Rm,p).

Next step in the Matching Approach is to impose continuity to the functions at $x_c = -p/\alpha$. The continuity in $R_0(x)$ fixes the value of x_0 in terms of R_m and p:

$$x_0 = x_c + \frac{\ln u_*}{\sqrt{-\mu_+ + p^2}}, \tag{8}$$

where

$$u_*^2 = -\frac{a}{\sqrt{b}} + \frac{2\sqrt{b}}{r_c^2} + \frac{2}{r_c^2}\sqrt{r_c^4 - ar_c^2 + b} \tag{9}$$

and $r_c = R_m - \varepsilon x_c^2$.

The continuity of $R_{0x}(x)$ at $x = x_c$ gives us the first relation between R_m and p:

$$f(Rm, p) \equiv \sqrt{-\frac{\gamma_+}{3}}r_c\sqrt{r_c^4 - ar_c^2 + b} + 2\varepsilon x_c = 0. \tag{10}$$

A second relation emerges from a condition of consistency, introduced in [3], obtained multiplying equation (4) by $R_0(x)$ and integrating the real axis. Since $R_0(x)$ is a symmetric function, the relation is reduced to:

$$g(R_m, p) \equiv \mu_- \int_{-\infty}^{0} R_0^2 dx - \beta_- \int_{-\infty}^{0} R_0^4 dx - \gamma_- \int_{-\infty}^{0} R_0^6 dx = 0. \tag{11}$$

The integrals can be evaluated and the result is:

$$\int_{-\infty}^{0} R_0^2 dx = \frac{1}{2}\sqrt{-\frac{3}{\gamma_+}} \ln\left|\frac{a+\sqrt{b}(u_*^2+2)}{a+\sqrt{b}(u_*^2-2)}\right| - R_m^2 x_c + \frac{2}{3}R_m \varepsilon x_c^3. \tag{12}$$

$$\int_{-\infty}^{0} R_0^4 dx = -\sqrt{-\frac{3}{\gamma_+}}\frac{(a^2-4b)\sqrt{b}+abu_*^2}{(-4b+(a+\sqrt{b}u_*^2)^2)} + \frac{a}{4}\sqrt{-\frac{3}{\gamma_+}}\ln\left|\frac{a+\sqrt{b}(u_*^2+2)}{a+\sqrt{b}(u_*^2-2)}\right| - R_m^4 x_c + \frac{4}{3}R_m^3 \varepsilon x_c^3. \tag{13}$$

$$\begin{aligned}\int_{-\infty}^{0} R_0^6 dx \;=\; & \frac{\sqrt{-\frac{3}{\gamma_+}}}{16(-4b+(a+\sqrt{b}u_*^2)^2)^2}\Big(-12a\sqrt{b}(a^2-4b)^2 - 4bu_*^2(a^2-4b)(9a^2+4b) - 12a(3a^2-4b)b^{\frac{3}{2}}u_*^4 \\ & +4b^2(-3a^2+4b)u_*^6 + (3a^2-4b)(-4b+(a+\sqrt{b}u_*^2)^2)^2 \ln\left|\frac{a+\sqrt{b}(u_*^2+2)}{a+\sqrt{b}(u_*^2-2)}\right|\Big) \\ & -R_m^6 x_c + 2R_m^5 \varepsilon x_c^3.\end{aligned} \tag{14}$$

The construction of relations $f(R_m, p) = 0$ and $g(R_m, p) = 0$ is essential because the existence of the pulses is conditioned by them: the pulse exist if the matching takes place. This occurs when the relations are simultaneously fulfilled and graphically implies the intersection between the curves that were defined in the (R_m, p) plane.

Through this procedure the mechanism of pulses appearance has been explained: there is a critical value μ_{c1}, so that for$\mu < \mu_{c1}$ the curves $f = 0$ and $g = 0$ do not intersect at any point suggesting that there are no pulses. For $\mu > \mu_{c1}$ the curves intersect in two points, that represent stable and an unstable pulses. This means that the appearance mechanism of pulses is a saddle-node bifurcation [11, 12].

MECHANISM OF STABLE PULSES DISAPPEARANCE

If μ is further increased, then the numerical observations [7] show that the stable pulse disappears leading to the emergence of fronts [13]. Therefore, there is a second interesting critical μ (μ_{c2}).

From the point of view of the Matching Approach the disappearance of pulses is related to the restrictions imposed by x_0 (obtained in (8)), since it should be real. It is only in this way that the expression (7) of $R_0(x)$ has sense.

The restriction in x_0 results from u_* which is the result of the evaluation of $\sqrt{u_*^2}$. Hence, u_*^2 should be positive. The condition $u_*^2 = 0$ can be rewritten as $a^2 = 4b$ (near the intersection of $f = 0$ and $g = 0$). Then, p can be fixed in a constant value p_{crit} through direct algebra. Next, x_0 is real in the region $p > p_{crit}$ (near the intersection).

$$p_{crit} = \sqrt{\frac{D_r\mu - \alpha_1 + \sqrt{\alpha_2 - \alpha_3\mu}}{D_r^2}}, \tag{15}$$

with

$$\alpha_1 = \frac{3}{16}\frac{(D_r\beta_r+D_i\beta_i)^2}{D_r\gamma_r+D_i\gamma_i}\left(1+\frac{D_i^2}{|D|^2}\right), \quad \alpha_2 = \left(\frac{3}{8}\frac{(D_r\beta_r+D_i\beta_i)^2}{D_r\gamma_r+D_i\gamma_i}\frac{D_i}{|D|}\right)^2, \quad \alpha_3 = \frac{3}{4}\frac{(D_r\beta_r+D_i\beta_i)^2}{D_r\gamma_r+D_i\gamma_i}\frac{D_rD_i^2}{|D|^2}. \tag{16}$$

Written in this way, it is clear that p_{crit} depends only on the parameters of (1), particularly of μ.

In terms of μ, μ_{c2} sets the existence of the pulse (Fig. 2). For $\mu < \mu_{c2}$ there is an intersection of $f = 0$ and $g = 0$ where $p > p_{crit}$. This means that there exists a stable pulse (Fig. (2a) and (2b)). For $\mu > \mu_{c2}$ the intersection of $f = 0$ and $g = 0$ is obtained in $p < p_{crit}$. In this case there is an absence of pulse (Fig. 2c). It is necessary to notice that μ_{c2} is such that the intersection occurs in $p = p_{crit}$.

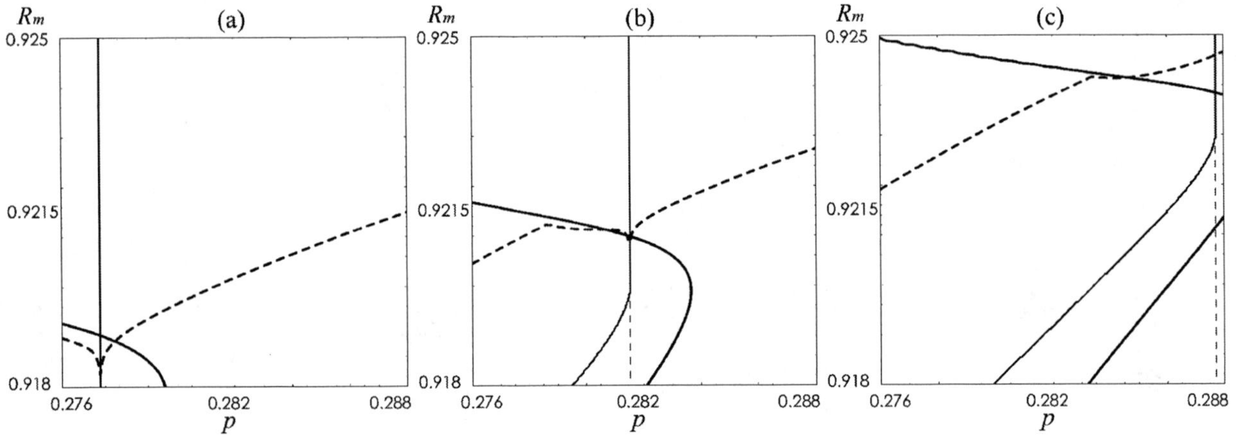

FIGURE 2. The (R_m, p) plane for different μ. The thick continuous line corresponds to $f = 0$. The thick dotted line corresponds to $g = 0$. The thin line determines the validity of x_0: in the right side x_0 is real. The thin dashed line corresponds to $p = p_{crit}$. If the thick constant and dotted lines are intersected with $p > p_{crit}$, then there is stable pulse. The parameters used in every figure are: $\beta_r = 1$, $\beta_i = 0.2$, $\gamma_r = -1$, $\gamma_i = 0.15$, $D_r = 1$, $D_i = -0.1$. (a) $\mu = -0.122 < \mu_{c2}$, the intersection of $f = 0$ and $g = 0$ is produced with $p = 0.277841 > p_{crit}(\mu) = 0.277313$. Therefore, there exists pulse. (b) $\mu = -0.120 < \mu_{c2}$, the intersection is produced with $p = 0.281575 > p_{crit}(\mu) = 0.281523$. There is pulse too. (c) $\mu = -0.117 > \mu_{c2}$, there is an absence of pulse because the intersection occurs at $p = 0.284637 < p_{crit}(\mu) = 0.287704$.

OBTENTION OF μ_{c2}

For the Matching Approach all the variables and functions are written in only two free parameters R_m and p. however, the study of bifurcations has always included μ playing the same role. This is so because in the physical interpretation μ is a control parameter. For instance, in an Rayleigh-Bérnard type experiment μ is proportional to the difference of temperatures ΔT among the plates of the fluid layer.

From this viewpoint, the functions f and g can be written as $f(R_m, p, \mu)$ and $g(R_m, p, \mu)$. On the other hand, it was already mentioned that the existence of pulses is determined by the condition $p = p_{crit}(\mu)$. This condition eliminates one of the dependencies, and let us obtain μ_{c2} through the intersection of the functions, now in the (R_m, μ) plane.

When the definition (11) of g, the integrals (12), (13), (14) and the conditions fixing p_{crit} ($u_*^2 = 0$ and $a^2 = 4b$) are observed, it is clear that the function g is undefined in p_{crit}. Thus, and keeping in mind that the dominant terms of g are logarithms, it approaches to:

$$g_{aprox}(R_m,\mu) \equiv 8\mu_- - 4\beta_- a - \gamma_-(3a^2 - 4b) = 0, \tag{17}$$

which is a function only of μ. Consequently, μ_{c2} is obtained explicitly:

$$\mu_{c2} = \frac{\alpha_2 - (\alpha_1 - D_r\alpha_4)^2}{\alpha_3}, \tag{18}$$

with $\alpha_4 = \frac{3}{4}\frac{D_r\beta_r + D_i\beta_i}{D_r\gamma_r + D_i\gamma_i}\left(\beta_r - \frac{3}{4}\gamma_r\frac{D_r\beta_r + D_i\beta_i}{D_r\gamma_r + D_i\gamma_i}\right)$.

In order to fix ideas we use (in Figs. 2 and 3) parameters [13] from which wide numerical results are known [7]. These parameters are $\beta_r = 1$, $\beta_i = 0.2$, $\gamma_r = -1$, $\gamma_i = 0.15$, $D_r = 1$, $D_i = -0.1$. With them can be obtained: $\mu_{c1} = -0.16776$ for the appearance of pulses and $\mu_{c2} = -0.117066$ for their disappearance. When comparing the numerical simulations it is found that the method's error is approximately 1% and 5% respectively.

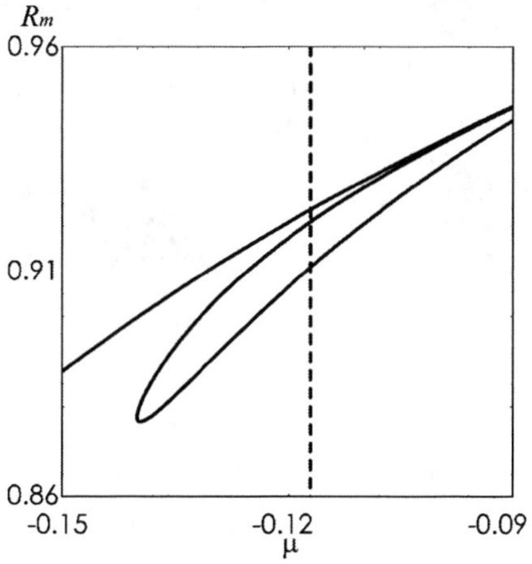

FIGURE 3. The (R_m,μ) plane. The thick continuous line corresponds to $f = 0$ and the thick dotted line corresponds to $g_{approx} = 0$, evaluated from (17). Using the same parameters than Fig. 2, $\mu_{c2} = -0.117066$ can be obtained.

CONCLUSIONS

In this article we have focused on pulses using the QCGLE as a model. This envelope equation arises generically near subcritical bifurcations for systems with oscillatory dynamics.

Firstly, we revised a quasi-analytical method to study pulses in a wide range of parameters. The method consists of approximating the shape of the pulse (its amplitude and phase gradient) separating the space into two regions: the core and outside the core, in order to impose continuity. With this scheme the appearance of pulses via saddle-node bifurcation was previously characterized, and one can obtain the minimum μ when the pulses exist.

Then, one looks for the maximum μ when the pulses exist. For that, when one increases μ, one must check the validity of the approximation that guarantee the pulse existence. In particular, one can see that when the approximated solution of the pulse is explicitly written, one of its variables (x_0) loses its physical sense. With this, we conclude that the pulse disappear leading to the emergence of fronts. The second μ of bifurcation (that is the maximum μ when the pulses exist) can be analytically approximated because the restrictions on x_0 simplify the matching conditions until μ_{c2} is fixed. When it is compared, μ_{c2} agree within 5% of recent numerical simulations.

As a perspective, it would be interesting to apply these results to the two dimensional QCGLE (trivial generalization), where a radial pulse solution was observed with a very different existence range although the same parameters are used.

ACKNOWLEDGMENTS

O.D. wishes to thank the support of FAI (Project No. ICIV-001-06, U. de los Andes), FONDECYT (Project No. 1070098) and Project Anillo en Ciencia y Tecnología ACT15. P.G. acknowledges support from Project Anillo en Ciencia y Tecnología ACT15 and FONDECYT (Project No. 1050660).

REFERENCES

1. P. Kolodner, D. Bensimon, and C. M. Surko, *Phys. Rev. Lett.* **60**, 1723–1726 (1988).
2. H. H. Rotermund, S. Jakubith, A. von Oertzen, and G. Ertl, *Phys. Rev. Lett.* **66**, 3083–3086 (1991).
3. O. Thual and S. Fauve, *J. Phys. France* **49**, 1829–1833 (1988).
4. B. A. Malomed and A. A. Nepomnyashchy, *Phys. Rev. A* **42**, 6009–6014 (1990).
5. V. V. Afanasjev, N. Akhmediev and J.M. Soto-Crespo, *Phys. Rev. E* **53** 1931–1939 (1996).
6. R. J. Deissler and H. R. Brand, *Phys. Rev. Lett.* **72** 478–481 (1994).
7. O. Descalzi and H. R. Brand, *Phys. Rev. E* **72**, 055202(R) (2005).
8. O. Descalzi, P. Gutiérrez and E. Tirapegui, *Int. J. Mod. Phys. C* **16**, 1909–1916 (2005).
9. V. Hakim and Y. Pomeau, *Eur. J. Mech. B/Fluids* **10**, 137–143 (1991).
10. S. Fauve and O. Thual, *Phys. Rev. Lett.* **64**, 282–284 (1990).
11. O. Descalzi, M. Argentina and E. Tirapegui, *Int. J. Bifurcation Chaos* **12**, 2459–2465 (2002).
12. O. Descalzi, M. Argentina and E. Tirapegui, *Phys. Rev. E* **67**, 015601(R) (2003).
13. W. van Saarloos and P. C. Hohenberg, *Phys. Rev. Lett.* **64**, 749–752 (1990).

Hole Solutions in the Cubic Complex Ginzburg-Landau Equation versus Holes in the Cubic-Quintic Complex Ginzburg-Landau Equation

Helmut R. Brand*, Orazio Descalzi†,* and Jaime Cisternas†

*Department of Physics, University of Bayreuth, 95440 Bayreuth, Germany
†Facultad de Ingeniería, Universidad de los Andes, Santiago, Chile

Abstract. In this contribution we compare the properties of hole solutions of the cubic complex Ginzburg-Landau equation with those of the cubic-quintic complex Ginzburg-Landau (CGL) equation in one spatial dimension. Both equations occur as prototype envelope equations near the onset of an oscillatory bifurcation to traveling waves.

While hole solutions of the cubic CGL equation have been discussed already in detail for about two decades and are known to be structurally unstable, the study of stable hole solutions for the cubic-quintic CGL equation has only attracted more attention rather recently. The hole solutions of the latter equation turn out to be structurally stable. In addition, several classes of moving and breathing hole solutions have been found.

Here we critically compare our current knowledge of hole solutions for both type of equations and also point out connections to hole solutions that have been found in other prototype equations describing pattern formation in dissipative systems.

Keywords: Localized Solutions, Holes, complex Ginzburg-Landau Equation, Traveling Waves, Pattern Formation, Stability
PACS: 05.70.Ln, 82.40.Bj, 42.65.Sf

INTRODUCTION

Throughout this note we discuss one-dimensional (1D) hole solutions as they arise for two types of prototype equations characterizing driven nonlinear systems with dissipation and dispersion (compare ref.[1] for a review of pattern formation in nonequilibrium systems). The first one is the cubic complex Ginzburg-Landau (CGL) equation, which arises as envelope equation near a forward bifurcation to traveling waves. This equation has been studied for over three decades and many features of its solutions have been reviewed. In the next section we just summarize the important features of hole solutions arising for this equation. The other equation of interest here is the cubic-quintic complex Ginzburg (CGL) equation, which arises near the onset of a weakly inverted bifurcation to traveling waves. For this equation hole solutions have been studied in more detail only recently. We therefore focus the bulk part of this article on the various classes of hole solutions as they arise for the cubic-quintic CGL equation. It seems important to note that both, the cubic CGL as well as the cubic-quintic CGL equation are envelope equations [2]-[4]. That is they are valid near the onset of an instability and the distance from onset serves as the expansion parameter for deriving all contributions in the resulting equations.

HOLES IN THE CUBIC COMPLEX GINZBURG-LANDAU EQUATION

The cubic complex Ginzburg-Landau equation reads:

$$\partial_t A = \mu A + (\beta_r + i\beta_i)|A|^2 A + (D_r + iD_i)\partial_{xx} A, \tag{1}$$

where $A(x,t) = R(x,t)e^{i\phi(x,t)}$ is a complex field. The linear growth term can be chosen to be real without loss of generality by going into the rotating frame. By rescaling amplitude, time and spatial coordinate the number of independent parameters is reduced to two.

Holes for the cubic CGL equation have been found first by Nozaki and Bekki [5, 6] about two decades ago. Subsequently numerical investigations by Sakaguchi [7] have shown, however, that the analytically found hole solutions are subject to several instabilities. Later Popp et al. [8, 9] (compare also the review [10] focusing predominantly on the

CP913, *Nonequilibrium Statistical Mechanics and Nonlinear Physics, XV Conference*
edited by O. Descalzi, O. A. Rosso, and H. A. Larrondo

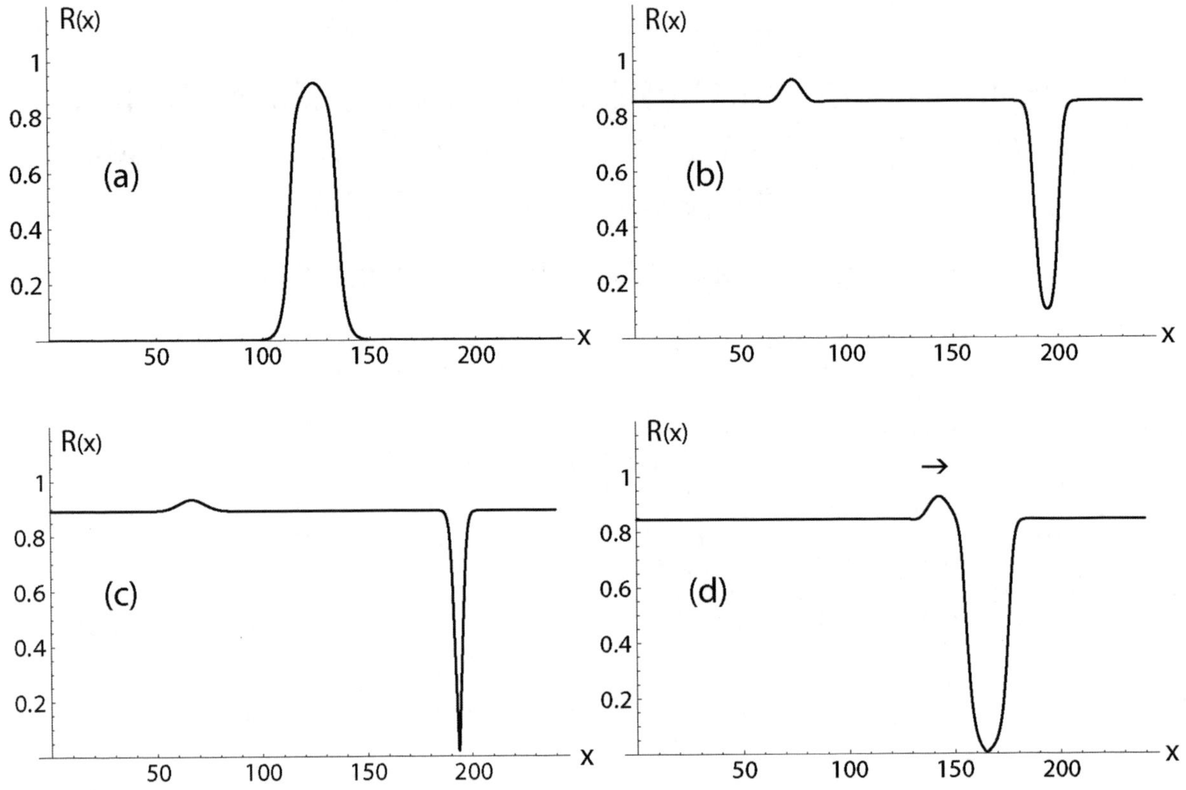

FIGURE 1. The modulus R is shown for different stable localized structures. (a) Stationary pulse for $\mu = -0.1200$. (b) Stationary 2π-hole for $\mu = -0.1080$. (c) Stationary π-hole for $\mu = -0.1080$. (d) Right-moving hole for $\mu = -0.1133$. (Adopted from ref.[12])

properties of the cubic CGL equation) demonstrated that the Nozaki-Bekki holes are structurally unstable. By this one means that even a small admixture of a quintic term renders the Nozaki-Bekki holes unstable [8]-[10]. We note that for a real physical systems such small admixtures are impossible to avoid, since the cubic CGL equation is derived as an envelope equation near the onset of an instability. Thus there are always higher order corrections including a quintic term. In view of our discussions in the next section we also note that there are only hole solutions with one type of fixed shape of the modulus for the cubic CGL equation.

HOLES IN THE CUBIC-QUINTIC COMPLEX GINZBURG-LANDAU EQUATION

The starting point of our investigations is the subcritical complex Ginzburg-Landau equation:

$$\partial_t A = \mu A + (\beta_r + i\beta_i)|A|^2 A + (\gamma_r + i\gamma_i)|A|^4 A + (D_r + iD_i)\partial_{xx} A, \tag{2}$$

where $A(x,t) = R(x,t)e^{i\phi(x,t)}$ is a complex field.

A numerical study of Eq.(2) has been carried out using the following parameter values $\beta_r = -\gamma_r = D_r = 1$, $\beta_i = 0.2$, $\gamma_i = 0.15$ and $D_i = -0.1$ [11] and periodic boundary conditions. Different stable localized solutions arise as a function of μ. In Fig.1 we have plotted - in addition to the pulse solutions of fixed shape known since 1988 [13] (Fig.1a) - the three known types of hole solutions with fixed shape: $2\pi-$ hole (Fig.1b), $\pi-$ hole (Fig.1c) and moving hole (Fig.1e).

In Fig.2 we have plotted the two types of stable breathing holes known for the cubic-quintic CGL equation, both as snapshots as well as in the form of $x-t$-plots.

As for the moving and moving breathing holes shown in Figs. 1 and 2 we emphasize that right- and left-moving hole solutions occur with equal probability as it must be the case for reasons of symmetry.

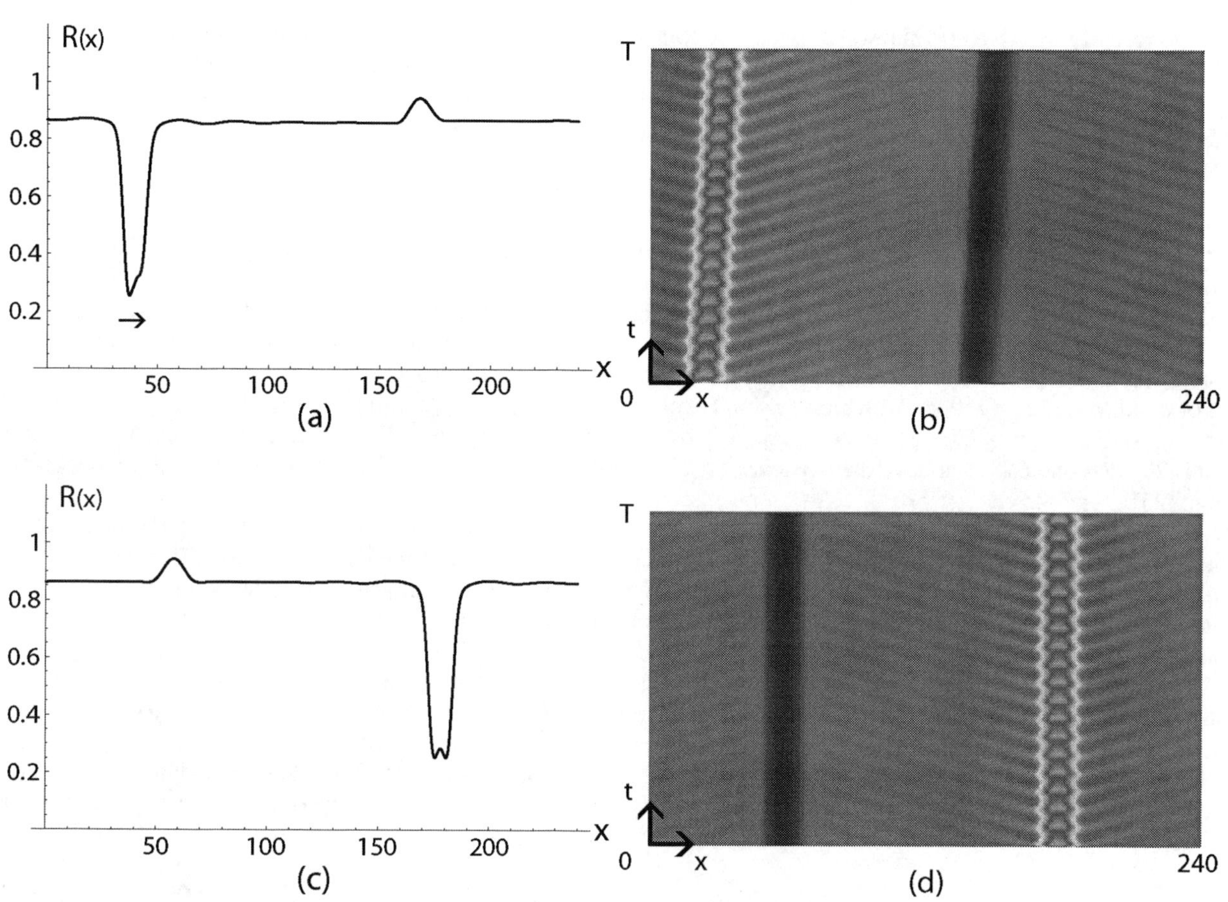

FIGURE 2. The modulus R and space-time plots for the modulus of breathing stable localized structures. (a) and (b) Breathing right-moving hole for $\mu = -0.08795$. (c) and (d) Breathing non-moving hole for $\mu = -0.08780$. In the $x-t$- plots (b) and (d) the time shown on the ordinate is $T = 10^3$ corresponding to 10^4 iterations. (adopted from ref.[12])

Concerning figures 1 and 2 we would like to point out first that - in contrast to the cubic CGL equation - the shapes of all types of hole solutions of the cubic-quintic CGL equation found up to now, are not known analytically.

Quite recently an analytic approximations scheme has been worked out for $2\pi-$ hole solutions [14] generalizing the method pioneered by Descalzi's group for pulse solutions [15, 16, 17].

We also emphasize that all types of hole solutions found until now for the cubic-quintic CGL equation are structurally stable in the sense that the admixture of a term higher order in the amplitude (for example, a septic term $\sim |A|^6A$) does not render the various classes of hole solutions discussed here to be unstable.

Historically the first two classes of hole solutions for the cubic-quintic CGL equation were reported by Sakaguchi [18]. For one class the modulus of A, $|A|$, touches zero in the center - this solution referred to conventionally as a $\pi-$ hole throughout this note - while for the other class the modulus stays finite in the center - referred to as a $2\pi-$ hole in this paper. The $2\pi-$ hole does therefore not show any singular behavior in contrast to the $\pi-$ hole. Sakaguchi has also shown numerically [18] that for the interactions between a $\pi-$ hole and a $2\pi-$ hole the $\pi-$ hole survives. We mention in passing also some early work on the influence of noise on holes and their interaction [19].

Over the last couple of years [12, 20] we have investigated in more detail the question of hole solutions for the cubic-quintic CGL equation and we have found various additional classes of stable hole solutions including especially breathing, non-moving holes and breathing moving holes (compare Fig.2 and the discussion above).

Thus there are at least five different classes of stable hole solutions for the cubic-quintic CGL equation: stationary $\pi-$ holes, stationary $2\pi-$ holes, moving holes, breathing holes and moving breathing holes. For all these classes of hole solution we have also checked that they are stable against noisy perturbations as long as the noise strength is sufficiently small.

Quite recently we have also shown [21], that for counter-propagating pulses arising as solutions from two coupled cubic-quintic CGL equations one can obtain propagating $2\pi-$ holes and propagating $\pi-$ holes after the collision in an annular geometry for small approach velocity of the original pulses.

DISCUSSION

Naturally there also other evolution equations characterizing dissipative pattern-forming nonequilibrium systems giving rise to hole solutions. Among them we briefly just highlight a few.

For the case of the optical bistability with dispersion and dissipation Brand and Deissler [22] have shown that stable solutions can arise. As a function of one parameter - the coefficient of nonlinear dispersion - one can obtain in this system stably first pulse solutions, then holes and afterwards particle solutions again.

For a rather simple reaction - diffusion system Hayase et al. have shown [23] that $\pi-$ holes and $2\pi-$ can not only stably exist, but that they even occur simultaneously over some parameter range. What is more, it has even been shown in ref.[23] that one can even have the existence of $\pi-$ holes, $2\pi-$ holes and pulse solutions over a finite parameter range [23].

We note that among all hole solutions found for the cubic-quintic CGL equation $\pi-$ holes are the most robust ones. They are less sensitive to large amplitude noise than $2\pi-$ holes and time-dependent (for the modulus) holes thus indicating that $\pi-$ holes have the largest domain of attraction. We also emphasize that the width of all classes of hole solutions presented here and known today is not related to the system size for large enough systems, but clearly a quantity set by the parameters of the equation.

We have investigated subsequently the same type of question for the cubic-quintic CGL equation and have indeed found the same type of behavior [12]. Our results are summarized in Fig.3. In addition, we have shown that moving pulses and moving holes can also stably exist simultaneously over a narrow parameter range [21].

This situation must be contrasted with that for the cubic CGL equation for which stable pulse solutions have never been found and for which hole solutions are structurally unstable. Thus question like a simultaneous existence for fixed values of the parameters in the equation cannot even be asked.

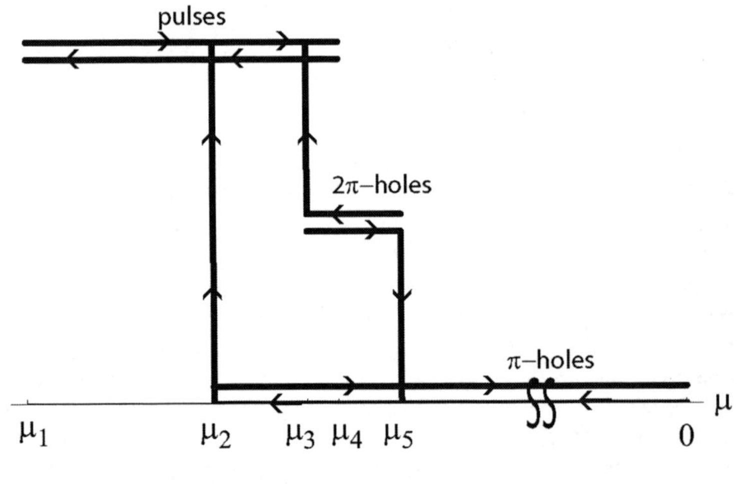

FIGURE 3. Phase-diagram of stationary pulses and holes. Pulses and holes coexist for $\mu_3 < \mu < \mu_4$. $\mu_1 = -0.16897$, $\mu_2 = -0.129$, $\mu_3 = -0.1126$, $\mu_4 = -0.1110$, $\mu_5 = -0.1045$. The $\mu-$axis is not to scale; regions of coexistence have been magnified (adopted from ref.[12]).

Another class of prototype equations are order parameter equations. They have been pioneered by Swift and Hohenberg [24] for Rayleigh-Bénard convection above onset to restore the symmetry in the plane of the convection cell, which is broken explicitly in the Newell-Whitehead-Segel equation [2] that has been derived for roll solutions. Order parameter equations, unlike envelope and phase equations [25], are not derived completely systematically, but contain a phenomenological aspect. They are constructed such that they reduce to the appropriate envelope equation near the onset of the instability and also reflect the correct symmetry of the underlying problem (for example, the isotropy in the plane for Bénard convection), In addition, one adds the nonlinearities, which are expected to be the most important ones for the problem at hand. Such equations are then quite often useful to describe pattern-forming nonequilibrium systems even far above the onset of the instability.

For the oscillatory instability near a weakly inverted bifurcation an order parameter equation of this type has been suggested in ref. [26]. In this paper various types of particle-like solutions in one and two spatial dimensions have been discussed. In addition, a compound state has been analyzed [26], which consists of a particle-like solution, but with a hole inside. The whole compound object was shown to be temporally chaotic. After having found various types of holes in the cubic-quintic CGL equation as well as for reaction-diffusion models, it would be a natural next step to investigate hole solutions for this order parameter equation derived near a weakly inverted bifurcation to traveling waves.

CONCLUSIONS AND PERSPECTIVE

For phase dynamics, the analogue of hydrodynamics for large aspect ratio pattern-forming nonequilibrium systems [25], the question of the existence of stable holes is clearly one, which has not been asked yet. The stable existence of stationary localized domains in wavelength, however, is well established, both theoretically [27, 28, 29] as well as experimentally for the case of slot convection [30]-[33]. From a modeling point of view also propagating particles have been predicted [34], but not yet experimentally observed.

By holes we clearly mean in this connection a hole in the wavelength, that is a smaller wavelength over part of an experimental cell. Clearly localized domains with a larger wavelength than the mean one correspond to holes in the wave vector as a function of space in a large aspect ratio pattern-forming nonequilibrium system.

We would also like to point out, that the question of stable hole-like solutions is not confined to envelope equations of the conventional type discussed here, for which the distance from onset is the expansion parameter, but could also be raised for Sivashinsky-type envelope equations. In this case one is interested in systems which show near onset a long wavelength instability thus giving rise to evolution equations of a rather different type. Indeed, envelope equations of Sivashinsky type share several features with phase equations.

In the same spirit the question arises to what extent hole solutions (as well as pulses) can arise when envelope equations are coupled to phase equations [35, 36, 37].

Clearly the biggest challenge arising from our investigations is the experimental observation of stable holes in a dissipative and dispersive pattern-forming system.

As a topic of future investigations naturally the question emerges whether hole solutions can also stably exist in two spatial dimensions for the cubic-quintic CGL equation. For pulse solutions this question has already been studied for the cubic-quintic CGL equation as well as for other prototype equations and various types of pulse solutions have been found to exist stably in two spatial dimensions [38, 39, 22, 40, 41, 42, 18]. Inspecting the literature, we are not aware, however, of a single stable hole solution in 2 D for a dissipative system. We mention here briefly, that the 2D cubic-quintic CGL equation offers a very rich scenario between the quasi-1D symmetry and rotational symmetry. Results will be published elsewhere [43]. Even in the cubic equation we have observed chains of holes never reported before [43].

There are several open questions closely related to the results discussed here.

One is related to dynamic evolution equations, which arise for a conserved order parameter such as equations of Cahn-Hilliard type. Here the question of holes seems to be completely open.

Another field for fruitful studies are most likely block-copolymers. While one has found many patterns including lamellar phases, cubic phases, double gyroid phases, hexagonal phases etc. in the bulk, the question of localized patterns, perhaps in the presence of an additional external field of a gradient in temperature and/or concentration does not seem to have been investigated as yet.

For the field of order parameter equations an open question is the efficiency of the trapping mechanism for hole solutions in general. Order parameter equations, which contain rapid as well as slow spatial variations, offer the possibility that walls associated with slow spatial variations are trapped by the rapid spatial variation of the underlying pattern. This mechanism has been pioneered for walls by Y. Pomeau [44] and then generalized to pulse-like solutions in refs.[26, 40]. Here we speculate that for hole-type solutions such a trapping mechanism should work as well.

ACKNOWLEDGMENTS

H.R.B. thanks the Deutsche Forschungsgemeinschaft for partial support of his work. O.D. wishes to acknowledge the support of FAI (Project No. ICIV-001-06), FONDECYT (Project No. 1070098) and Project Anillo en Ciencia y Tecnología ACT 15. J.C. thank FONDECYT (Project No. 1050660) for financial support.

REFERENCES

1. M. C. Cross, and P. C. Hohenberg, *Rev. Mod. Phys.* **65**, 851–1112 (1993).
2. A. C. Newell, and J. A. Whitehead, *J. Fluid Mech.* **38**, 279–303 (1969).
3. H. R. Brand, P. S. Lomdahl, and A. C. Newell, *Phys. Lett. A* **118**, 67–73 (1986).
4. H. R. Brand, P. S. Lomdahl, and A. C. Newell, *Physica D* **23**, 345–361 (1986).
5. K. Nozaki, and N. Bekki, *J. Phys. Soc. Japan* **53**, 1581–1582 (1984).
6. K. Nozaki, and N. Bekki, *Phys. Lett. A* **110**, 133–135 (1985).
7. H. Sakaguchi, *Prog. Theor. Phys.* **85**, 417–421 (1991).
8. S. Popp, O. Stiller, I. S. Aranson, A. Weber, and L. Kramer, *Phys. Rev. Lett.* **70**, 3880–3883 (1993).
9. S. Popp, O. Stiller, I. S. Aranson, and L. Kramer, *Physica D* **84**, 398–423 (1995).
10. I. Aranson, and L. Kramer, *Rev. Mod. Phys.* **74**, 99–143 (2002).
11. W. van Saarloos, and P. C. Hohenberg, *Phys. Rev. Lett.* **64**, 749–752 (1990).
12. O. Descalzi, and H. R. Brand, *Phys. Rev. E* **72**, R055202 (2005).
13. O. Thual, and S. Fauve, *J. Phys. (Paris)* **49**, 1829–1833 (1988).
14. O. Descalzi, G. Düring, and E. Tirapegui, *Physica A* **356**, 66–71 (2005).
15. O. Descalzi, M. Argentina, and E. Tirapegui, *Phys. Rev. E* **67**, 015601 (2003).
16. O. Descalzi, Y. Hayase, and H. R. Brand, *Phys. Rev. E* **69**, 026121 (2004).
17. O. Descalzi, Y. Hayase, and H. R. Brand, *Int. J. Bif. Chaos* **14**, 4097–4104 (2004).
18. H. Sakaguchi, *Prog. Theor. Phys.* **86**, 7–12 (1991).
19. R. J. Deissler, and H. R. Brand, unpublished (1999).
20. O. Descalzi, H. R. Brand, and J. Cisternas, *Physica A* **371**, 41–45 (2006).
21. O. Descalzi, J. Cisternas, and H. R. Brand, *Phys. Rev. E* **74**, R065201 (2006).
22. H. R. Brand, and R. J. Deissler, *Physica A* **216**, 288–298 (1995).
23. Y. Hayase, O. Descalzi, and H. R. Brand, *Phys. Rev. E* **69**, R065201 (2004).
24. J. Swift, and P. C. Hohenberg, *Phys. Rev. A* **15**, 319–328 (1977).
25. H. R. Brand, "Phase Dynamics - The Concept and some recent Developments", in *Patterns, Defects and Materials Instabilities* edited by D. Walgraef, and N. M. Ghoniem, NATO ASI Series E: Applied Sciences - Vol. 183, Kluwer, Dordrecht, 1990, pp. 21–34.
26. H. Sakaguchi, and H. R. Brand, *Physica D* **117**, 95–105 (1998).
27. H. R. Brand, and R. J. Deissler, *Phys. Rev. Lett.* **63**, 508–511 (1989).
28. H. R. Brand, and R. J. Deissler, *Phys. Rev. A* **41**, 5478–5481 (1990).
29. R. J. Deissler, Y. C. Lee, and H. R. Brand, *Phys. Rev. A* **42**, 2101–2113 (1990).
30. M. Dubois, R. DaSilva, F. Daviaud, P. Bergé, and A. Petrov, *Europhys. Lett.* **8**, 135–139 (1989).
31. F. Daviaud, P. Bergé, and M. Dubois, *Europhys. Lett.* **9**, 441–446 (1989).
32. M. Dubois, P. Bergé, and A. Petrov, "Properties of quasi 1-D Rayleigh-Bénard convection" in *The Geometry of Nonequilibrium* edited by P. Coullet and P. Huerre, NATO ASI Series - Vol. 237, Plenum, New York, 1990, pp. 227–235.
33. J. Hegseth, J. M. Vince, M. Dubois, and P. Bergé, *Europhys. Lett.* **17**, 413–418 (1992).
34. H. R. Brand, and R. J. Deissler, *Phys. Rev. A* **46**, 888–892 (1992).
35. P. Coullet, anbd G. Iooss, *Phys. Rev. Lett.* **64**, 866–869 (1990).
36. H. Sakaguchi, *Prog. Theor. Phys.* **87**, 1049–1053 (1992).
37. H. Sakaguchi, *Prog. Theor. Phys.* **89**, 1123–1146 (1993).
38. R. J. Deissler, and H. R. Brand, *Phys. Rev. A* **44**, R3411-3414 (1991).
39. R. J. Deissler, and H. R. Brand, *Phys. Rev. E* **51**, R852-855 (1995).
40. H. Sakaguchi, and H. R. Brand, *Physica D* **97**, 274–285 (1996).
41. H. Sakaguchi, and H. R. Brand, *Europhys. Lett.* **38**, 341–346 (1997).
42. H. Sakaguchi, and H. R. Brand, *J. Phys. II (Paris)* **7**, 1325–1330 (1997).
43. O. Descalzi, J. Cisternas, and H. R. Brand, to be published (2007).
44. Y. Pomeau, *Physica D* **23**, 3–11 (1986).

Zipping Characterization of Chaotic Sequences Used in Spread Spectrum Communication Systems

L. De Micco, C. M. Arizmendi and H. A. Larrondo

Facultad de Ingeniería, Universidad de Mar del Plata (UNMDP).
Juan B. Justo 4302 (7600) Mar del Plata, Argentina.

Abstract. In this paper finite sequences, known as Pseudo Noise codes (PN-codes) are studied. PN-codes are used in Spread Spectrum Digital Systems (SSDS), a widely employed wireless communication technique. The selection of the best "set of PN-codes" (PN-family) is an important issue in SSDS. This selection is based on a very involved statistical analysis. This paper presents three contributions in this field: the first contribution is the definition of new global quantifiers instead of those used in the engineering literature. The second contribution is to show that a quantifier based on the Zipping Complexity Measure may be a global optimization quantifier of the whole system. Finally a comparison between usual PN-families and chaotic generated PN-families is presented, showing that chaotic systems are competitive in this application field.

Keywords: chaos, zipping, spread spectrum, PN-codes
PACS: 05.45+b

INTRODUCTION

Since the introduction of chaos synchronization [1] the application of chaos to communication systems has attracted attention and many different systems has been designed. The particular case considered here, SSDS is widely employed nowadays in wireless communications (cell phones, wireless hubs, etc.). As in any other communication system a signal (message) is sent from a transmitter to a receiver. In SSDS the transmitter encode the message (text, voice, image, etc.) as a binary signal. Then a finite length binary sequence (PN-code), that is independent of the encoded message, is employed as a modulation waveform to produce a modulated signal to be transmitted.

Each user of the same communication system, has a different PN-code. The set of all the PN-codes employed in the same communication system is known as the PN-family. When a user (the transmitter) wants to send a message to another user (the receiver) it must employ the PN-code assigned to the receiver (pre-agreed code).

Each receiver continuously evaluates the correlation between the received signals and the particular PN-code assigned to it. When the correlation exceeds a threshold an incoming call is supposed to exist (this is the *detection step*) and the receiver looks for the maximum of the correlation between the incoming signal and the receiver PN-code, making a circular shift of it (this is the *synchronization step*). Then the demodulation and decoding processes start (this is the *decoding step*). In this way many users can transmit simultaneously sharing the same frequency band but with different PN-codes according to which receiver they want to contact with, and that particular receiver picks up the signal.

The name "spread spectrum" comes from the fact that the modulated signal has a spectrum that covers a whole frequency band larger that the one covered by the original encoded message. This is a requirement common to all digital systems in order to reduce the Electromagnetic Interference (EMI) they produce.

The influence of the statistical properties of the PN-family on the overall performance of the system is of upmost importance in the design process [2, 3, 4] and different PN-families were proposed as is explained in the next section. More recently PN-codes generated by chaotic maps were also considered [5, 6, 7, 8]. Such chaotic systems present a potential advantage over conventional SSDS communication in the opportunity they provide to simultaneously coding and spreading the information signal and they can provide families with a high number of PN-codes.

CHAOTIC AND NON CHAOTIC FAMILIES

A PN-code is a N-bit binary string and then there exist 2^N number of possible different strings. This number increases exponentially with the number of bits but most families have a considerable lower number of usable PN-codes (family

CP913, *Nonequilibrium Statistical Mechanics and Nonlinear Physics, XV Conference*
edited by O. Descalzi, O. A. Rosso, and H. A. Larrondo

members) because of the restrictions imposed by the following optimization requirements:

1. All the members of a given PN-family must be generated using only one algorithm: this is a hardware restrictions and there are several strategies to produce PN-families reported in the engineering literature. Each strategy gives rise to a different family. Chaotic maps are good candidates because they are very simple to be implemented.
2. Availability of a large number of elements: The number of members depends on the family generation method but usually it is $p << 2^N$. this number determines the maximum number of users sharing the spectrum. Again chaotic maps are good candidates because they can provide a high p.
3. The autocorrelation of each PN-code must be a delta function: this makes each PN-code easily distinguishable from a circular time-shifted version of itself to make the *synchronization step* faster. This property is also important for reducing the effect of multipath propagation.
4. The cross correlation between any pair of PN-codes must be a null function: each sequence in the family must be easily distinguishable from (a possibly circular time-shifted version of) every other signal in the same family. This requirement is important during the *detection step* for minimizing the multiuser interference.
5. Spread spectrum: the modulated signal must have a uniform spectrum over the assigned bandwidth. This is a requirement common to all digital systems in order to reduce the Electromagnetic Interference (EMI).

The problem is that the global optimization is not easy. For example, the auto-correlation optimization (item 3) is usually attained at the expense of a poorer cross correlation performance (item 4) and vice versa [2]. A similar situation occurs with the spreading properties of the sequence (item 5). Furthermore any optimization reduces the number p of usable sequences in the family (item 2). That is in fact the reason why several families coexist, each family using a different generating algorithm that partially optimizes over some of the requirements mentioned above.

Chaotic families studied in this paper have two important advantages: the big number of sequences generated by a chaotic map and the ease of their generation. In this paper they are compared with the commonly used families in the engineering literature. A brief description of each studied family follows and their main characteristics (see section for a detailed comparison between them):

1. **Non chaotic PN-codes** studied in this paper are
   (a) **M-sequences**: They are the most widely studied PN-codes and they are the basis for other PN-families. Each family is generated by a m-stage shift register with linear feedback according to a primitive polynomial. The length of each PN-code is $N = 2^m - 1$. The spreading and autocorrelation properties are almost ideal but the cross-correlation between any pair of codes is a periodic function with high peaks. The number of members p of each family is small.
   (b) **Gold-codes**: Gold PN-codes are obtained by means of an exclusive OR operation between two *preferred M-sequences* of the same family. Preferred M-sequences are pairs with minimal cross correlation. Including the preferred pair, a total of $p = 2^m + 1$ Gold Codes can be produced from any m-stage feedback shift register. The cross-correlation between any pair of Gold PN-codes is uniform and bounded. Gold PN-families have a large number of members. The spread spectrum properties are not optimal.
   (c) **Walsh-Hadamard codes**: These PN-codes have zero cross-correlation for zero circular time-shift, but for a shift different from zero their cross-correlation is very much dependent on the particular pair of codes used. For synchronous systems (where shift is always zero) the correlation properties of these codes are optimal. The spread spectrum does not cover the whole bandwidth, but only a number of discrete frequency components. This family has $p = N$ members.
2. **Chaotic PN-codes** studied in this paper are generated by iterating two chaotic maps: the Three-Way Bernoulli Map (TWBM) and the Four-Way Tailed Shift Map (FWTSM)[5]. The sequences generated by each map are transformed into a binary code by means of the usual symbolics dynamics rule ($[0, 0.5) \to 0, (0.5, 1] \to 1$). The number p of family members is limited by the periodicity of the pseudo chaotic digital sequence. Autocorrelation, cross correlation and Spreading Spectrum properties are studied below.

GLOBAL PERFORMANCE QUANTIFIERS

In this section three new quantifiers are proposed so as to compare the performance of the considered PN-families. The first one C is measure of both the autocorrelation (3) and the cross correlation (4) quality. In view of the decoding procedure the linear correlation properties of each family of PN sequences play a major role in the system efficiency,

TABLE 1. $\tilde{n}_{corr}$ as a function of threshold for different families with $N = 127$

			$\tilde{n}_{corr}$		
c_{th}	M-seq	Gold	FWTSM	TWBM	Walsh
1	0	0	0	0	0.0004
0.9	0	0	0	0	0.0005
0.8	0	0	0	0	0.0010
0.7	0	0	0	0	0.0019
0.6	0	0	0	0	0.0029
0.5	0	0	0	0	0.0057
0.4	0	0	0	0.0001	0.0090
0.3	0.0028	0	0.0024	0.0023	0.0169
0.2	0.0028	0	0.0380	0.0384	0.0310
0.1	0.4164	0.4961	0.3401	0.3412	0.0703
0.0	1.0000	1.0000	1.0000	1.0000	1.0000

since they determine the level of multiple access interference, the self-interference due to multipath propagation, and the time required for code synchronization. The second proposed quantifier S measures the spreading of the spectrum over the whole possible frequencies. The third proposed global quantifier Z is proposed as a global parameter of the whole system.

All these quantifiers are global, to compare the different families as a whole, it means they include all the members of each family instead of the more traditional approaches that use randomly selected representative members. Comparisons are made between families with members of the same length N. Furthermore the definition of global quantifiers is a first step towards a global optimization process.

- **Global Correlation Quantifier** C: this quantifier includes the auto correlation and cross correlation properties of each family. For a perfect detection and synchronization the autocorrelation is required to be a delta function and the cross correlation between the different members of the family must be zero as pointed in section . This is an ideal case but actually a peaked autocorrelation and low cross correlations are the best one can obtain. For this reason it is considered that a *good family* is one that has a cross correlation lower than the threshold used by the receiver in the *detection step*. This threshold is a percentage of the autocorrelation value at zero shift. The proposed global quantifier C is calculated as follows: 1) Let $\{x^{(i)},\ (i = 1, ...p)\}$ be the members of a given family, each member being a N-bit binary vector. Calculate correlations between any pair of them, for all possible circular time shifts.

$$c^{ij}(s) = \begin{cases} \sum_{k=0}^{N-|s|-1} x_k^i x_{k+s}^j & s \geq 0 \\ c^{ij}(-s) & s < 0 \end{cases} \tag{1}$$

The number of values is $N_{corr} = 2\,p^2\,N$.
2) Choose a threshold level c_{th} defined as a fraction of the maximum autocorrelation located at zero shift A_0.
3) Count the number of elements higher than the threshold level n_{corr} and normalize it:

$$\tilde{n}_{corr} = n_{corr}/N_{corr} \tag{2}$$

Table 1 shows that $\tilde{n}_{corr}$ is a decreasing function of c_{th}. 4) Finally C is the minimum threshold c_{th} giving $\tilde{n}_{corr} = 0$. In Fig. 1 for the studied families. On the x-axis families are ordered from best (left) to worst (right).

- **Global Spectrum Quantifier** S: the PN-codes are periodic and they have always a discrete spectrum but if the modulated signal energy is concentrated over a small number of discrete frequency components the spreading is not very efficient. The ideal case would be a constant spectrum over the whole band. Therefore the proposed quantifier S is calculated as follows:
1) Evaluate the magnitude of the Fast Fourier Transform (FFT) for each PN-code (see Fig. 2a).
2) Normalize each FFT magnitude by dividing it by its mean value.
3) Calculate the variance σ_i of each normalized FFT vector. magnitude (see the label on each panel of Fig. 2a). 4) The spectrum quantifier S is $S = < \sigma_i >$, the mean value over all the members of the family. Each subplot of Fig. 2b shows σ_i for a given family, as a function of the $N/2$ discrete frequencies. The order of the families is from top

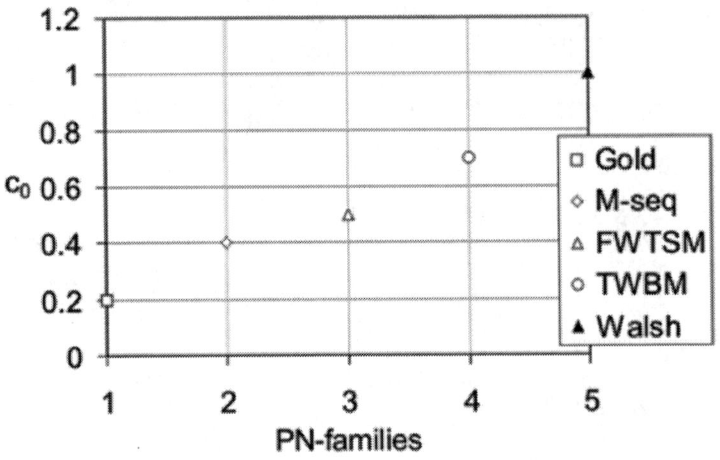

FIGURE 1. (a) Global Correlation Quantifier C as a function of the threshold c_{th} for several PN-families with $N \simeq 127$

to bottom: M-sequences, Gold, FWSTM, TWBM, and Walsh. S is shown as a label and also as an horizontal line over each subplot. The M-sequence family has the lowest S and then this family has the most uniform spectrum. Gold, TWTSM and FWBSM have quite similar values. Finally the Walsh family has the highest S showing that the spectrum is spread over a small number of frequency components. For each family it is possible to delete those codes with high σ_i's in order to improve the performance of the whole family. Of course this decision diminishes p, the number of allowed simultaneous users.

- **Zipping complexity quantifier** Z. For a string of characters the algorithmic complexity is defined as the length in bits of the smallest program that produces the string as output [9]. The problem with this definition is that it is impossible, even in principle, to find such a program. Nevertheless, the zippers or file compressors are algorithms conceived to do that job at least approximately. The Lempel and Ziv algorithm is used for most zippers and it is one of the best known file compressors [10, 11].
For a given finite length string the zipping complexity may be defined as l_Z/l where l, is the string length and l_Z is the zipped string length. The problem is that the compressing ability of a zipper depends on the string length and also on the order of the PN-codes inside the string. Then for our specific application it is necessary to construct surrogate series to obtain a parameter independent of these two factors. Then in this paper the zipping complexity quantifier Z is then evaluated as follows:
1) Construct a binary string consisting of all the members of a given family one after the other. The PN-family with highest p defines the size of the string $l = N\,p$. The string corresponding to a family with a lower p is completed by repeating the family PN-codes.
2) Generate $p!$ surrogates by changing the order of the PN-codes inside the original string in all the possible ways.
3) Zip each surrogate to give z_i. The global quantifier Z is given by $Z = < z_i >$, the mean value for all the surrogates.
Table 2 shows the results obtained for non chaotic and chaotic families. This parameter is not completely independent on the two previously defined because:
  - The string of a PN-family with high cross correlations between its PN-codes is more compressible.
  - A PN-family with a colored spectrum has periodicities and then the string will be more compressible.
  - A family with a few members must be repeated many times to complete the string, and then the string will be more compressible.

Then Z decreases if correlation increases, spreading is not flat and the number of members decreases as Table 3 shows.

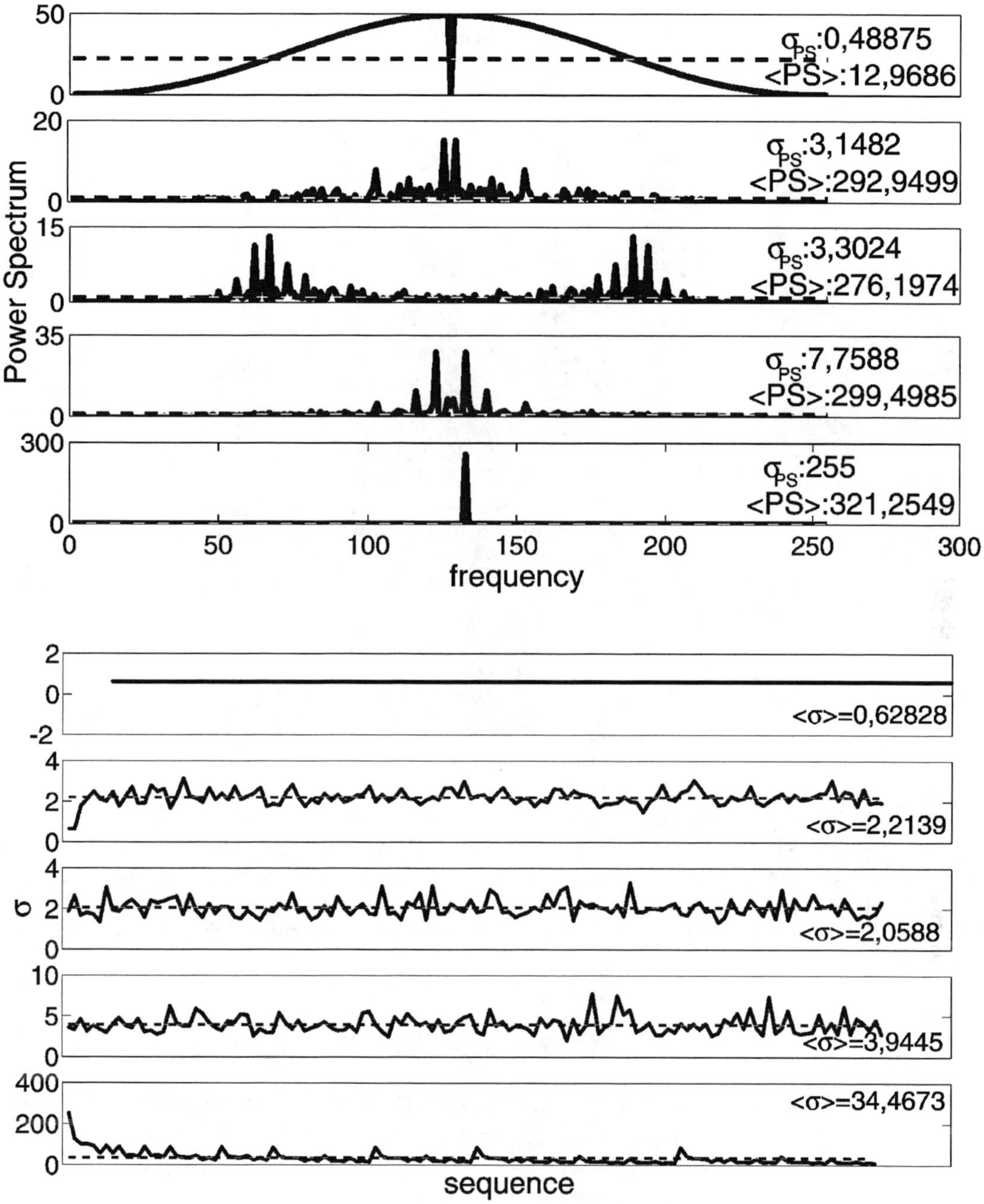

FIGURE 2. (a) Typical spectrum of a PN-code for the different families the mean value and the variance σ_i are shown as labels on each subplot; b) Variances σ_i for the PN-codes of the different families. The value of S is shown as a label on each subplot

CONCLUSIONS

Three global quantifiers were defined. They are the correlation quantifier C, the spreading quantifier S and the zipping quantifier Z. These quantifiers are global in the sense there is one value for each whole PN-family. The zipping

TABLE 2. Global Quantifier Z for different families and different lengths

family	N	p	l	Z
Gold	31	33	2,112	0.1416
FWTSM	31	33	2,112	0.1330
TWBM	31	33	2,112	0.1311
Walsh	32	32	2,112	0.0971
M-seq	63	6	768	0.1575
Gold	63	65	8,320	0.1174
FWTSM	63	520	66,560	0.1111
TWBM	63	520	66,560	0.1113
Walsh	64	64	8,320	0.0534
M-seq	127	18	4,608	0.1220
Gold	127	129	33,024	0.1096
FWTSM	127	129	33,024	0.1075
TWBM	127	129	33,024	0.1075
Walsh	128	128	33,024	0.0032

TABLE 3. Proposed global quantifiers for different classical and chaotic PN-families

family	N	p	C	S	Z
Gold	127	129	0.2	3.1148	0.1174
FWTSM	127	129	0.4	3.3024	0.1109
TWBM	127	129	0.5	7.7588	0.1104
M-seq	127	19	0.4	0.6202	0.0072
Walsh	128	128	1	255.00	0.0363

quantifier appears as a possible quality measure to compare different families, with the highest Z corresponding to the best family, because Z is higher for strings consisting of correlated PN-codes and it also decreases if the spectrum is not flat. Using the proposed quantifiers classical and chaotic PN-families were studied. The results presented show that chaotic families can have a higher number of members than conventional ones, with equivalent correlation and spreading properties. Furthermore they can be easily implemented and consequently they are good candidates to be used in real systems.

ACKNOWLEDGMENTS

This work was supported by Consejo Nacional de Investigaciones Científicas y Técnicas (CONICET) (PIP 5569/04), Agencia Nacional de Promoción Cientifica y Tecnológica (PICT 21409/04 and PICTO 04-495) and University of Mar del Plata, Argentina. HAL is a CONICET researcher.

REFERENCES

1. L. M. Pecora and T. L. Carroll, *Phys. Rev. Lett.* **64(8)** 821-824 (1990).
2. L. P. Welch, L.R. *IEEE. Trans. Inform. Theory* **20(3)**, 397-399 (1974).
3. D. V. Sarwate and M. B. Pursley *Proc. of the IEEE.* **68(5)**, 593-619, (1980).
4. M. B. Pursley *IEEE. Trans. Commun.* **25(6)**, 795-799 (1977).
5. G. Mazzini, G. Setti and R. Rovatti, *IEEE Trans. Circuits Sys. 1* **44(10)**, 937–947 (1997).
6. R. Rovatti, G. Mazzini and G. Setti, *IEEE Trans. Circuits Sys. 1* **51(7)**, 1336–1347 (2004).
7. R. Rovatti, G. Mazzini and G. Setti, *IEEE Trans. Circuits Sys. 1* **51(7)**, 1348–1364 (2004).
8. M. Sushchik, L. S. Tsimring and A. R. Volkovskii, *IEEE Trans. Circuits Sys. 1* **47(12)**, 1684–1691 (2000).
9. J. R. Sanchez, F. Family, C. M. Arizmendi, *Phys. Lett. A* **249**, 281–285 (1998).
10. J. Ziv and A. Lempel, *IEEE Trans. Inf. Theory* **22**, 75–81 (1976).
11. J. Ziv and A. Lempel, *IEEE Trans. Inf. Theory* **23**, 337–343 (1977).

Heterologous Infection During Chagas' Disease

G. J. Sibona*, C. A. Condat*,† and S. Cossi Isasi**

*CONICET, ARGENTINA
†FAMAF, Universidad Nacional de Córdoba, Ciudad Universitaria, 5000 Cordoba, ARGENTINA
**Fac. de Cs Médicas, Universidad Nacional de Córdoba, Ciudad Universitaria, 5000 Cordoba, ARGENTINA

Abstract. Human populations are often infected with more than one parasite strain. This is frequently the case with ChagasŠ disease, which is endemic to large regions of Latin America. In the present work we study the dynamics of the heterologous infection for this disease, using a model for the interaction between the trypanosoma cruzi parasite and the immune system. We find the dependence of the nature of the post-acute stage on the parameters characterizing the inoculated infectious strains.

Keywords: Chagas, competition, antibodies, heterologous infection, phase portrait, chronic disease
PACS: 87.17Aa, 87.19.Xx

INTRODUCTION

Chagas' disease is endemic in large regions of South America. The WHO estimates that 90 - 100 million people are at risk of infection and that 10-20 million are actually infected [1]. About 50,000 annual deaths result from chronic cardiomyopathy. The causative agent of the disease is the protozoan parasite Trypanosoma cruzi. This parasite, which is transmitted by certain bloodsucking insects, such as Triatoma infestans and Rhodnius Robustus, presents 3 main morphological stages: replicating epimastigotes, found in the insect vector, intracellular, reproductive amastigotes and blood-circulating, infective, trypomastigotes. The last two stages are identified in the vertebrate host. Upon invading the host cells, the trypomastigotes differentiate into amastigotes and initiate the replication until the cell bursts.

There are large genetic variations among *T. cruzi* parasites [2],[3]. According to genetic markers, there are two major phylogenetic lineages: *T. cruzi* I and *T. cruzi* II [4]. Lineage I is observed in regions where the parasite circulates in the sylvatic cycle, while lineage II is most frequent in the American Southern Cone, where *T. cruzi* develops a domestic cycle. It is remarkable that, in those regions where both lineages circulate, there are almost no cases of simultaneous infections with both lineages, even if the hosts may be assumed to have been repeatedly bitten. Reinfections with different parasite strains have been experimentally studied by Machado and co-workers [5], showing a mild cross-immunity to reinfection among strains. In this paper we extend a parasite-antibody competition model to investigate the effects of the simultaneous presence of two parasite strains in the vertebrate host. The model considered was previously used to successfully reproduce experimental data of the parasite/antibody populations during the acute phase of the infection, the cardiac damage induced through cell destruction [6] and the amastigote duplication time [7].

THE ANTIBODY-PARASITE COMPETITION MODEL

Recently, we developed a mathematical model to describe the interactions between the parasites and the immune system in the mammal host, during the acute stage of the disease [6]. We studied the evolution of the populations of trypomastigotes (n), antibody molecules (a), and infected cells (r). We now consider the possibility that more than one parasite species can coexist in the same host. Labeling the different species with the index i, the evolution equations for the various populations are:

$$\dot{n}_i = \eta_i N_i r_i - \alpha_{ii} a_i n_i - \sum_{j \neq i} \alpha_{ij} a_j n_i - \zeta_i n_i \tag{1}$$

$$\dot{r}_i = \zeta_i n_i - \eta_i r_i \tag{2}$$

CP913, *Nonequilibrium Statistical Mechanics and Nonlinear Physics, XV Conference*
edited by O. Descalzi, O. A. Rosso, and H. A. Larrondo

$$\dot{a}_i = (\gamma_{ii} - \alpha_{ii}a_i)\, n_i + \sum_{j \neq i} (\gamma_{ij} - \alpha_{ji}a_i)\, n_j - \frac{1}{\tau_i}(a_i - a_{i0}) \tag{3}$$

Here η_i is the cytotoxicity, ζ_i the infectivity, and N_i the number of emerging trypomastigotes per ruptured cell. The meaning of the other coefficients is as follows: a_{i0} is the initial antibody population, α_{ij} determines the removal efficiency of antibody species j with respect to parasite strain i, γ_{ij} indicates the strength of the activation of the sources for antibody i due to the presence of the $j-th$ parasite strain, and τ_i is the mean lifetime of the $i-th$ species antibody molecules. The first term in the sum in Eq. 3 corresponds to the cross immunity, while the second term in that sum corresponds to cross removal.

In writing Eqs. (1) to (3) we made the following assumptions: (i) Terms of the form $\eta_{ij}N_i r_i$ with $j \neq i$ were neglected in the equation for n_i: the presence of one strain does not directly induce the production of members of a different strain. (ii) Each parasite invades a different cell: we neglect the number of cells invaded by more than one parasite. (iii) The presence of one strain influences the activation of antibodies specific to the other strains (cross-immunity). We also observe that there is no reason to assume that cross-effects are symmetric. Thus, generally, $\gamma_{ij} \neq \gamma_{ji}$ and $\alpha_{ij} \neq \alpha_{ji}$.

SIMPLE INFECTION: STEADY STATES

To understand the model dynamics it is convenient to start by considering the simple case of one parasite and one antibody species. Under these conditions, defining the reproduction rate $\kappa = \zeta\,(N_r - 1)$, and studying the $t \to \infty$ limit, we find three possible asymptotic outcomes [8] (see Fig. 1):

- Case I: **Chronic Disease.** If $\alpha a_0 < \kappa < \gamma$, the high parasite production combined with the low annihilation rate gives a non-zero parasite population at long times.
- Case II: **Healing.** If $\kappa < \gamma$, but $\alpha a_0 > \kappa$, the high stimulation of the immune system and the high annihilation rate combine to eliminate the parasite infection ($n \to 0$).
- Case III: **Death.** If $\gamma < \kappa$, the antibody production is not strong enough to control the parasite population ($n \to \infty$), leading to the host death.

HETEROLOGOUS INFECTION: STEADY STATES AND PHASE PORTRAITS

Asymptotic Behavior of the Two-Strain System

Next we consider the analytical steady-state solutions for the two-parasite, two-antibody species problem with cross reactivity ($\gamma_{ij} \neq 0$, $\alpha_{ij} \neq 0$). There are four possibilities:

- Case Ia. **Chronic disease with two surviving strains.** An equilibrium may emerge where both strains survive at the end of the infection, coexisting with both antibody species. Defining $\Delta = \alpha_{11}\alpha_{22}$ - $\alpha_{12}\alpha_{21}$, and $k_i = (N_i - 1)\zeta_i$, the steady-state values of the antibody populations are given by,

$$a_{1s} = \Delta^{-1}(k_1\alpha_{22} - k_2\alpha_{12}) \tag{4}$$

$$a_{2s} = \Delta^{-1}(k_2\alpha_{11} - k_1\alpha_{21}) \tag{5}$$

Once these values are known, we can compute the steady state numbers of circulating trypomastigotes. Using the auxiliary quantities

$$\Gamma_{ij} = \gamma_{ij} - \alpha_{ji}a_{is} \tag{6}$$

and

$$\Theta_i = \frac{a_{is} - a_{i0}}{\tau_i} \tag{7}$$

we get

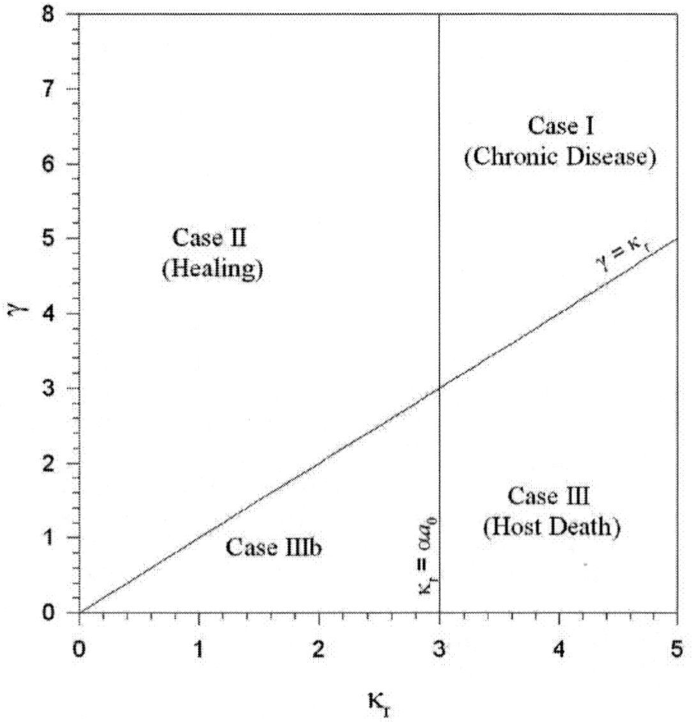

FIGURE 1. Phase diagram for the one-parasite, one antibody case

$$n_{1s} = \frac{\Gamma_{22}\Theta_1 - \Gamma_{12}\Theta_2}{\Gamma_{11}\Gamma_{22} - \Gamma_{12}\Gamma_{21}} \tag{8}$$

and

$$n_{1s} = \frac{\Gamma_{11}\Theta_2 - \Gamma_{21}\Theta_1}{\Gamma_{11}\Gamma_{22} - \Gamma_{12}\Gamma_{21}} \tag{9}$$

- Case Ib. **Chronic disease with one surviving strain**. If we assume that $n_{1s} \neq 0$, but $n_{2s} = 0$, i. e. $r_{2s} = 0$, we find,

$$a_{2s} = \frac{k_1 - \alpha_{11}a_{1s}}{\alpha_{12}} \tag{10}$$

Somewhat surprisingly, $a_{2s} \neq a_{20}$: although parasite strain 2 has been eliminated, the population of the corresponding antibody species does not return to its original value. This conclusion will be confirmed by the simulations presented below. On the other hand, the steady state population of antibody species 1, a_{1s}, can be obtained as the solution of the quadratic equation,

$$\alpha_{11}\left(\alpha_{11}\tau_1 - \alpha_{12}\tau_2\right)a_{1s}^2 - Ba_{1s} + \gamma_{11}\tau_1 k_1 - \alpha_{12}\tau_2 k_1 a_{10} + \alpha_{12}\gamma_{21}\tau_2 a_{10} - \alpha_{12}\gamma_{11}\tau_1 a_{20} = 0 \tag{11}$$

with

$$B = \alpha_{11}\gamma_{11}\tau_1 + \left(\alpha_{11}\tau_1 - \alpha_{12}\tau_2\right)k_1 - \alpha_{11}\alpha_{12}\left(\tau_2 a_{10} + \tau_1 a_{20}\right) + \alpha_{12}\gamma_{21}\tau_2 \tag{12}$$

- Case II. **Healing.** Under certain conditions, the immune system prevails and both parasite strains are eliminated. The health attractor is defined by $(n_{1s}, n_{2s}, a_{1s}, a_{2s}) = (0, 0, a_{10}, a_{20})$.
- Case III. **Host death.** It occurs if the immune system cannot control one of the strains. In this case both strains end up growing exponentially.

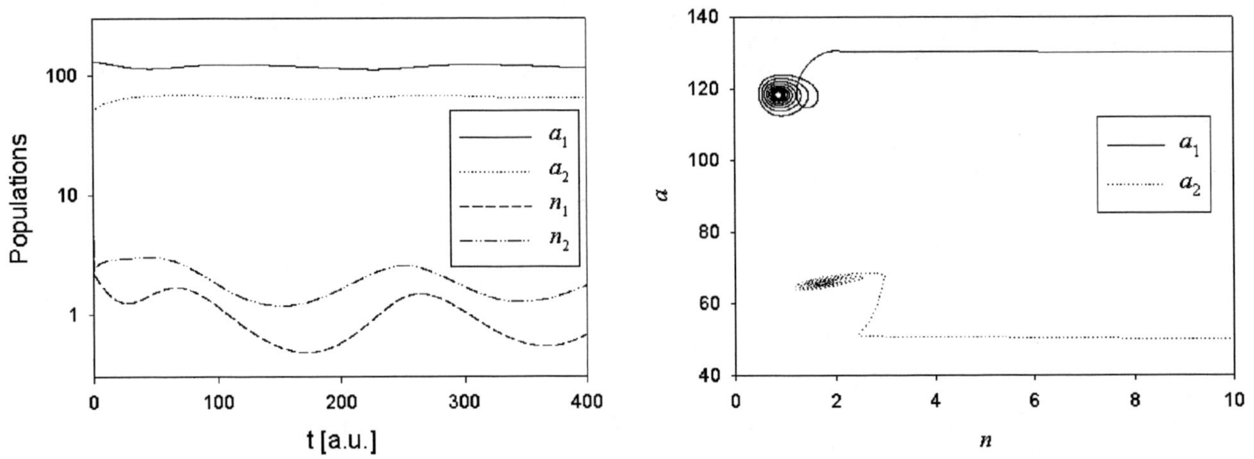

FIGURE 2. Case Ia. **(left)** Time evolution of the populations. **(right)** Phase portrait. Note the very slow attenuation of the oscillations near the coexistence attractor. Here the parameters considered are: $N_i = 2$, $\zeta_i = 12.5$, $\eta_i = 7.5$, $\tau_i = 8.3$, $\alpha_{11} = \alpha_{22} = 0.1$, $\alpha_{12} = 0.05$, $\alpha_{21} = 0.01$, $\gamma_{11} = 7.5$, $\gamma_{22} = 20$, $\gamma_{12} = \gamma_{21} = 1$; $n_1(t=0) = n_2(t=0) = 10$, $a_{10} = 130$, $a_{20} = 50$.

Now that we have determined the possible infection outcomes, we will use phase portraits to investigate the system dynamics.

PHASE PORTRAITS

We obtain numerical time-dependent solutions to equations (1) to (3) using a standard Euler method. A particularly informative way of presenting these solutions is to build phase portraits: once $a_i(t)$, $n_i(t)$ and $r_i(t)$ are calculated for any given parameter set, we assign to each value of the time a point in the (a_i, n_i) plane. By joining these points we then draw a trajectory describing the system evolution for each set of initial conditions. We can see an example of this mechanism in Fig. 2 for a typical evolution of the chronic case (Ia).

For the chronic case with two coexisting parasites (Ia) or for the one surviving parasite (Ib), each population oscillates toward the steady state values predicted by Eqs. (4) to (9) (see Figs. 2 and 3). In the Healing case (II) the parasite populations are eliminated, often after a transient peak (see Fig. 3). For the Death case (III) the antibodies can temporarily reduce the number of parasites, but the immune reaction is finally overcome and the parasite populations grow exponentially (see Fig 3).

It is interesting to analyze in some detail the interaction between the two strains of the parasite. Does the introduction of a second strain help the original strain to survive the immune attack, or is one of the strains eliminated through competition? To answer this question we have performed simulations varying the parameters characterizing the parasite strains and the corresponding antibodies. We have found outcomes ranging from strain synergy to single-survivor competition, depending on the size of the interaction of each strain with the non-specific antibodies. As an example we show in Fig. 3 the effect of the introduction of a second strain (solid line) in an infection corresponding to case II (healing) while infecting only with the original strain (dotted line). If the reproduction rate is low enough, the two antibody species completely control the original and new strains, leading to their annihilation (Fig. 3, upper-left). If the second strain has a higher κ_2, it can survive the immune system response, leading to an equilibrium between the populations (Fig. 3, upper right). Is interesting to note that the interaction between the second strain and the non-specific antibodies can weaken the immune system, allowing the first strain to survive too (Fig. 3, bottom-left). Finally if the reproduction rate is high enough, the immune system cannot control the infection and both parasite populations growth unbound.

We note that the evolution of the immune system in the different cases observed before. For the healing case the asymptotic limit is straightforward: the antibodies populations return to the original values a_{i0}. This situation changes in both chronic cases where the immune system reaches an equilibrium with the parasite population/s. Now the asymptotic limit can be higher or lower than the initial value, according to the parameters involved, as is observed in Fig. 3 (bottom-left). While for the original strain the specific antibody ends with a population lower than the initial

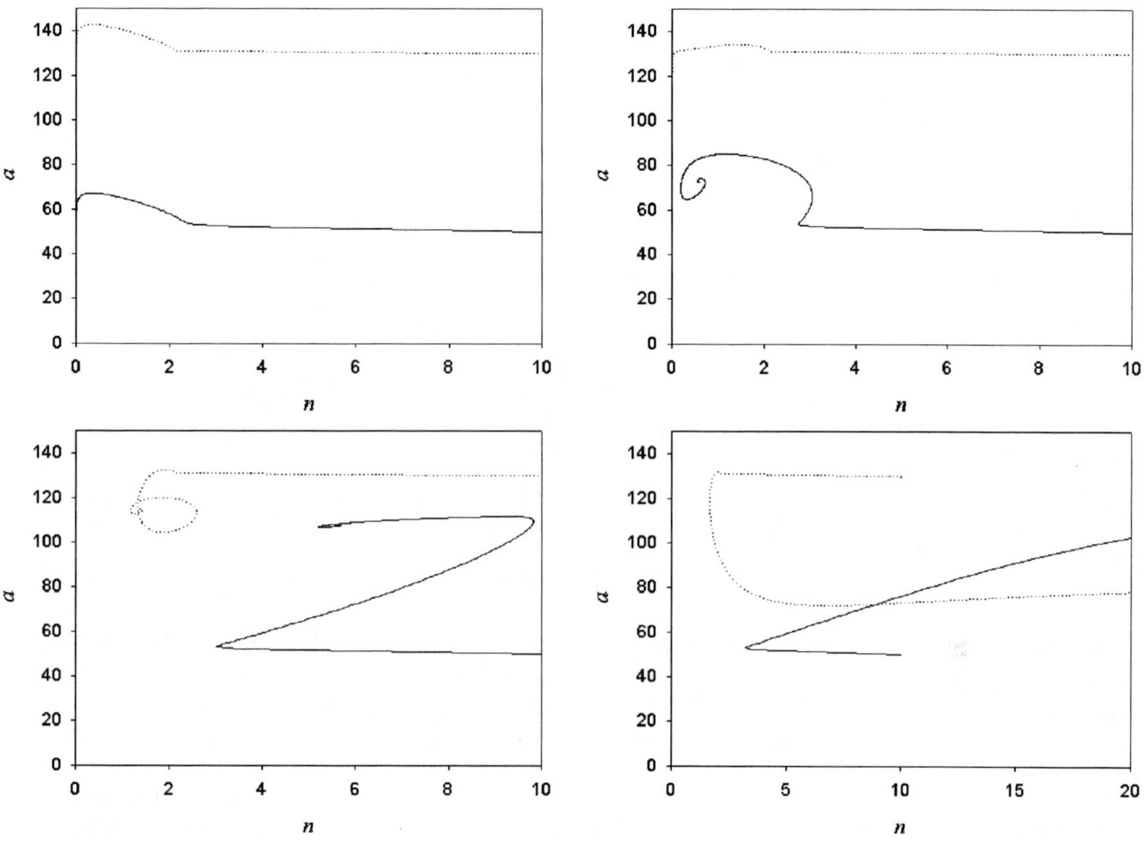

FIGURE 3. Phase portraits for the two parasites case varying the number of emerging parasites per ruptured cell of the second strain (solid line). The original strain (dotted line) corresponds to case II while infected alone. Here the parameters are, $\zeta_i = 12.5$, $\eta_i = 7.5$, $\tau_i = 8.3$, $N_1 = 2$, $a_{11} = a_{22} = 0.1$, $a_{12} = 0.03$, $a_{21} = 0.01$, $\gamma_{11} = 12$, $\gamma_{22} = 20$, $\gamma_{12} = \gamma_{21} = 1$; $n_1(t=0) = n_2(t=0) = 10$, $a_{10} = 130$, $a_{20} = 50$; (upper-left) $N_2 = 1.6$ case II, (upper-right) $N_2 = 1.87$ case Ib, (bottom-left) $N_2 = 2.13$ case Ia, (bottom-right) $N_2 = 2.4$ case III.

one, for the second strain the higher antibody induced generation leads to a higher final value. Finally, for the death case (III), the antibodies populations tend to an asymptotic finite value.

Nevertheless, the example presented in Fig. 3 is not the only possibility as we mentioned before. Could it occur that a parasite, that is in equilibrium with the immune system while inoculated alone (case II), is eliminated due the competition of a second strain (case Ib)?. An example of this behavior is observed in Fig. 4. Here the initial parasite (solid line) tends oscillatorily to the equilibrium value (left), but with the introduction of the second strain, the non-specific antibody is overstimulated (note the higher value for a_1 in the right figure) breaking the previous equilibrium and allowing the immune system to control the original strain. Now is the second strain which is in equilibrium with the antibody populations.

CONCLUSIONS

In this work we have extended our parasite - immune system competition model to describe an heterologous infection during Chagas' disease. In particular, we considered two parasite strains and two antibody species, obtaining the analytical form of the equilibrium and describing the system evolution with the use of phase portraits. We concluded that, if two parasite strains are present in the mammalian host, four results are possible: healing, chronic disease with one strain, chronic disease with both strains, and death. Thus, by modeling the parasite - immune system interaction we can describe all observed clinical cases.

According to our results, the situation observed in nature, with the existence of two different lineages of *T. cruzi*,

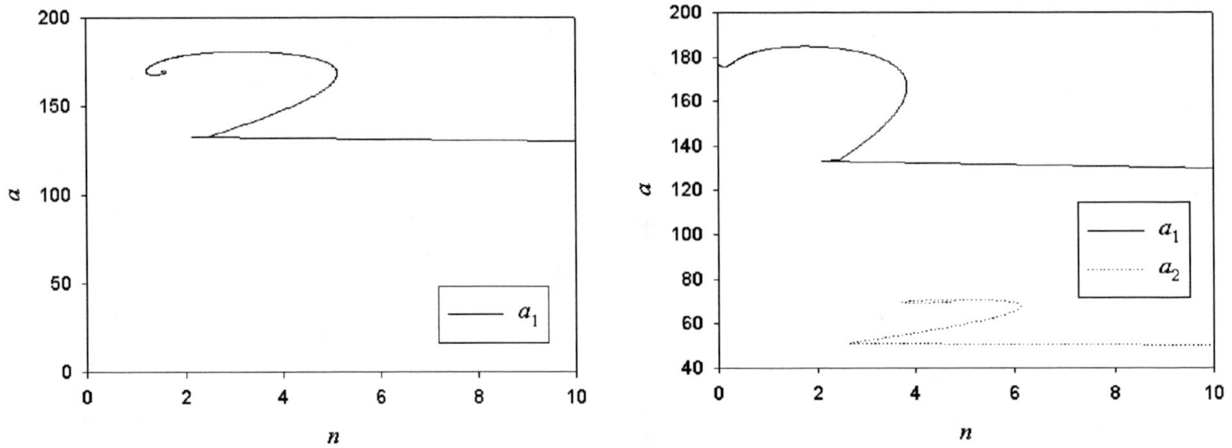

FIGURE 4. Phase portraits for the two parasites case. Here the parameters are $\zeta_i = 12.5$, $\eta_i = 7.5$, $\tau_i = 8.3$, $N_1 = 2.2$, $N_2 = 2.27$, $a_{11} = a_{22} = 0.1$, $a_{12} = 0.05$, $a_{21} = 0.01$, $\gamma_{11} = 7.5$, $\gamma_{22} = 20$, $\gamma_{12} = 10$, $\gamma_{21} = 1$; $n_1(t=0) = 10$, $a_{10} = 130$, $a_{20} = 50$; (left) $n_2(t=0) = 0$; (right) $n_2(t=0) = 10$.

can be explained by considering that the interaction of both types with the human immune system favors the survival of the domestic lineage of the parasite. The situation is reversed if we consider the interaction of both lineages with the immune systems of wild mammals existing in the south American forests, where the sylvatic lineage prevail.

In a future publication we will present the system phase diagram, which is currently under construction, we will discuss the implications of the model results for reinfection experiments, and we will investigate the influence of an heterologous infection on the efficiency of a possible therapeutic course. Using this model we can also predict the extent of the cell damage inflicted by a trypanosoma cruzi reinfection (with the same or a different strain).

ACKNOWLEDGMENTS

This work was supported by CONICET (Argentina) through Grant PIP 6311/05 and SECyT of the Universidad Nacional de Córdoba.

REFERENCES

1. Report of a WHO Expert Committee, Control of Chagas Disease. WHO Technical Report Series Vol. 811 (WHO, Geneva, 1991).
2. R.W. Finley and J.A.Dvorak, J. Protozool. **34**, 409-415 (1987).
3. M. Carneiro, A.J. Romanha, and E. Chiari, Mem. Inst. Oswaldo Cruz, **86**, 387-393 (1991).
4. N. Añez et al., Tropical Medicine and International Health **9**, 1319-1326 (2004).
5. E.V. Machado et al., Am. J. Trop. Med. Hyg. **65**(6), 958-965 (2001).
6. G.J. Sibona, C.A. Condat and S. Cossy Isasi, Phys. Rev E **71**, 020901(R) 1-4 (2005).
7. G.J. Sibona, C.A. Condat and S. Cossy Isasi, "Relevance of Intracellular Replication to the evolution of Chagas Disease" in Mathematical Modeling of Biological Systems, Volume I, edited by A. Deutsch, L. Brusch, H. Byrne, G. de Vries and H.-P. Herzel. Birkhauser, Boston, 2007, pp. 299-308.
8. S. Cossy Isasi, G.J. Sibona, and C.A. Condat, J. Theor. Biol. **208**, 1-13 (2001).

Reduced Electric Linear Dichroism of DNA Fragments in Aqueous Solutions

J. A. Bertolotto*, G. B. Roston*, G. M. Corral* and M. E. Ascheri†

*Departamento de Física, Facultad de Ciencias Exactas y Naturales, Universidad Nacional de La Pampa, Uruguay 151, (6300) Santa Rosa (LP), Argentina
†Departamento de Matemática, Facultad de Ciencias Exactas y Naturales, Universidad Nacional de La Pampa, Uruguay 151, (6300) Santa Rosa (LP), Argentina

Abstract. We calculate the reduced electric linear dichroism at the steady state for DNA fragments in aqueous solutions modeled as flexible broken rod molecules (BRM). The electrical and optical properties of the BRM are characterized by the corresponding tensors and by a charge q with a uniform linear distribution. Each rod has N monomers that are chromophore groups with an average transition probability tensor. The reduced electric linear dichroism of BRM solution is calculated from the anisotropy in the absorption of each molecule, integrating over all the orientations, employing the steady-state orientational distribution function of the system. We first obtain the analytical solution of the Fokker-Planck equation for the system by using a perturbation method where the orientational distribution function is expanded in powers of the perturbation parameter until the fourth order. Then we solve the differential equation for each order of perturbation applying the Fourier method for expanding series. Finally, we compare the results of this theory with experimental curves of reduced electric linear dichroism as a function of the electric field for aqueous solutions of DNA fragments.

Keywords: Electric Linear Dichroism, DNA, Free Electrophoresis, Flexibility
PACS: 05.20.-y, 33.55.-b, 87.14.Gg

INTRODUCTION

The electrooptical techniques of electric birefringence and dichroism are very useful for analyzing the structure and dynamics of macromolecules and colloids in solution [1], [2]. The electric linear dichroism (electric birefringence) is defined as the difference $\Delta A = A_{//} - A_{\perp}$ ($\Delta n = n_{//} - n_{\perp}$), where $A_{//}$ and $A_{\perp}$ ($n_{//}$ and $n_{\perp}$) are the absorbances of the sample (refraction indexes) for light with its polarization plane, parallel and perpendicular to the applied electric field **E**, respectively.

The reduced electric linear dichroism is $\Delta A/A$, where A is the absorbance of the solution in the absence of an applied electric field.

Porschke [3] found that fragments of DNA in buffers with Na ions (ionic strengths between 1mM and 12mM) and chain lengths between 399bp (base pairs) and 859 bp have positive reduced electric dichroism for low electric field strength. For smaller and larger lengths the dichroism is negative for any electric field strengths. The electric dichroism changed to the usual negative value at higher electric fields. They also observed a positive electric birefringence for low electric fields and a negative one for high electric fields. They explained this behavior by introducing a permanent "effective" electric dipole moment, originated by the asymmetric distribution charge of the system with respect to the center of diffusion of the bent rod that has a linear uniformly distributed charge.

On the other hand, in a previous work [4] we explained this phenomenon by modeling the DNA fragments as broken rod molecules (BRM) that, under the action of an applied electric field, move by electrophoresis originating a positive birefringence. This is due to the fact that the translation rotation coupling orientates the BRM with its bisector parallel to the applied electric field. Employing adequate parameters to characterize the DNA fragments, we obtained a positive electric birefringence at low electric field strengths, even when we took into account the orientation due to the interaction between **E** and the induced dipole moment which contributes with a negative sign to the birefringence.

In order to explain the sign reversal of the electric birefringence, in Bertolotto et al. [5], we considered a solution of molecules represented by charged BRM, polarizable but not polar. The orientational distribution function, that is solution of the Fokker-Planck equation of the system at the steady state, is given in powers of E until the second order for the terms that depend on the electric charge q and until the forth order for the ones involving the electric polarizability. We found that the electric birefringence has a positive value at low electric fields and changes to a negative one when the field **E** increases. The order of magnitude of the reversal field was also corroborated with the

CP913, *Nonequilibrium Statistical Mechanics and Nonlinear Physics, XV Conference*
edited by O. Descalzi, O. A. Rosso, and H. A. Larrondo

experimental data.

Considering that DNA fragments that show a positive (negative) dichroism in the ultraviolet spectrum (depending on the orienting electric field strength) have also associated a positive (negative) birefringence in the visible spectrum, respectively [1], the theoretical results obtained for the electric birefringence are also valid, at least in a qualitative way, for the electric dichroism.

Porschke and Antosiewicz [6] analyzed the behavior of a solution of charged and polarizable rigid bent rods by means of a Brownian dynamics simulation taking into account the translation rotation coupling due to the translation movement of the molecules. They found that the electric dichroism both at the transient and at the steady state, strongly depend on this coupling over a wide range of electric field strengths and concluded that a "permanent" dipole moment is essential for observation of a positive electric dichroism.

On the other hand, Bertolotto [7] described in detail how to obtain the Fokker-Planck equation for a system composed of a solution of BRM (or bent rod) with a charge q polarizable but not polar. He concluded that it is not necessary to introduce "permanent" dipole moment for the molecules but the translation rotation coupling of molecules that move by electrophoresis gives rise to a positive dichroism.

Corral and Bertolotto [8] calculated the linear electric dichroism of charged and polarizable BRM based on the quantum theory for interaction between molecules and light developed by Schellman [9], who discussed, among others topics, the linear dichroism. The proposed model predicts positive values for the reduced dichroism at low electric field strength and negative ones when field E increases.

Although the theoretical curves obtained from our previous works show, in a qualitative way, the same behavior that the experimental curves, in order to reduce the approximation error in our results, it is necessary to extend the expansion in series of the orientational distribution function until the fourth order for the term involving the electric charge q.

In this work, on the basis of the theory developed in previous works [4], [5], [7], [8] we calculate the electric dichroism of DNA fragments modeled as flexible charged and polarizable BRM. The orientational distribution function of the molecules is obtained from the Fokker-Planck equation of the system by means of a perturbation method until the fourth order of field E both for the terms involving q and for the ones depending on the electric polarizability. By extending our perturbation method, we will be able to improve the accuracy in obtaining the orientational distribution function and adjust the experimental curves versus the applied electric field to the experimental data for DNA fragments in aqueous solution in Ref. [3], in order to get the DNA characteristic electrooptical parameters.

THEORY

DNA double helices are modeled as BRM formed by two rodlike arms of length L, having an angle χ between them with a bending potential at the hinge $V = (1/2)\varepsilon(\chi-\pi)^2$, where ε is the spring stiffness constant and π is the bending angle with minimum energy configuration. Each arm is composed by N monomers, with an average transition probability tensor per monomer $[T_i]$ and has an electric charge q uniformly distributed and an electric polarizability tensor $[\alpha^E]$.

The principal terms of the electric polarizability tensor, a_i for $i = 1, 2, 3$, were calculated in Ref. [5] and Ref. [10]. In these works we assumed that the parallel electric polarizability for a rodlike macroion resulted directly proportional to the square of the length L, $\alpha_{//} = KL^2$ where K is the electric polarizability constant. The simple model proposed there verified this dependence and is used to calculate the principal terms of the electric polarizability tensor for the BRM. These parameters resulted proportional to K and depend on L, a and χ with a being the transversal distance to the symmetry axis of each rod, where the counterions move.

The characteristic hydrodynamical parameters for the BRM are the rotational diffusion tensor $[R]$ and the translation-rotation coupling diffusion tensor $[P]$.

Let be $OXYZ$ the laboratory coordinate system. The orienting electric field $\mathbf{E}$ is applied along the Z positive axis. We assume that the BRM is placed in a molecule coordinate system $O'X'Y'Z'$ with its origin fixed to the center of mass of the molecule. The broken rod system lies on the $X'Z'$ plane and the X' axis coincides with the bisector of the angle χ. We also assume that the transition dipole moment is perpendicular to each arm's axis and lies on the $X''Y''$ plane with respect to a local coordinate system $O''X''Y''Z''$. The Z'' axis is parallel to the longitudinal axis of each arm.

The plane polarized light propagates along the Y positive axis. The electric fields of the incident plane polarized light, parallel to Z axis (or perpendicular to X axis) applies to a molecule with any orientation given by the Euler angles and interacts with the transition dipole moments which we suppose are located in the DNA basis plane.

In order to calculate the linear dichroism we need the transition probability tensor $[T_i]$ with N_i the number of chromophores per molecule. This tensor can be expressed as follows

$$[T_i] = \begin{bmatrix} \mu^2 & 0 & 0 \\ 0 & \mu^2 & 0 \\ 0 & 0 & 0 \end{bmatrix} \tag{1}$$

where μ is the transition dipole moment.

The reduced electric linear dichroism (RELD) is defined in terms of a difference between the absorbed energy per monomer, per second, for two perpendicular orientations, $w_{//}$ and $w_{\perp}$ with respect to w, the average absorbed energy, per monomer, for randomly oriented molecules and is written as follows

$$\frac{\Delta A}{A} = \frac{w_{//} - w_{\perp}}{w} \tag{2}$$

After some calculations we arrive to the following expression

$$\frac{\Delta A}{A} = \frac{1}{w} \int_0^{2\pi} \int_0^{2\pi} \int_0^{\pi} (Z - X) f(\theta, \psi) \sin\theta d\theta d\phi d\psi \tag{3}$$

where X is the interaction energy of the incident linearly polarized light parallel to the X axis (this axis is perpendicular to the orienting electric field **E**) and is given by

$$X = 2\mu^2 \left[(-\cos\theta\cos\psi\sin\phi - \cos\phi\sin\psi)^2 + \sin^2\left(\frac{\chi}{2}\right)(\cos\theta\cos\psi - \cos\theta\sin\phi\sin\psi)^2 \right] \tag{4}$$

whereas Z is the interaction energy of the incident linearly polarized light parallel to the Z axis (this axis is parallel to the orienting electric field **E**) and is represented by the following formula

$$Z = 2\mu^2 \sin^2\theta \left[\cos^2\psi + \sin^2\psi \sin^2\left(\frac{\chi}{2}\right) \right] \tag{5}$$

The average absorbed energy w can also be expressed as a function of Z and X

$$w = \frac{1}{8\pi^2} \int_0^{2\pi} \int_0^{2\pi} \int_0^{\pi} (Z + 2X) \sin\theta d\theta d\phi d\psi \tag{6}$$

For the case of randomly oriented BRM, we can easily calculate this energy resulting in

$$w = \mu^2 (3 - \cos\chi) \tag{7}$$

By substituting Equation (7) into Equation (3) we obtain an expression for RELD that becomes independent of μ^2 and only depends on the Euler angles, the conformational angle χ and the orientational distribution function $f(\theta, \psi)$.

We solve the Fokker-Planck differential equation for the system of BRM, in the steady state, for a small electric field E, in order to get the steady-state orientational distribution function $f(\theta, \psi)$. We obtain an analytical solution by using a perturbation method where the function $f(\theta, \psi)$ is expanded in powers of the perturbation parameter until the fourth order. We solve the differential equation for each order of perturbation applying the Fourier method for expanding series. The details of calculation were described in Ref. [11]. Thus, we can express the orientational distribution function in the fashion

$$\begin{aligned} f(\theta, \psi) = {} & \tfrac{1}{8\pi^2} + [h_q(\theta, \psi)]E + [s_q(\theta, \psi) + s_{a_i}(\theta, \psi)]E^2 + [p_q(\theta, \psi) + p_{q,a_i}(\theta, \psi)]E^3 + \\ & [r_q(\theta, \psi) + r_{a_i}(\theta, \psi) + r_{q,a_i}(\theta, \psi)]E^4 \end{aligned} \tag{8}$$

In this expression we observe that the coefficients of the different powers of E are functions that depend on the electric charge q (denoted with a sub index q), on the principal terms of the electric polarizability tensor, a_i for $i = 1, 2, 3$ (denoted with a sub index a_i) and on both parameters, q and a_i (denoted with a sub index q, a_i). Except for the functions with sub index a_i, the parameters of all the other functions have terms with the rotational diffusion tensor R_{ii} for $i = 1, 2, 3$, and the translation-rotation coupling diffusion tensor P_{23} and P_{32}. These hydrodynamical parameters were calculated by means of the Hydrox program by García de La Torre and Bloomfield [11], [12].

The orientational distribution function is normalized in such a way that, when you integrate it all over the space, the first addend of Equation(8) is equal to one, while the corresponding integrals of the remaining terms are equal to zero.

The flexibility of the molecules is included considering the solution of flexible BRM as a mixture of rigid BRM with a conformational distribution function $G(\chi)$ [5].

From Equation (3), we first calculate the electric dichroism for the system of BRM with a fixed conformational angle χ (denoted as $(\Delta A/A)_\chi$), using the orientational distribution function given by Equation (8). Then we find the RELD of the system averaging $(\Delta A/A)_\chi$over all the different angles of conformation χ, employing the conformational distribution function $G(\chi)$

$$\frac{\Delta A}{A} = \frac{\int_0^\pi \left(\frac{\Delta A}{A}\right)_\chi G(\chi)\,d\chi}{\int_0^\pi G(\chi)\,d\chi} \tag{9}$$

For low electric field strengths, it is valid to neglect the deformation originated by **E**, so that, we can write the function $G(\chi)$ as a Gaussian distribution centered at $\chi = \pi$, which reads

$$G(\chi) = \sqrt{\frac{\varepsilon}{2\pi kT}}\exp\left(-\frac{\varepsilon(\chi-\pi)^2}{2kT}\right) \tag{10}$$

RESULTS AND DISCUSSION

Applying the developed theory, we first calculate the RELD for a system of BRM in aqueous solution, using convenient parameters to characterize the DNA fragments. Then we plot the theoretical curves showing the RELD dependence on the applied electric field E and adjust them to the experimental data from Porschke [3]. He obtains the stationary values of RELD versus E for DNA fragments with 587 bp in buffers (1mM NaCl, 1mM Na-cacodylate pH 7.0 and 0.2 mM EDTA) of different salt concentrations. In our case the length of each arm of the BRM is $L = 294$ bp.

The parameters employed to get the theoretical curves are (see Fig. 1):

Fig 1. (a): $q = 0.160Q$, $K = 8.5\times10^{-6}$ cm, $\varepsilon = 3.26\times10^{-13}$ g cm^2s^{-2}
Fig 1. (b): $q = 0.150Q$, $K = 6.7\times10^{-6}$ cm, $\varepsilon = 3.27\times10^{-13}$ g cm^2s^{-2}
Fig 1. (c): $q = 0.071Q$, $K = 4.0\times10^{-6}$ cm, $\varepsilon = 1.22\times10^{-13}$ g cm^2s^{-2}
Fig 1. (d): $q = 0.062Q$, $K = 3.5\times10^{-6}$ cm, $\varepsilon = 1.09\times10^{-13}$ g cm^2s^{-2}

The effective charge of the BRM is q and $Q = 5.64\times10^{-7}$ statC represents the BRM charge when the phosphate groups are completely ionized.

In Fig.1 the experimental data are represented by: triangles for a ionic strength $I = 1.2$ mM (Fig.1(a)); circles for a ionic strength $I = 2.4$ mM (Fig.1(b)); squares for a ionic strength $I = 7.4$ mM (Fig.1(c)) and diamonds for a ionic strength $I = 12.4$ mM (Fig.1(d)). The samples from Fig.1(c) and (d) have NaCl of 5mM and 10mM added, respectively.

Let us observe that the three parameters needed for the fitting, q, K and ε, in general, decrease when the ionic strength increases. Moreover, the polarizability constant K, obtained from the fitting is a decreasing linear function of the logarithm of the ionic strength. This dependence coincides with Elias and Eden findings [13] for DNA fragments in dilute solutions of 145 bp over a wide range of ionic strengths.

The decreasing of the charge and the stiffness constant with the increasing of the ionic strength is more irregular but show similar behavior between them.

We can see that the decreasing of the above mentioned quantities is more evident in the samples from Fig.1(c) and (d) that have NaCl added. This behavior for q and ε may be explained from the fact that when the charge of the molecule decreases the repulsion between the neighboring phosphate groups will decrease so that the stretching and the stiffness constant will also decrease.

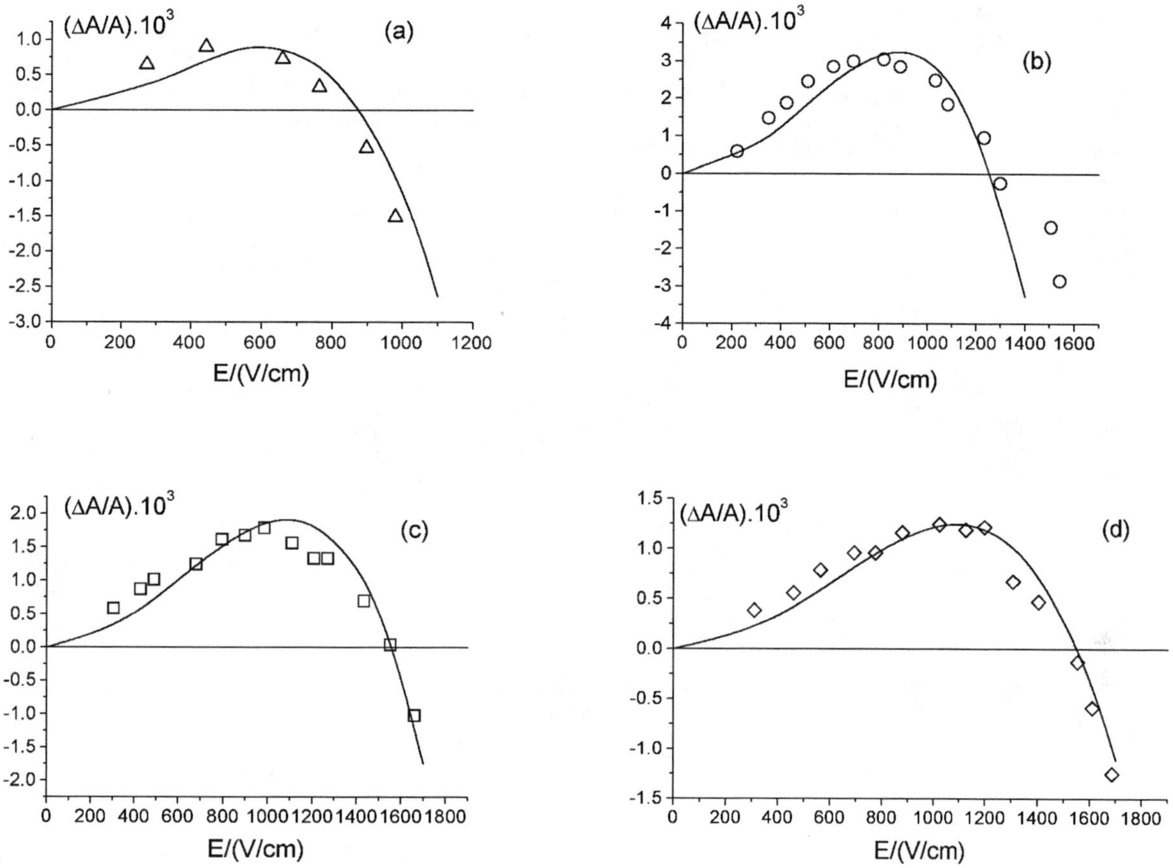

FIGURE 1. Theoretical curves fitting to experimental data using results from the developed theory.

Though the charged and polarizable HBR model is quite simple and we do not consider more complex possible conformations it describes the steady state electrooptical behavior of DNA fragments. It is observed that the theoretical results derived from this model adequately fit to the experimental curves. Moreover, the parameters obtained from the fitting show the expected dependence on the solution ionic strength.

These results reinforce the idea that in DNA fragments should not be considered any electric permanent dipole moment and, for low electric field strengths, the behavior of the dichroism curves versus the electric field can be explained considering two orientation mechanisms: the effect of the translation rotation coupling of the molecules when they move in the presence of electric field and the interaction between the electric induced dipole moment and the applied electric field.

REFERENCES

1. E. Fredericq, C. Houssier, *Electric dichroism and electric birefringence*, Clarendon Press, Oxford, 1973.
2. S. P. Stoylov, *Colloid Electrooptics*, Academic Press, London, 1991.
3. D. Porschke, *Biophys. Chem.* **49**,127–139 (1994).
4. J. Bertolotto, G. B. Roston, M. E. Ascheri and M. Campo, *Physica A* **327**, 185–189 (2003).
5. J. A. Bertolotto, G. B. Roston and M. E. Ascheri, *Progr. Colloid.Polym. Sci.* **128**, 25–29 (2004).
6. D. Porschke and J. M. Antosiewicz, *J. Phys. Chem. B* **109**, 1034–1038 (2005).
7. J. A. Bertolotto, *Physica A* **373**, 29–42 (2007).
8. G. M. Corral, and J. A. Bertolotto, "Cálculo del Dicroísmo Lineal Eléctrico de Soluciones Diluidas de Fragmentos de ADN," in *89 Reunión de la Asociación Física Argentina 2004*, edited by R. Gratton et al., AFA Conference Proceedings 16, Asociación Física Argentina, Bahía Blanca, 2004, pp. 185–188.

9. J. A. Schellman, *Chemical Reviews* **75**, 323-331 (1975).
10. J. A. Bertolotto, M. G. Campo, G. B. Roston and M. E. Ascheri*J. Colloid Surf. A*, **203** 167–183 (2002).
11. J. A. Bertolotto, G. B. Roston, M. E. Ascheri and G. M. Corral, "Resolución de la Ecuación Diferencial de Smoluchowski por Métodos de Perturbación," in *XV Congreso sobre Métodos Numéricos y sus Aplicaciones ENIEF-2006*, edited by A. Cardona, et al., ENIEF Conference Proceedings 25, Asociación Argentina de Mecánica Computacional, Santa Fe, 2006, pp. 1033–1038.
12. J. García de la Torre and V. A. Bloomfield, *Q. Rev. Biophys.* **14**, 81–139 (1981).
13. J. G. Elias and D. Eden, *Macromolecules* **14**, 410-419 (1981).

Constructing the Average Natural History of HIV-1 Infection

L. Diambra*, A. Capurro† and C.P. Malta**

*Instituto de Física, Universidad Nacional de La Plata,
C.C. 67, (1900) La Plata, Argentina
†Bernstein Center for Computational Neuroscience, Freiburg Albert-Ludwigs-University,
Freiburg Hansastr. 9A, 79104 Freiburg, Germany
**Instituto de Física, Universidade de São Paulo,
CP 66318, 05315-970, São Paulo, SP, Brasil

Abstract. Many aspects of the natural course of the HIV-1 infection remains unclear, despite important efforts towards understanding its long-term dynamics. Using a scaling approach that places progression markers (viral load, $CD4^+$, $CD8^+$) of many individuals on a single average natural course of disease progression, we introduce the concept of inter-individual scaling and time scaling. Our quantitative assessment of the natural course of HIV-1 infection indicates that the dynamics of the evolution for the individual that developed AIDS (opportunistic infections) is different from that of the individual that did not develop AIDS. This means that the rate of progression is not relevant for the infection evolution.

Keywords: natural history; CD4+ cell count; AIDS Prognosis; Longitudinal viral load; HIV
PACS: 87.19.Xx; 7.18.-h; 87.18.Bb

INTRODUCTION

The quantification of plasma viral load and $CD4^+$ T-cell count in human immunodeficiency virus type 1 (HIV-1) infection have proved to be powerful markers for diagnostic assessment of disease progression and responsiveness to antiviral therapy [1, 2, 3]. The infection is marked by a progressive decline in the number of circulating $CD4^+$ T-cells, that over a period of years leads to death from immune failure and opportunistic infection. The most longitudinal studies of viral load have focused on long-term trends from time series (TS) generated by measurements made in individuals, with the goal of explaining the pathogenesis or defining prognostic indicators. These TS may contain few points, due to many circumstances. In additiom the rate of disease progression among persons infected with HIV-1 varies greatly. This poses difficulties and limitations in studying the time evolution of the process related to that time series. So we have devised a method to put together the data from many individuals, thus generating a single TS corresponding to an "average individual". Our method involves a time scaling besides a normalization of the quantities object of measurements.

The dataset used to illustrate our method was provided by the Transfusion Safety Studies (TSS) and is composed of reports from plasma samples relative to 106 non-treated patients infected with HIV-1 during the early eighties [4]. 42 of them, called donors, have unkown date and agent of infection. The remaining 64 patients were infected by transfusion (IT) from the donors. When available, the data contain information about the patient ID, infection date, AIDS date, and the levels of log-transformed viral load (VL), $CD4^+$ and $CD8^+$ cells, with their respective collecting date. There was no regularity in collecting plasma samples, and the amount of data relative to $CD4^+$, $CD8^+$ cells and VL levels were insufficient to consider an individual analysis.

We shall consider only samples from the IT group where the AIDS dates are available. Furthermore not all the recipients developed AIDS during the follow-up, so we have splitted the IT patients in two groups: (i) AIDS group, with 23 patients, containing samples from individuals that developed AIDS (clinically affected) during the study; (ii) non-AIDS group, with 41 patients, containing samples from individuals that did not develop AIDS (clinically unaffected) during the follow-up program. Each group will hereafter be treated independently. VL is the most scarce marker, we will consider only the cases for which there exist at least 3 VL samples. This requirement reduces the AIDS and non-AIDS groups to 17 and 26 patients, respectively.

The differences between individual patterns of HIV-1 evolution, and markers level, are expected to be smoothed out by the scaling procedures adopted. After the scaling procedures, we will be referring to the resulting TS as the natural evolution of infection in an "average individual". In the next section we present the scaling used to build the "average individual" time series. In the section 3 we analyze and discuss the time evolution of these "average individuals".

CP913, *Nonequilibrium Statistical Mechanics and Nonlinear Physics, XV Conference*
edited by O. Descalzi, O. A. Rosso, and H. A. Larrondo

TABLE 1. Parameters value used in the time scaling based on a critical CD4$^+$ level, L^C_{CD4}, so that $\delta^{REF}(x) = \delta^C(x)$.

	AIDS	non-AIDS
number of patients	17	26
L^C_{CD4}	177.1	264.5
number of CD4$^+$ (VL) draws	156 (51)	191 (58)
Δ (mean $\pm$ sd)	53$\pm$14	56$\pm$24
CD4$^+$ sampling points [min,max]	[4,15]	[4,20]

THE SCALINGS

Inter-individual scaling

In order to take into account inter-individual differences in the levels of CD4$^+$, CD8$^+$ cells and VL we normalize the level $y_k(x,t)$ of the marker k =CD4, CD8 and VL of each patient x according to

$$\lambda_k(x,t) = \frac{y_k(x,t) - \mu_k(x)}{\sigma_k(x)} \quad \text{with } k = \{\text{CD4}, \text{CD8}, \text{VL}\}, \tag{1}$$

where $\mu_k(x)$ and $\sigma_k^2(x)$ are the mean and variance of the marker level k in the patient x. In order to obtain clinical reference values for each marker level, we scale the normalized data λ_k to the same extreme values of the original time series. This scaling is performed by the following transformation

$$L_k(t) = \frac{\left(\lambda_k(t,x) - \lambda_k^{min}\right)\left(y_k^{max} - y_k^{min}\right)}{\lambda_k^{max} - \lambda_k^{min}} + y_k^{min}, \tag{2}$$

where λ_k^{max} and λ_k^{min} are the maximum and minimum values, respectively, of the normalized level of the marker k, and y_k^{max} and y_k^{min} are the respective maximum and minimum values of the original level of the marker k. We include in our study only patients having at least 3 VL samples (for VL is the most scarce marker).

Time scaling

Even after these inter-individual scaling of our data points, we need to take into account the individual patterns of HIV-1 time evolution, since the immune response of each individual to the viral spreading occurs at different rates [5]. Thus, the data were time scaled through the following transformation:

$$t = \frac{\delta(x) - \delta^{INF}(x)}{\delta^{REF}(x) - \delta^{INF}(x)}, \tag{3}$$

where $\delta(x)$ and $\delta^{INF}(x)$ represent, respectively, the drawing date of the sample, and the infection date for patient x. $\delta^{REF}(x)$ is a reference date that will correspond to scaled time $t^{REF} = 1.0$ ($t^{INF} = 0$).

For patients belonging to the AIDS group the natural choice for the reference date is the AIDS date. The patients belonging to the non-AIDS group did not develop AIDS during the follow up program, so we need an alternative criterion to define a reference date. This is necessary in order to allow comparison of the evolution of the two groups. The criterion that we have used for choosing the reference date, which is applicable for both AIDS and non-AIDS patients, is based on the definition of a threshold for the CD4$^+$ level. The critical date, $\delta^C(x)$, for each individual, corresponds to the date at which the CD4$^+$ level crosses this threshold.

We have used the following procedure in order to establish the critical value, L^C_{CD4}, of the CD4$^+$ level. For the patients belonging to the AIDS group, we use (3) to make the following time scale by taking $\delta^{REF}(x) = \delta^{AIDS}(x)$:

$$\tau = \frac{\delta(x) - \delta^{INF}(x)}{\delta^{AIDS}(x) - \delta^{INF}(x)}, \tag{4}$$

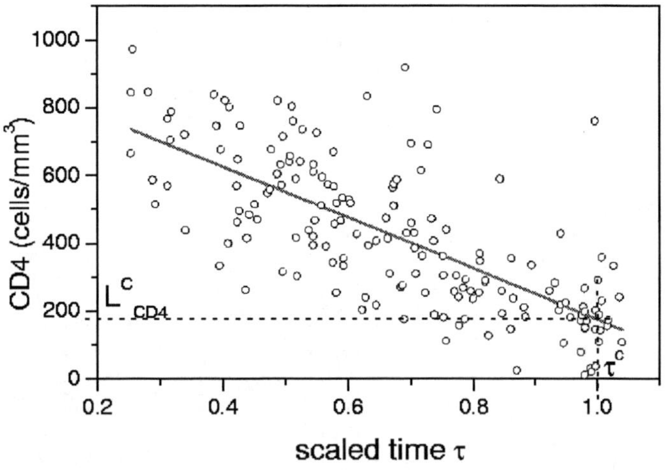

FIGURE 1. Linear fitting of data from the AIDS group patients with at least 3 VL sample. As explained in the text, the CD4$^+$ marker time has been scaled using $\delta^{REF}(x) = \delta^{AIDS}(x)$. The critical value for the AIDS group, L^C_{CD4}, is the CD4$^+$ level at the critical time, $\tau^C = 1$, on the line fitting the data.

so that $\tau^{AIDS} = 1.0$.

After the scaling procedures (2) and (4), it is possible to put together all the data points of each marker k for the patients of the AIDS group. The next step is to make a linear fitting of the AIDS group CD4$^+$ level with the scaled time $\tau \in [0, 1.0]$. This linear fitting for the AIDS data is displayed in the Fig. 1. The critical value of the CD4$^+$ level, L^C_{CD4}, corresponds to its value at the critical time $\tau^C = 1$ on the line fitting the AIDS group scaled data (see Fig. 1).

After establishing the critical value L^C_{CD4}, the critical date for each patient x is the date $\delta^C(x)$ at which $L_{CD4}(\delta^C(x)) = L^C_{CD4}$ (in general an interpolation is required, and we use a simple linear interpolation). The reference date for each patient is taken as $\delta^{REF}(x) = \delta^C(x)$, and equation (3) becomes

$$t = \frac{\delta(x) - \delta^{INF}(x)}{\delta^C(x) - \delta^{INF}(x)} \,. \tag{5}$$

Equation (5) gives the time t scaled in terms of the critical date, so that $t^C = 1$.

The time scaling (5) requires only the determination of the critical date $\delta^C(x)$, therefore it can also be used in the case of the patients belonging to the non-AIDS group. The non-AIDS patients have the basal level different from that of the AIDS patients, so the critical CD4$^+$ level for non-AIDS patients is obtained by scaling L^C_{CD4} (the AIDS patients critical CD4$^+$ level) using the maximum and the minimum values, y^{max}_{CD4} and y^{min}_{CD4}. The date at which this scaled critical level of CD4$^+$ is crossed (if required, linear interpolation is used) corresponds to the critical date for the non-AIDS patients, to be used in equation (5). In table 1 we give a summary of the quantities involved in the time scaling procedure described above, the variable Δ being defined as: $\Delta \equiv \delta^C(x) - \delta^{INF}(x)$.

The average individual

After applying the scaling procedures (2) and (5), the data of all the patients belonging to the AIDS group will constitute a single time series that describes the natural evolution of an "average individual" that developed AIDS. Likewise, for the non-AIDS group, the time series resulting from the application of the scaling procedures (2) and (5), describes the natural evolution of an "average individual" of the non-AIDS group after being infected by the HIV-1. It is important to stress that the time scaling (5) guarantees the use of the same time scale for both the "average AIDS individual", and the "average non-AIDS individual". This is absolutely essential for comparing the infection evolution of AIDS and non-AIDS patients.

TABLE 2. Sliding-point linear regression results for individuals with at least 3 VL samples. For all parameters we give the *mean* value $\pm sd$ (standard deviation). $\bar{t}$ is the value of t that minimizes the chi-square of the overall fitting using (6). The markers slopes are in cells per mm^3/months. As neither $CD8^+$ nor VL give support to two linear regimens, the parameters shown are for a single linear fitting.

		$\delta^{REF} = \delta^{AIDS}$	**AIDS**	**non-AIDS**
$CD4^+$				
	$\bar{t}$	-	0.98	0.99
	a_E	933.1 ± 39.9	891.87 ± 78.0	988.55 ± 80.9
	b_E	-14.1 ± 1.1	-11.2 ± 2.1	-8.3 ± 2.2
	a_L	-	273.5 ± 295.4	148.64 ± 81.9
	b_L	-	-1.9 ± 2.0	3.7 ± 0.8
$CD8^+$				
	a	1092.2 ± 2.74	1073.19 ± 95.19	1011.31 ± 69.17
	b	-8.5 ± 148.6	-395.11 ± 115.76	129.37 ± 56.35
VL				
	a	7.0 ± 0.6	7.15 ± 0.59	7.12 ± 0.60
	b	3.9 ± 0.9	3.14 ± 0.70	0.21 ± 0.45

Hereafter the label x will be dropped since we shall be working with the "average individual" series obtained by applying the scaling procedures (2) and (5). It should be emphasized that the scaling procedures were applied separately to the AIDS ("average AIDS patient") and non-AIDS ("average non-AIDS patient") groups.

RESULTS

In this section we analyze the results obtained from the "average individual" time series, in order to test its usefulness. As mentioned in the introduction, our method aims at generating a TS with a larger number of points, instead of a series having at most 15 points (20) for the AIDS (non-AIDS) group.

Fitting the "average individual" time series to a linear function would be the simplest possible model. Since the data points are rather spread, we have considered the next simplest model that assumes two linear segments. To this end the time series for each marker k was divided in two segments by a point t', and each segment was fitted, independently, to a linear expression:

$$\begin{aligned} L_k(t) &= a_E + b_E t \quad \forall t < t' , \\ L_k(t) &= a_L + b_L t \quad \forall t \geq t' . \end{aligned} \tag{6}$$

For each "average individual" time series we find the time point $t' = \bar{t}_k$ that minimizes the chi-square between measured and fitted values for the two segments together. If the point $\bar{t}_k$ has associated with it significantly different parameters a_E, a_L, b_E and b_L, we can infer that the marker k admits two different linear regimens.

In the case of $CD8^+$ and VL markers we have used the simplest linear fitting since the fitting using (6) produced two linear segments having practically the same slope (for both AIDS and non-AIDS groups). Only the $CD4^+$ marker series have been fitted with the two linear segments (6). The parameters that produce the best linear fitting of the data are summarized in table 2. The column with the heading $\delta^{REF} = \delta^{AIDS}$ corresponds to the results for the AIDS group using the time scaling (4), and we can see that the linear fitting of the average individual time series generated using this time scaling produces results similar to those obtained using the time scaling (5) (column with the heading AIDS).

In the Fig. 2 we display the results for the AIDS (left) and non-AIDS (right) groups.

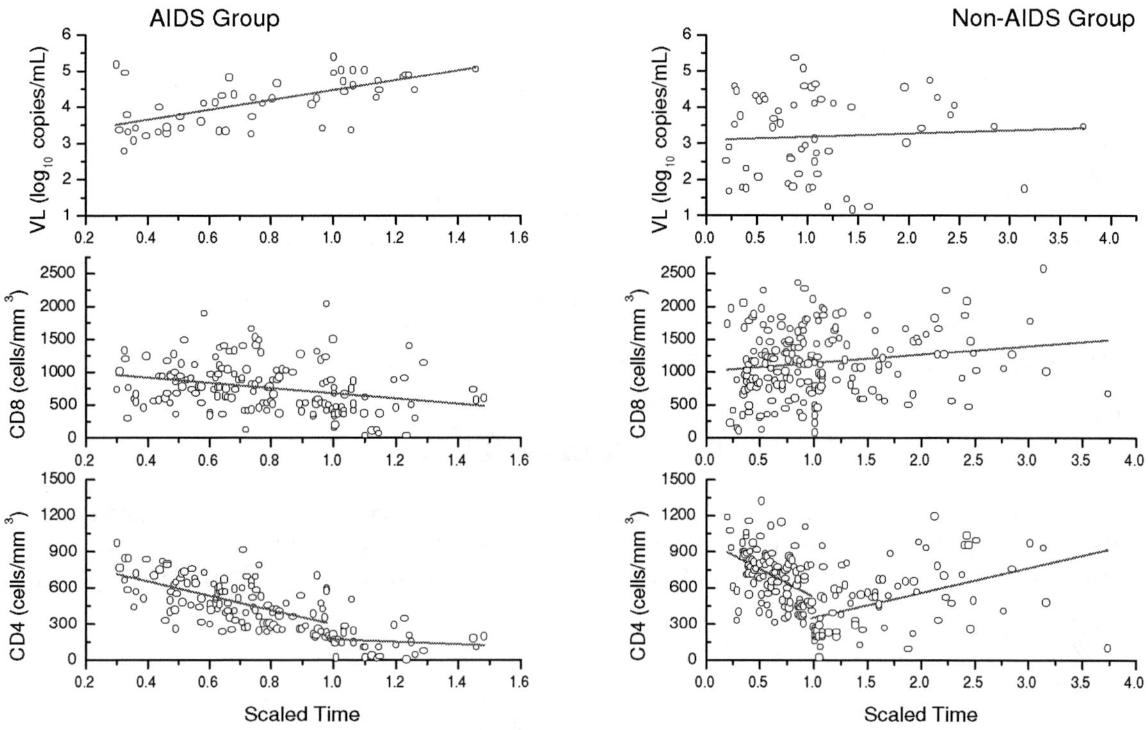

FIGURE 2. Scaled (equation(2)) clinical markers from the patients of the AIDS group (left) and non-AIDS group (right), with at least 3 VL samples. The time has been scaled according to (5), so that $\delta^C(x)$ is the reference date (see the text).

Long-term trends in the AIDS group

The patients from this group developed first AIDS symptoms in average at 1621 days (54± 19 months) after seroconversion. When we use the date at which the CD4$^+$ level crosses the threshold 177 cells/mm^3 as reference date, we obtain 53± 14 months (see table 1). These values are compatible with 66 months reported in [7]. In the first column of table 2 ($\delta^{ref} = \delta^{AIDS}$), we report the long-term trends in VL and markers obtained when we use the AIDS date as reference date. We found that the log of VL increases at a rate of 0.072 units per months, i.e. increases a decade each 13.8±4.8 months. While the copies of virus increase exponentially, the data corresponding to CD4$^+$ marker admits splitting in two linear stage: early and late. The decline rate for the early stage of this markers was 11.6 cells/mm^3 per months, while CD4$^+$ marker admits declines at a similar rate (10.3 cells/mm^3 per months).

AIDS and non-AIDS groups comparison

Figure 2 shows that the progression of the markers of the AIDS group is different from the progression of the markers of the non-AIDS group. The fitted slopes from non-AIDS and AIDS groups are different with a p−value $< 10^{-4}$. In the case of VL, the non-AIDS group (Fig. 2, right) has a slower rate of progression than the AIDS group (Fig. 2, left).

Regarding CD8$^+$ level, we obtained an increasing trend (positive slope) for the non-AIDS group (Fig. 2, right) while for the AIDS group (Fig. 2, left) we obtained a decreasing trend (negative slope), indicating that the CD8$^+$ level evolves differently for AIDS and non-AIDS patients, according to the "average individual" time series.

For the CD4$^+$ marker, there were two significantly different linear regimens for both groups. It should be stressed that it is difficult to establish a trend for AIDS group for $t > 1$, due to the small amount of data points. For $t < 1$ the non-AIDS group has a slower progression rate for CD4 than the AIDS group, but the difference is not significant. The important finding is that the level of this marker exhibits an abrupt change at $t = 1$ for the non-AIDS group. For $t > 1$ the slope is 205±47 for the non-AIDS group while the for the AIDS group it is -101±263 (see table 2). Thus, we have a positive trend in the CD4 level for patients belonging to the non-AIDS group rather than a negative trend as found in the AIDS individuals. The slopes are different with a p−value $< 10^{-4}$. These results are in agreement with the experimental observations regarding AIDS evolution [6].

The above analysis could not be made for each individual separately, due to the small number of points for most patients. In conclusion, our method of generating an "average individual" time series is useful when the the number of data points for each individual is insufficient for analysing the process time evolution. We have illustrated the method using the TSS data for HIV infected individuals that did not receive treatment. Nevertheless the method can be adapted to other types of measurements that generate time series.

ACKNOWLEDGMENTS

LD is researcher of CONICET (Argentina). CPM acknowledges partial financial support from CNPq and FAPESP. AC is Supported by CNPq/CLAF (process number 152426/2005-5), and BMBF 01GQ0420 to BCCN-Freiburg and thank to Prof. Ad Aertsen (BCCN Freiburg) and Prof. Andreas Schulze-Bonhage (University Clinic Freiburg).

REFERENCES

1. H. W. Sheppard, W. Lang, M. S. Ascher , E. Vittinghoff, and W. Winkelstein, *AIDS* **7**, 1159–1166 (1993).
2. C. J. Pitcher, C. Quittner, D. M. Peterson, M. Connors, R. A. Koup, V. C. Maino, and L. J. Picker *Nature Medicine* **5**, 518– 525 (1999).
3. J. W. Mellors, A. Munoz, J. V. Giorgi, J. B. Margolick, C. J. Tassoni, P. Gupta, L. A. Kingsley, J. A. Todd, A. J. Saah, R. Detels, J. P. Phair, and C. R. Rinaldo Jr., *Annals of internal medicine* **126**, 946–954 (1997).
4. E. A. Operskalsky, M. P. Bush, J. W. Mosley, and D. O. Stram, *Journal of Acquired Immune Deficiency Syndromes and Human Retrovirology* **15**, 145–150 (1997).
5. S. L. Liu, T. Schacker, L. Musey, D. Shriner, M. J. McElrath, L. Corey and J. I. Mullins, *J. Virol.* **71**, 4284–4295 (1997).
6. R. H. Lyles, A. Muños, T. E. Yamashita, H. Bazmi, R. Detels, C. R. Rinaldo, J. B. Margolick, J. P. Phair, and J. W. Mellors, *JID* **181**, 872–880 (2000).
7. S. Jurriaans, B. Van Gemen, G. J. Weverling, D. Van Strup, P. Nara, R. Coutinho, M. Koot, H. Schuitemaker, and J. Goudsmit. *Virology* **204**, 223–233 (1994).

Modeling the Transmission of *Piscirickettsia salmonis* in Farmed Salmon

Jaime Cisternas and Adolfo Moreno

Facultad de Ingeniería, Universidad de los Andes, San Carlos de Apoquindo 2200, Santiago, Chile

Abstract. Farming Atlantic salmon is an economic activity of growing relevance in the southern regions of Chile. The need to increase efficiency and reach production goals, as well as restrictions on the use of water resources, had led in recent years to certain practices that proved prone to bacterial infections among the fish.

Our study focuses on the impact of rickettsial bacteria in farmed salmon, and the possibility of controlling its incidence once it is established along the salmon life cicle. We used compartmental models to separate fish in their maturation stages and health status. The mathematical analysis will involve differential equations with and without delays, and linear stability principles. Our goal was to build a simple model that explains the basic mechanisms at work and provides predictions on the outcome of different control strategies.

Keywords: aquaculture, salmon, Piscirickettsia salmonis, compartmental models, delay differential equations
PACS: 87.17Aa, 8719.Xx

INTRODUCTION

The rise of salmon farming to an industrial scale has coincided with the emergence of bacterial infections and other diseases in the fish. The endemic presence of *Piscirickettsia salmonis* in Chilean farms is probably the most severe case of such infections. This disease made its public appearance in 1989 when a large number of farmed salmon mortalities [1] was linked to a new rickettsia bacteria [2, 3, 4]. Only in 1994, infections caused by *P. salmonis* have caused $50 million in losses to the Chilean industry, dedicated mostly to Atlantic salmon (*Salmo salar*) [5]. In recent years, *P. salmonis* has found its way into North America, Scotland, Ireland, and Norway, affecting almost every salmonid: Coho salmon, steelhead trout, chinook salmon, and Atlantic salmon.

Further study has shown that rickettsiae can be present along the whole *anadromous* life cycle of farmed salmon: egg, sac fry, Parr, smolt, until spawning or butchery; though only smolts develop symptoms of *P. salmonis*. Infected fish can be recognized by a lethargic behavior and for being oriented against the water current. Among the pathological changes we can mention pale gill filaments, liver and kidney hemorrhages, ulcerations in the fins, and other signs consistent with septicemia [6].

The main mechanisms of horizontal transmission are faeces and urine of infected fish that enter other fish through the gill or wounds in the skin [7]. Vertical transmission also exists [8].

Succesful control strategies are based on vaccination and oral therapy, though some of the drugs are not approved in all countries and this may induce the creation of commercial barriers. Other ways of preventing the disease are raising husbandry standards and reducing stocking densities [9].

Though a more or less complete picture of *P. salmonis* can be made up from the literature, it is not well understood how it operates at the level of the population *i.e.* how its incidence evolves over time. This article describes two mathematical models for the population dynamics of salmon rickettsiosis in a farm, that take into account the transmission mechanisms, the disease progression in individual fish, and the production levels of the farm, but leaves many other interesting processes under the rug. These compartmental models were built using differential equations following standard methodology (see for instance [10, 11]).

This article was structured as follows: in the next section, we pose a basic compartmental model, and study the properties of the equilibria that it predicts. The following section considers a reduced model that considers the time delay induced by the fresh-water phases of salmon when the infection is latent. This model can also be analyzed mathematically, but its predicted behavior is somehow different from the previous model. The implications of both models and their use as tools for policy making are assessed in the last section.

CP913, *Nonequilibrium Statistical Mechanics and Nonlinear Physics, XV Conference*
edited by O. Descalzi, O. A. Rosso, and H. A. Larrondo

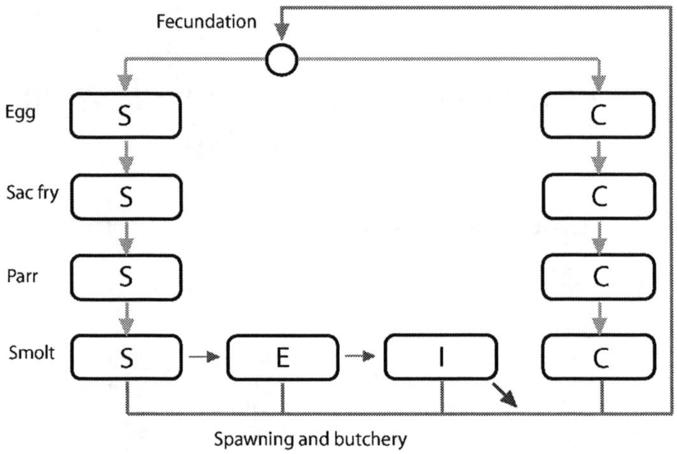

FIGURE 1. Basic model of 10 compartments.

BASIC COMPARTMENTAL MODEL

Let's begin this section with a more detailed description of the life cycle of farmed salmon, from egg to adult fish:

1. incubation of the eggs, from fecundation to eclosion, lasts 2 months;
2. sac fry (small alevin carrying a sac with nutrients), lasts 3 months;
3. Parr (fish with camouflaging vertical stripes), 6 months;
4. smolt, begins once the fish is able to survive in salt water, lasts 12 months until spawning and butchery.

Each phase takes place in a separate place, either a pool or a net cage in a body of water. As it was mentioned in the Introduction, during the fresh water phases, fish can be either susceptible or infected, but no changes of status are allowed. Infected fish are latent carriers that don't develop symptoms of the disease or suffer from higher mortality than healthy individuals [7].

Vertical transmission takes place with a small probability, when one of the gametes comes from an infected parent.

Horizontal transmission between smolts sharing the same cage has the peculiarity that newly infected smolts, after an incubation period, develop internal as well as external signs of the disease, and are more infectious than carrier smolts that don't develop signs. Infected smolts also have a higher mortality than carriers. The survival time of infected smolts is considerably shorter than the time allowed for maturation [12].

As it is explained in Fig. 1, fish can be distinguished by their four maturity stages (egg, sac fry, Parr or smolt) and four health status (susceptible, exposed, infected and carrier).

The main parameters of the model are the mean residence time of every compartment and the infectivity β in the smolt phase. For the infection rate we use the standard mass-action formula [10]:

$$\text{new infections per unit time} \;=\; \beta \text{ susceptibles} \, \frac{\text{infectious}}{\text{total}} \, ,$$

where by 'total' we mean the total number of fish sharing the same cage. Infectious fish are the sum of carriers and infecteds. The parameter β, with units of inverse time, is a summary of different effects: presence of bacteria in faeces and urine, number of fish sharing the same water (stocking density, water temperature and quality), susceptibility to infection (coverage and effectiveness of vaccination), among others.

The equations for the populations of susceptible S_i, carriers C_i, exposed E_i, and infecteds C_i, of each stage $i \in \{1,2,3,4\}$, will take a similar form. For the eggs:

$$\frac{dS_1}{dt} = (1-q)\dot{T}_{\text{eggs}} - \delta_1 S_1 \, , \tag{1}$$

$$\frac{dC_1}{dt} = q\dot{T}_{\text{eggs}} - \delta_1 C_1 \, , \tag{2}$$

where q is the fraction of infected eggs, and $1/\delta_1$ is the time before eclosion takes place. The quantity of incubated eggs $\dot{T}_{\text{eggs}}$ depends on the farmer, as it is going to be explained later.

For the sac fry and Parr populations:

$$\frac{dS_2}{dt} = \delta_1 S_1 - \delta_2 S_2 , \tag{3}$$

$$\frac{dC_2}{dt} = \delta_1 C_1 - \delta_2 C_2 , \tag{4}$$

$$\frac{dS_3}{dt} = \delta_2 S_2 - \delta_3 S_3 , \tag{5}$$

$$\frac{dC_3}{dt} = \delta_2 C_2 - \delta_3 C_3 , \tag{6}$$

where $1/\delta_2, 1/\delta_3$ are the times salmon live as fry sac and Parr, respectively.

The differential equations for the Smolt populations, including the contributions of infections by carriers and infected fish, and disease specific mortality:

$$\frac{dS_4}{dt} = \delta_3 S_3 - \beta S_4 \left(\frac{C_4 + f I_4}{S_4 + E_4 + I_4 + C_4} \right) - \delta_4 S_4 + \Omega I_4 , \tag{7}$$

$$\frac{dC_4}{dt} = \delta_3 C_3 - \delta_4 C_4 , \tag{8}$$

$$\frac{dE_4}{dt} = \beta S_4 \left(\frac{C_4 + f I_4}{S_4 + E_4 + I_4 + C_4} \right) - \delta_4 E_4 - \gamma E_4 , \tag{9}$$

$$\frac{dI_4}{dt} = \gamma E_4 - \alpha I_4 - \delta_4 I_4 - \Omega I_4 , \tag{10}$$

where $1/\delta_4$ is the duration of the smolt phase, and α, Ω are the mortality rate due to *P. salmonis* and the recovery rate. $1/\gamma$ is the duration of the incubation period of the disease. Nondimensional parameter $f > 1$ was introduced to reflect the fact that salmons with external signs are a lot more infectious than carriers. Parameter α could also include the removal of fish with visible signs of infection.

The symbol q that appears in the first two equations, is not a constant but the proportion of infected eggs, that depends on the proportion of infected adult salmon used for spawning:

$$q = v(2p - p^2) \quad \text{where} \quad p = \frac{C_4 + E_4}{S_4 + C_4 + E_4} , \tag{11}$$

where parameter v is the probability for an egg of being infected if one of its parents has *P. salmonis*.

The cycle gets closed by the farmer, who is interested in reaching a production goal $\dot{T}$ to satisfy the demand, and controls the number of fish eggs that get into the cycle (per unit of time):

$$\dot{T}_{\text{eggs}} = \dot{T} + K\left(\dot{T} - \delta_4(S_4 + E_4 + C_4)\right) , \tag{12}$$

parameter K measures the agressivenes of the farmer *i.e.* the way he/she reacts to a production smaller or bigger than his/her goals.

For parameters values that represent a realistic farm, the trajectories of the differential system built from equations (1)–(12) can be separated into two groups. Initial conditions where all individuals are in the susceptible compartments, lead to an equilibrium state for the production cycle of the form:

$$S_1 = \dot{T}/\delta_1 , \; S_2 = \dot{T}/\delta_2 , \; S_3 = \dot{T}/\delta_3 , \; S_4 = \dot{T}/\delta_4 ,$$

all other compartments remain empty. On the other side, initial conditions that contain a few infecteds or carriers, no matter how few, develop into an equilibrium state where all compartments are being occupied. This state is called endemic equilibrium.

As it can be appreciated in Fig. 2, the disease-free and the endemic equilibria have different stabilities under small perturbations. The stability of equilibrium states of the production cycle can be studied using the Jacobian of the

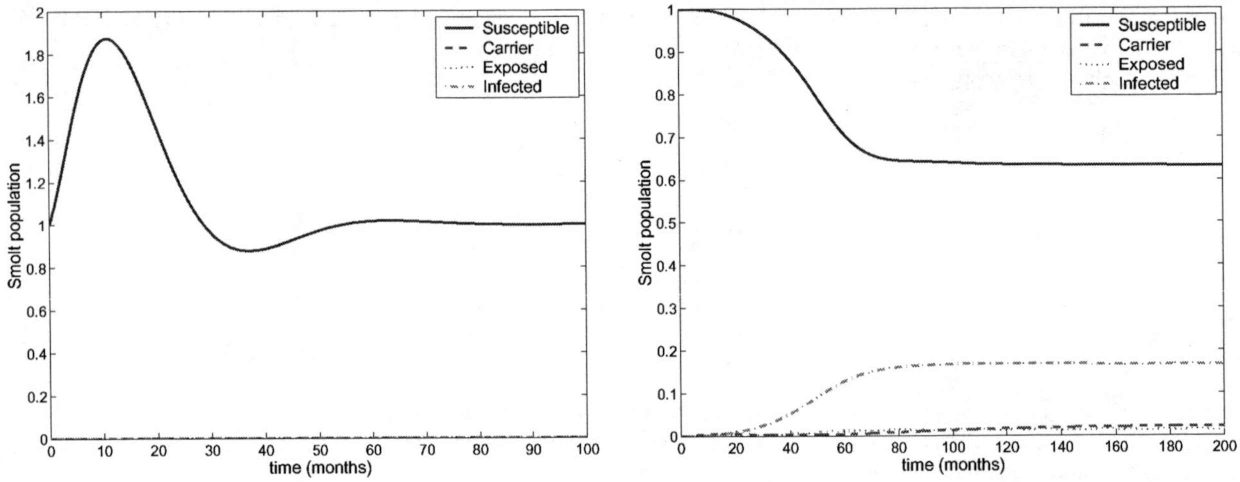

(a) (b)

FIGURE 2. (a) Solution of the dynamic model for an initial condition that uses only the susceptible compartments. (b) After perturbing the previous equilibrium, the system approaches an endemic equilibrium. The following parameter values were used: $\alpha = 1/4, \Omega = 0, f = 5, \nu = 0.16, \delta_1 = 1/2, \delta_2 = 1/3, \delta_3 = 1/6, \delta_4 = 1/12, \gamma = 1/0.25, \beta = 0.117, K = 1, \dot{T} = 1/12$.

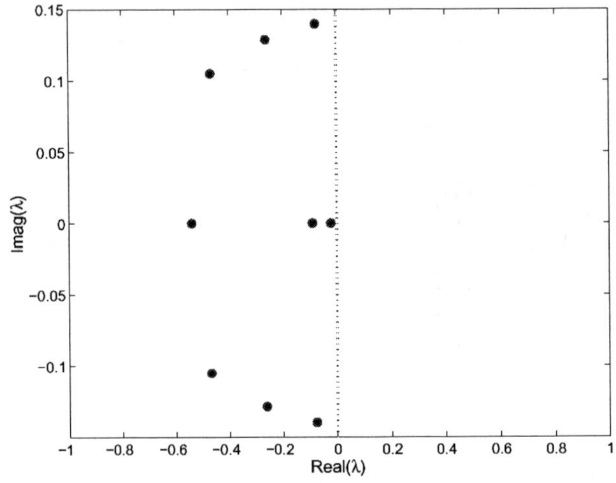

FIGURE 3. Complex eigenvalues of the Jacobian that results from linearizing our 10 dimensional system around the endemic equilibrium point. Same parameter values as in Fig. 2.

differential system and its spectrum. In Fig. 3 we plot the eigenvalues of the 10×10 matrix that results from the linearization of the differential equations (1)–(10) around the endemic equilibrium. As all eigenvalues have negative real parts, we can conclude that this state is stable to small perturbations. The same procedure could be repeated for the disease-free state to show it is unstable.

This stability picture can be changed if one is allowed to modify some parameters. For instance reducing β below 0.084 (let's say improving the quality of the water) one can make the endemic equilibrium unstable and the disease-free state stable. This phenomena could be studied using the concepts of bifurcation theory, although analytical results will be elusive given the high dimensionality of the system.

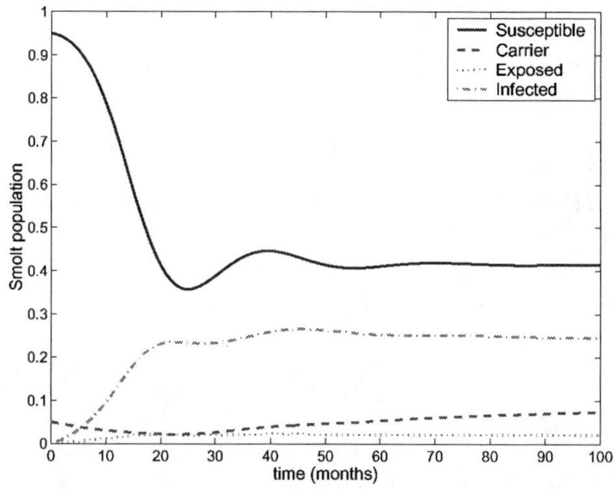

(a)

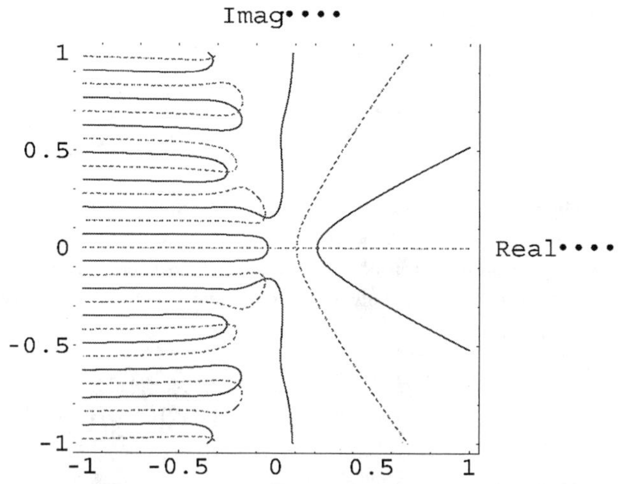

(b)

FIGURE 4. (a) Trajectory of the DDE using same initial conditions as in 2(a). (b) Blue curve shows the points of the complex plane were $\mathrm{Re}(h(\lambda)) = 0$, and segmented red curve shows where $\mathrm{Im}(h(\lambda)) = 0$; intersections corresponds to eigenvalues λ of the DDE around the disease-free state. As expected, there is one (real) eigenvalue λ with positive real part so the state is unstable, despite the presence of an infinite number of intersections in the left semiplane. Here we use a time delay of $\tau = 11$ months, other parameters as in Fig. 2.

REDUCED MODEL

A simplification of the previous model can be obtained observing that in the early stages of salmon growth there are no important transfers between compartments, besides the natural growth of fish. Equations (1)–(6) provide a *time delay* equal to the average time it takes for a salmon to become a smolt. If one would use that new smolts come from eggs obtained 11 moths ago, and using the convenient notation $x^{\mathrm{lag}} = x(t-\tau)$:

$$\frac{\mathrm{d}S}{\mathrm{d}t} = (1-q^{\mathrm{lag}})\dot{T}^{\mathrm{lag}}_{\mathrm{eggs}} - \beta S\left(\frac{C+fI}{S+E+I+C}\right) - \delta_4 S + \Omega I\,, \tag{13}$$

$$\frac{\mathrm{d}C}{\mathrm{d}t} = q^{\mathrm{lag}}\dot{T}^{\mathrm{lag}}_{\mathrm{eggs}} - \delta_4 C\,, \tag{14}$$

$$\frac{\mathrm{d}E}{\mathrm{d}t} = \beta S\left(\frac{C+fI}{S+E+I+C}\right) - \gamma E - \delta_4 E\,. \tag{15}$$

$$\frac{\mathrm{d}I}{\mathrm{d}t} = \gamma E - \alpha I - \delta_4 I - \Omega I\,. \tag{16}$$

The resulting system is a Delay Differential Equation (DDE), with properties far different from the ones of ordinary differential equations. Trajectories (computed with `dde23.m`, see [13]) show a richer structure, including limit cycles and bifurcation phenomena.

To study the stability of the disease-free state (or the endemic equilibrium) for the DDE system (13)–(16) we impose a single exponential for all variables:

$$X(t) = X_0 + \xi e^{\lambda t}\ ,\ X^{\mathrm{lag}}(t) = X(t-\tau) = X_0 + \xi e^{\lambda(t-\tau)}\,.$$

Now we get two Jacobian matrices:

$$J_{ij} = \frac{\partial f_i}{\partial x_j}\ ,\ J^{\mathrm{lag}}_{ij} = \frac{\partial f_i}{\partial x^{\mathrm{lag}}_j}\,.$$

The *characteristic equation* should give the values of λ for which there are nontrivial solutions ξ:

$$h(\lambda) = \det\left(J + e^{-\lambda\tau}J^{\mathrm{lag}} - \lambda\,\mathrm{Id}\right) = 0\,,$$

takes the following form:

$$h(\lambda) = -(\delta(1+Ke^{-\lambda\tau})+\lambda)\Big(\beta f\gamma(\delta+\lambda-2\delta\nu e^{-\lambda\tau})-$$
$$e^{-\lambda\tau}(\alpha+\delta+\lambda+\Omega)(e^{\lambda\tau}\lambda(\gamma+\lambda)+\delta^2(e^{\lambda\tau}-2\nu)+\delta(e^{\lambda\tau}(\gamma+2\lambda)-2\nu(\beta+\gamma+\lambda)))\Big)\ .$$

The previous function, separated in its real and imaginary parts, was used to compute the eigenvalues of the disease-free state in Fig. 4(b). As it was expected, there is one eigenvalue with positive real part and the state is unstable. The same procedure applied to the endemic equilibrium would show its stability, and would allow us to find the smallest modification of parameters that can turn the endemic equilibria unstable and make *P. salmonis* disappear slowly from the population.

DISCUSSION AND PERSPECTIVE

Our preliminary results show that the mathematical modeling of the dynamics of *Piscirickettsia salmonis* in farmed fish is indeed possible, despite all the uncertainties and the missing mechanisms. This work has been developed having *P. salmonis* in mind, but most of the work can be generalized to other diseases that affect aquaculture.

We built the model identifying the stages of the production cycle and the health status of salmon, as well as the basic processes that make fish transfer from a given compartment to other compartment of different status or age.

There are other complexities that this model doesn't include, for instance the distinction of severe and mild infections. Other missing mechanisms are the effect of water pollution and temperature, and the interaction with wild fish and other aquatic species. Nevertheless we are confident that the models presented here capture the basic intra-farm mechanisms. Even if this simple methodology doesn't have the power to make quantitative predictions, it does show the relevance of the different processes and parameters.

The basic model explains the existence of two equilibria: in the first the disease is absent, and in the second the infection is endemic. For typical parameters, the disease-free state is unstable and the endemic state is stable. This was explained by means of linear analysis and eigenvalues of a Jacobian matrix. If one or few fish get infected, let's say by contact with feral fish or presence of the bacteria in the water, or the use of contaminated eggs for the incubation, some adult salmons will become carriers and infectious, and the disease will propagate through all the production cycle until certain equilibria is reached, and susceptible fish will coexist with carriers and infectious fish.

The second model, based on delay differential equations, shares most of the above mentioned features and predict roughly the same stability of the endemic equilibria. This time the characteristic equation is a trascendental equation that can be analyzed in the complex plane.

A change of this scenario, an interchange of stability between the disease-free and the endemic states, could be achieved (within the limits of this simplistic models) by reducing the infectivity: either by reducing the stocking density, improving the quality of water, or using more effective vaccines. Better results could also be obtained by improving husbandry standards, and surveillance to eliminate infectious smolts.

Our findings confirm the notion that it is possible to remove *P. salmonis* from the population by only improving practices at the aquaculture centers. This may be more expensive and slower than relying on medication, but it's safer in the long run as commercial barriers for drug-treated fish are being raised.

Among directions for future research we can mention the addition of environmental factors such as temperature and coinfection with other diseases such as sea lice [14, 15]. But more relevant advance could be obtained introducing some aspects that are peculiar to salmon farming, for instance impulsive effects, as salmons are transfered not one by one but by large quantities according to a yearly schedule. Other interesting issue is the use of antibiotics. There are indications that the heavy use of antibiotics in the salmon industry is leading to the same sort of resistance that take place in human hospitals [16]. To make things worse, resistant bacteria could propagate into the human population.

As we add more compartments and processes, new parameters are going to be needed. Their uncertainties may conspire and ruin the explanatory power of our models. An interesting direction would be the use of a Bayesian framework [17], that would allow us to formally take into account uncertainty in all parameters, in particular the infectivity. This methodology would add more robustness to the dynamic models and provide a clearer picture of the policies needed to make salmon farming sustainable in the long run.

ACKNOWLEDGMENTS

We thank María Teresa Valenzuela, Marcela Cárcamo and Alejandro Tardel for their comments and suggestions. J.C. wishes to acknowledge the support of FONDECYT (Project No. 1070098).

REFERENCES

1. E. Branson, and D. Nieto, *J. Fish. Dis.* **14**, 147–156 (1991).
2. J. Fryer, C. Lannan, L. Garcés, J. Larenas, and P. Smith, *Fish. Pathol.* **25**, 107–114 (1990).
3. J. Cvitanich, O. Garate, and C. Smith, *J. Fish. Dis.* **14**, 121–145 (1991).
4. J. Fryer, C. Lannan, S. Giovannoni, and N. Wood, *Int. J. Sys. Bacteriol.* **42**, 120–126 (1992).
5. P. Smith, J. Contreras, and L. Garcés, *J. Aquat. Anim. Health* **8**, 130–134 (1996).
6. R. Cusack, D. Groman, and S. Jones, *Can. Vet. J.* **43**, 435–440 (2002).
7. P. Smith, P. Pizarro, P. Ojeda, J. Contreras, S. Oyanedel, and J. Larenas, *Dis. Aquat. Org.* **29**, 165–172 (1999).
8. J. Larenas, C. Astorga, J. Contreras, and P. Smith, *Arch. Med. Vet.* **29**, 161–166 (1996).
9. J. Larenas, J. Contreras, S. Oyanedel, M. Morales, and P. Smith, *Arch. Med. Vet.* **29**, 113–119 (1997).
10. R. Anderson, and R. May, *Infectious Diseases of Humans: Dynamics and Control*, Oxford University Press, Oxford, 1991.
11. F. Brauer, and C. Castillo-Chavez, *Mathematical Models in Population Biology and Epidemiology*, Springer, New York, 2000.
12. F. Almendras, I. Fuentealba, S. Jones, F. Markham, and E. Spangler, *J. Fish. Dis.* **20**, 409–418 (1997).
13. L. Shampine, and S. Thompson, *Appl. Numer. Math.* **37**, 441–458 (2001).
14. P. Reno, *J. Aquat. Anim. Health* **10**, 160–171 (1998).
15. M. Krkošek, M. Lewis, A. Morton, L. Frazer, and J. Volpe, *PNAS* **103**, 15506–15510 (2006).
16. G. Webb, E. D'Agata, P. Magal, and S. Ruan, *PNAS* **102**, 13343–13348 (2005).
17. B. Elderd, V. Dukic, and G. Dwyer, *PNAS* **103**, 15693–15697 (2006).

Neuron Bifurcations in an Analog Electronic Burster

Guillermo V. Savino* and Carlos M. Formigli†

*Laboratoy for Nonlinear Dynamic Systems, DEEC Department, Facultad de Ciencias Exactas y Tecnología, Universidad Nacional de Tucumán, Av. Independencia 1800, San Miguel de Tucumán,(4000), Argentina.
†Laboratory for Nonlinear Dynamic Systems, DEEC Department, Facultad de Ciencias Exactas y Tecnología, Universidad Nacional de Tucumán, Av. Independencia 1800, San Miguel de Tucumán,(4000), Argentina.

Abstract. Although bursting electrical activity is typical in some brain neurons and biological excitable systems, its functions and mechanisms of generation are yet unknown. In modeling such complex oscillations, analog electronic models are faster than mathematical ones, whether phenomenologically or theoretically based. We show experimentally that bursting oscillator circuits can be greatly simplified by using the nonlinear characteristics of two bipolar transistors. Since our circuit qualitatively mimics Hodgkin and Huxley model neurons bursting activity, and bifurcations originating neuro-computational properties, it is not only a caricature but a realistic model.

Keywords: electronic neuron model, burster, bifurcation, nonlinear dynamics
PACS: 87.19.La, 87.10.+e, 05.45.- a

INTRODUCTION

Some brain neurons and other biological excitable cells have bursting electrical activity. Mammalian neurons fire bursts to increase the reliability of synaptic transmission [1], and important information may be carried in the intraburst frequency [2]. Nevertheless, bursting functions and generation mechanism are as yet unknown.

In modeling such complex oscillation, hardware analog electronic models [3] have relevant advantages compared to their purely mathematical counterpart [4, 5, 6] solved digitally: 1) They can provide real time processing. If they are feasible for very large scale integration (VLSI) circuits [7], they may be suited for studying a large-scale neural network dynamics when many of them are connected with each other; 2) Physical interpretation of each parameter of a circuit is possible, and perhaps associated with electrophysiological quantities; 3) Circuit dynamics can be described geometrically on the phase-plane and by its mathematical equation; 4) Circuits are particularly appropriate for studying noise mediated behaviors, such as stochastic resonance [8], coherent resonance [9], and synchronization [10].

Although many electronic devices are bursting oscillators, not all of them can be considered a neuron model. Since real neuron dynamics are yet unknown, at least the circuit must preserve bifurcation dynamics of the largest accepted ionic channel-based Hodgkin and Huxley (HH) model [4]. Although many efforts have been addressed to HH equations hardware electronic implementations, the circuits have always resulted complicated [11].

In studying neuron excitability, there are theoretical methods for reduction of multi-dimensional models, e.g., the HH model to 2-dimensional systems, and all of them can be described with only two nullclines: a cubic parabola (flipped N-shape) and a sigmoidea shaped [12]. But to get bursting feasible, at least a three dimensional system is necessary, since bursting occurs due to the interplay of fast currents responsible for spiking activity, and slow currents that modulate the activity [13]. Until now a simple burster physical realization, conduction change-based and preserving bifurcation dynamics of the reduced HH model has not been reported.

In this paper we propose an analog electronic self-oscillating burster. Experimentally all bifurcation of the HH reduced model and some burster types are reproduced. The circuit is particularly suitable to identify and study the fast and slow currents interplay to understand bursting generation.

RESULTS

Figures 1a and 1b show the burster circuit and the waves at four circuit points. The circuit has two blocks with very different characteristic times: the fast subsystem or spikes **Generator** and the slow subsystem or **Modulator**, and they are linked only by the modulating current $I(t)$. Using resistor and capacitor values of Fig. 1a circuit, it self-oscillates as a burster with the waves of Fig. 1b, and period $T = T_r + T_d$. Experimental results show that burst duration T_r is

CP913, *Nonequilibrium Statistical Mechanics and Nonlinear Physics, XV Conference*
edited by O. Descalzi, O. A. Rosso, and H. A. Larrondo

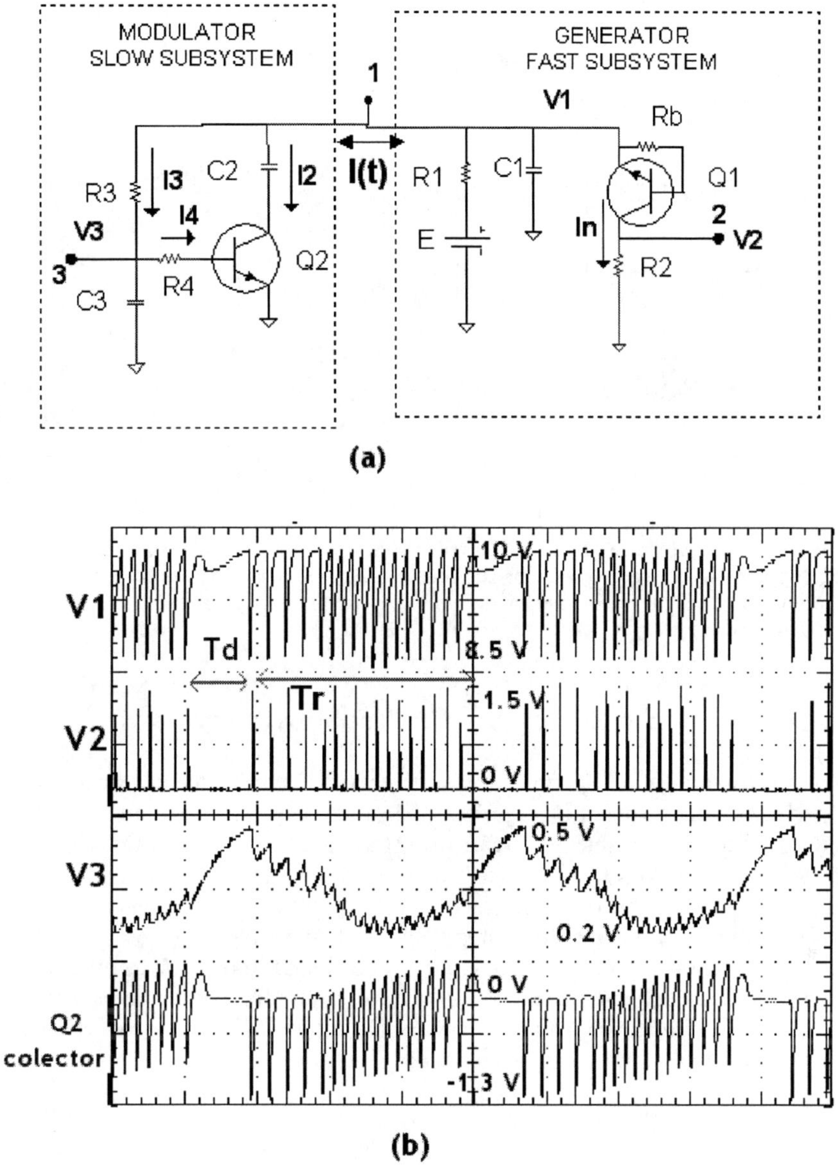

FIGURE 1. (a) Burster circuit. Current $I(t)$ links the fast and slow subsystems. Q1 and Q2 transistors 2N2222, $E = 10$ volt, capacitors in $[\mu\text{F}]$ and resistors in $[\text{K}\Omega]$; $R_1 = 3.3$, $R_2 = 0.1$, $R_b = 20$, $R_3 = 5$, $R_4 = 2.2$, $C_1 = 1$, $C_2 = 2$, $C_3 = 2$. (b) Bursting typical waveform at terminals 1, 2, 3 and Q2 transistor collector. Picture to fixed height

mainly determined by the current $I(t)$ through resistor R_3 in the range of 10 KΩ $< R_3 <$ 400 KΩ. With $R_3 >$ 400 KΩ, $I(t)$ intensity is small enough to have no modulating effect on the **Generator**, it oscillates with its own relaxation period given by the small time constant $C_1R_1 + C_1(R_1 + R_2)$, as compared with bursting period T. As R_3 decreases, $I(t)$ increases, and random many spikes long bursts appear, and then becomes periodic with duration T_r. With subsequent R_3 decrease, T_r becomes shorter and spikes number smaller, until the *burst* has only one spike. Finally bursting stops and the circuit becomes excitable (infinite period oscillation); it has a single spike in response to an excitatory external pulse. On the contrary, pause duration T_d minimally changes with R_3; it is mainly determined by C_2 capacity.

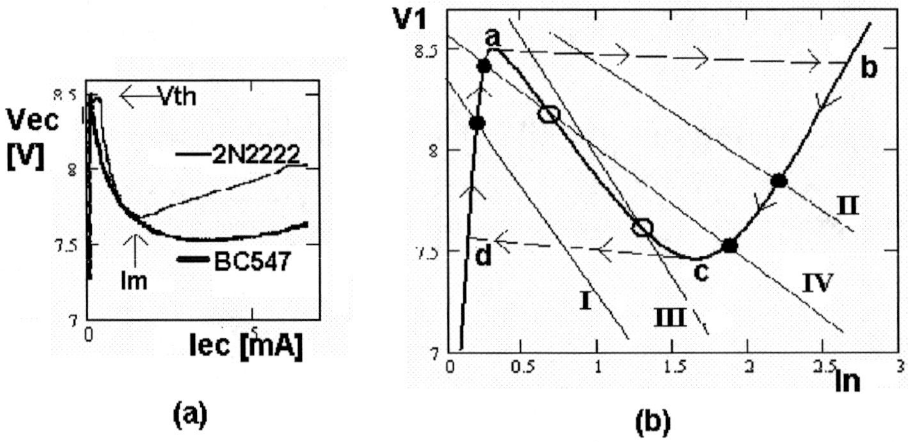

FIGURE 2. Generator 2-dimensional phase-plane. (a) Current-voltage $I_n - V_{ec}$ characteristic measured for transistors 2N2222 and BC547 showing negative resistance, due to reverse biased semiconductor junction avalanche conduction, the threshold voltage V_{th} and the minimum current I_m to sustain avalanche conduction. (b) Circuit draw phase-plane, black nullcline intersections are stable equilibriums and white unstable equilibriums.

The Generator Or Fast Subsystem

The Generator isolated is regarded as an electronic implementation of many 2-dimensional systems modeling neuron excitability [14, 15], and from Electronics it is an ordinary negative-resistance oscillator. Since Q1 is reverse biased by voltage V_1, an oscillation occurs each time V_1 reaches the Q1 avalanche conduction threshold V_{th}. Thus the negative resistance is implemented by transistor Q1 emitter-collector current-voltage characteristic curve $V_{ec} = V_{ec}(I_n)$. This curve has been measured in two ordinary transistors used in our experimental setup and is shown in Fig. 2a.

Only to write circuit equations for a geometrical approach, we assume $R_b = \infty$, $R_2 = 0$, static Q1 current-voltage $V_{ec} = V_{ec}(I_n) = V_1$ characteristic curve and the inductance L is included into the I_n current loop. Although L is not included physically in the experimental circuit of Fig. 1a, it is always present, since it takes into account instantaneous change impossibility of currents, such as I_n, due to unavoidable inductive effects or inertia in all physical system, including biological ones. Generator equation then becoming:

$$L\frac{dI_n}{dt} = V_1 - V_{ec}(I_n); \qquad C_1\frac{dV_1}{dt} = \frac{(E - V_1)}{R_1} - I_n \tag{1}$$

where the I_n nullcline is $V_1 = V_{ec}(I_n)$ and the V_1 nullcline is $V_1 = E - R_1V_1$. Nullcline I_n may be associated with the *activation* cubic parabola (flipped N-shape) and V_1 with the *restitution* function of 2-dimensional neuron models, same as in the FitzHugh–Nagumo model. The shape and the position of nullclines are controlled by circuit values, and they determine the circuit dynamics, i.e. the number of equilibria, their stability, the existence of limit cycle attractors, etc. Thus in Fig. 2b, the circuit becomes excitable for lines I and II, oscillatory on the closed trajectory **abcd** for line III and bistable for line IV. It is evident that as straight nullcline V_1 approaches one of the flipped-N shape I_n nullcline knee, a bifurcation will occur. By changing up and down only one parameter (I_{ex}) of the circuit, different bifurcation types are detected during the Generator transition from excitable (equilibrium) to oscillatory and viceversa, for particular and constant values of all other circuit parameters. The first case, parameter values as in the Fig. 1a, is shown by Fig. 3 that the circuit switches directly from quiescent on the stable equilibrium to oscillatory by a saddle invariant circle bifurcation, and viceversa when I_{ex} parameter is reversed. Fig. 3 shows the *fingerprint* or the typical scenery of a saddle-node on invariant circle bifurcation; the equilibrium is located on the circle trajectory and the oscillation frequency changes with parameter I_{ex}.

For the second case in Fig. 4, capacitances C_1 and C_2 are about ten times smaller than values in Fig. 1. That changes the V_1, I_n dynamical variables velocities field, in such way that the circuit transition from the quiescent equilibrium to oscillatory is not direct, as in the previous case, instead in a parameter I_{ex} range, the original stable equilibrium (EQ) coexists with a stable limit cycle (CL), that is a bistable behavior, by a saddle-homoclinic orbit bifurcation.

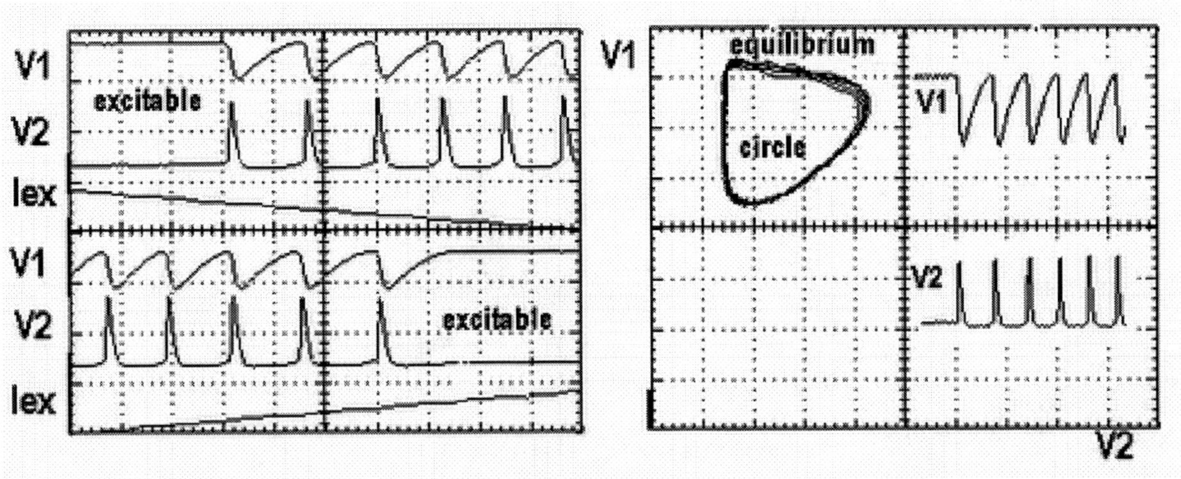

FIGURE 3. Saddle-Node on invariant circle bifurcation. Spikes frequency changes with the parameter I_{ex} as a Class 1 excitability.

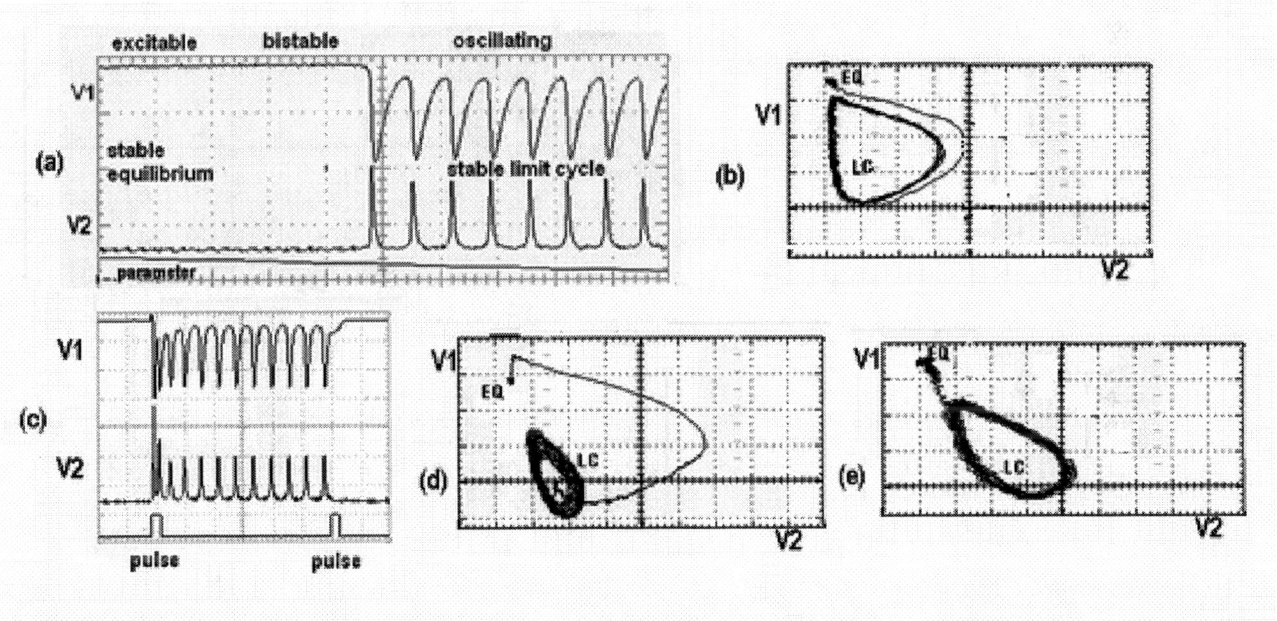

FIGURE 4. Saddle-Homoclinic orbit bifurcation. (a) and (b) From excitable on the stable equilibrium EQ to oscillatory on the stable limit cycle CL. (c), (d), (e) bistability: a pulse switches the oscillator between the EQ and LC, and viceversa.

With further same direction I_{ex} changes, bistability disappears by a saddle-node bifurcation and only the limit cycle oscillation persists.

Two different behaviors are shown in Fig. 5 and 6, both have small oscillations prior to and after a big one settles down or disappears. They take place by a supercritical and a subcritical Andronov-Hopf (A-H) bifurcation respectively. Both are obtained with a resistor R_1 higher than that used for the bifurcations of Fig. 3 and 4. As R_1 increases V_1 nullcline becomes steeper, the excitable state equilibrium switches directly to a smooth increasing limit cycle amplitude by a supercritical A-H bifurcation (Fig. 5). In a different (R_1, E) range, the equilibrium first changes to the bistable behavior clearly shown by Fig. 6 by a fold-limit cycle bifurcation. As I_{ex} changes, the stable equilibrium disappears and three limit cycles are created: a small stable, a medium unstable and a big stable. By later same direction parameter I_{ex} changes, bistability disappears by a subcritical A-H bifurcation and only the big limit cycle persists. The inverse

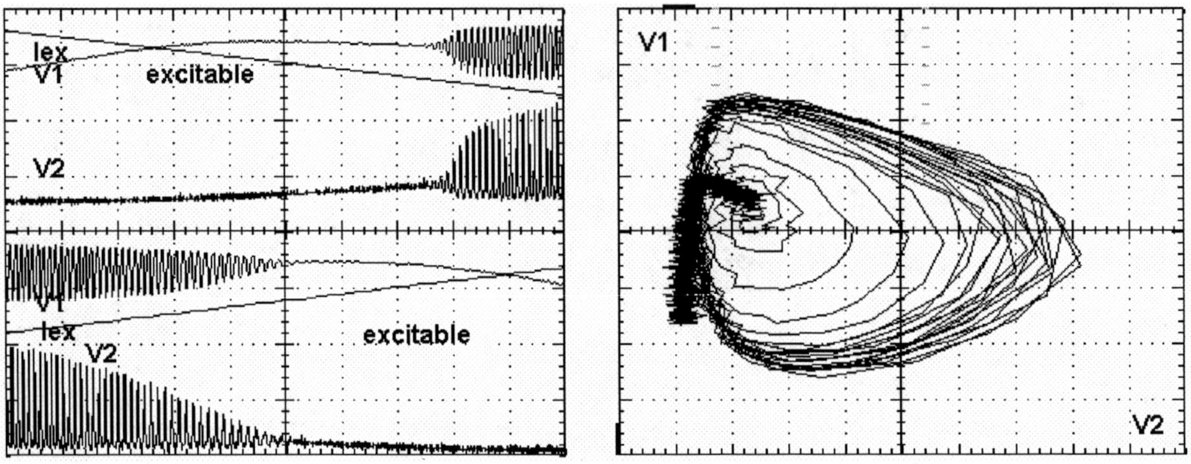

FIGURE 5. Supercritical Andronov-Hopf bifurcation. The circuit switches directly from excitable to oscillatory and viceversa.

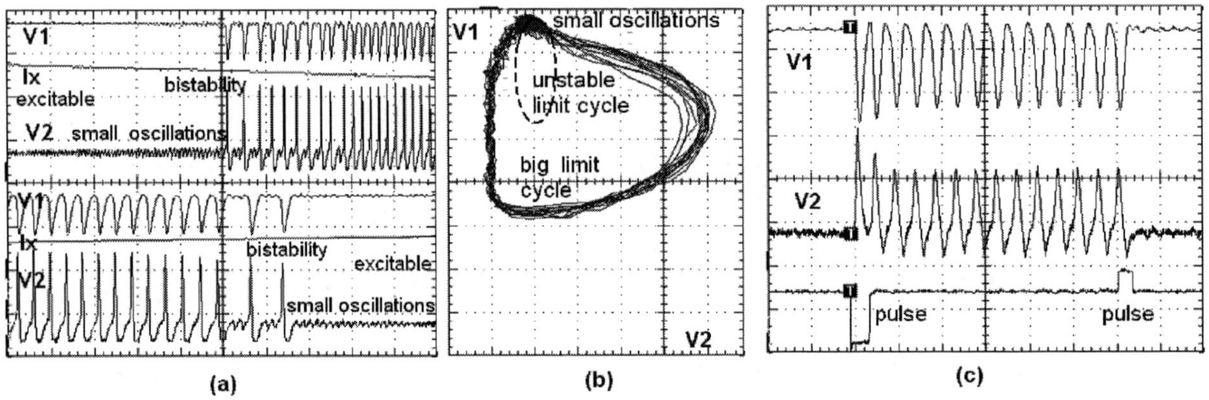

FIGURE 6. Subcritical Andronov-Hopf bifurcation. (a) From excitable (equilibrium) to bistable by a fold-limit cycle bifurcation and then from bistable to oscillatory by a subcritical A-H bifurcation, small oscillations suddenly jump to the maximum amplitude and viceversa (b) Two stable limit cycles and one unstable coexist, the unstable one is indicated by the dashed ellipse LC. (c) bistability: an external pulses switch the oscillator between small and big oscillations.

bifurcations sequence happens if the parameter I_{ex} changes in the opposite direction.

Interestingly, although the Generator circuit I_n nullcline of Fig. 2a looks like a distorted cubic parabola and the V_1 nullcline is just a straight line, our experimental results show that the Generator reproduces all the bifurcation types [12] of the HH reduced model with cubic parabola (flipped N-shape) and sigmoidea nullclines.

The Modulator Or Slow Subsystem

The circuit slow modulating conductance is implemented by the transistor Q2, and currents I_2 and I_3. They are not directly involved in the spike-generation mechanism, but rather modulate it by $I(t) = I_2 + I_3$ to get bursting oscillation. Bursting is characterized by: burst duration T_r, spikes number, pause duration T_d, adaptation pattern [16], and starting and ending bifurcation type. From our experimental results we conclude that: the duration T_r and spikes number are mainly controlled by resistor R_3, the pause T_d by C_2 capacity, and the adaptation patterns of Fig. 8 by R_3 and R_4, but the starting and ending bifurcation type is determined only by the Generator.

Through waveforms of Fig. 1b we can understand bursting occurs due to the interplay between the Generator fast currents I_1 and I_n responsible for spiking activity and the slow currents I_3 and I_2 that modulate the activity. In the

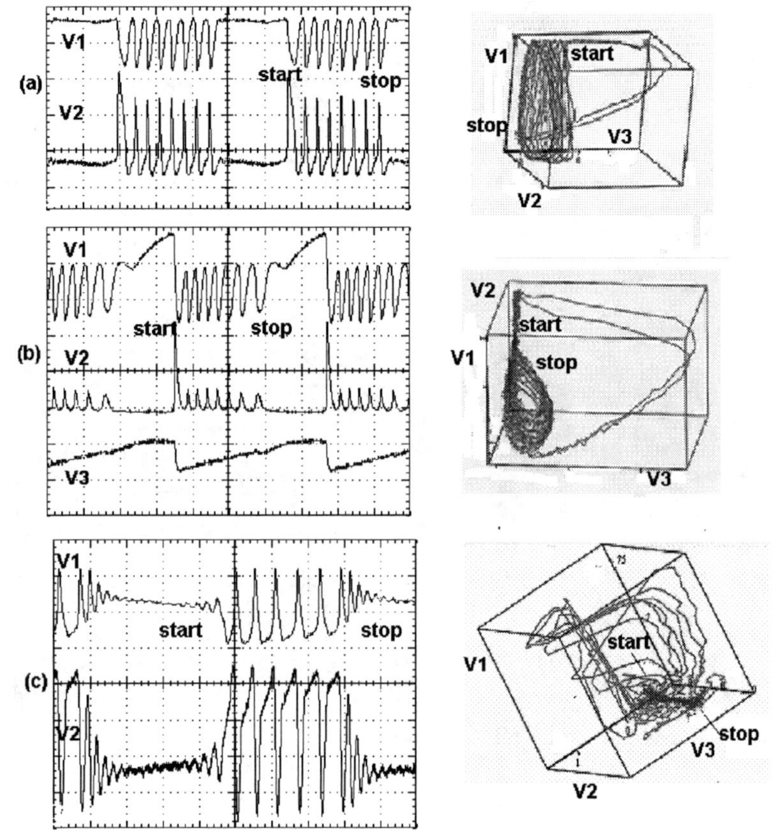

FIGURE 7. Bursters (a) Circle/Circle, (b) Saddle-Homoclinic/Saddle-Homoclinic, (c) subcritical Hopf/supercritical Hopf

Generator works a *voltage-dependant mechanism*, that is, a spike triggers only if V_1 reaches the Q1 threshold V_{th}. Notice, it is truthful for the first burst spike whatever starting bifurcation it is. The Modulator puts at work a *time-dependent mechanism*, it ends a burst whatever ending bifurcation it is, only by switching Q2 from cut to saturate. Notice that Q2 escapes from saturation just with the first burst spike, and it will return to saturation after the time T_r, spent by voltage V_3 -the stepped waveform V_3 in Fig. 1b- grows enough to supply the saturation Q2 base current. And this is a *time-dependent mechanism*, since voltage V_3 results from the balance of: a charging C_3 current I_3 and a discharging C_3 current I4, e.g. for constant R_4, burst duration T_r decreases as R_3 decreases since I_3 increases charging C_3 faster.

The three bursters of Fig. 7 were obtained with our circuit and named following the complete of sixteen (4x4) bursters classification provided by Izhikevich [12]. Their names come from the bifurcation type that initiates and terminates the burst. Notice that each starting or ending bifurcation type of any Fig. 7 bursters was already detected in the Generator. Nevertheless only three of sixteen possible bifurcation combinations were detected,the others may be are not present in the circuit dynamic or they are hidden by noise been experimentally undetectable.

Bursters also have been named [12, 17] according to their particular burst spike-frequency or spike-interval adaptation patterns, or adaptation pattern. The four types of Fig. 8 have been obtained with our circuit. Since the patterns are stable, each interspike interval was measured in ten consecutive bursts and averaged before adjusting the tendency lines for consecutive interval and instantaneous frequency of Fig. 8. Figure 1 circuit equations are:

$$L\frac{dI_n}{dt} = V_1 - V_{\text{ec}}(I_n); \qquad C_1\frac{dV_1}{dt} = \frac{(E-V_1)}{R_1} - I_n; \qquad C_2\frac{dV_{C2}}{dt} = I_2; \qquad C_3\frac{dV_3}{dt} = I_3 \tag{2}$$

where $I(t) = I_2 + I_3$,and C_2 capacitor voltage V_{C2} is about zero during the pause T_d and $V_{C2} = V_1 - V_3$ during the bursting T_r. Two equations have been added to equation (1), each one for dynamical variables $V_{C2}(t)$ and $V_3(t)$. Thus

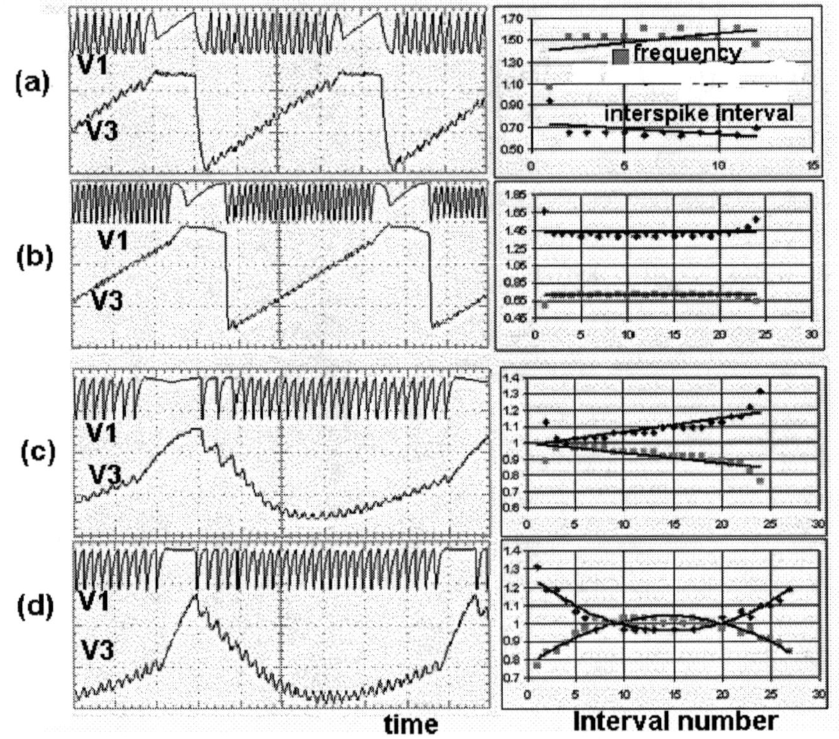

FIGURE 8. Adaptation patterns. (a) circle/circle or *accelerate triangular* burster with increasing interspike frequency. (b) saddle-homoclinic/saddle-homoclinic or *square* burster with constant frequency, (c) circle/circle or *decelerating* burster with diminishing frequency, (d) circle/circle or *parabolic* burster

the circuit becomes a 4-dimensional nonlinear dynamical system, able to perform complex oscillations such as bursting with different adaptation patterns.

Our electronic circuit is the first physical realization of all HH reduced models, since experimentally we have shown it reproduces all HH reduced model bifurcation types. In spite of the nullcline differences, it looks like a *dynamics-preserving model* perhaps with some neurocomputational properties [13]. Nevertheless additional experiments and analytical studies are necessary.

The simple hardware circuit intrinsic burster can be dissected in two parts: 1) a *voltage-dependent Generator mechanism* responsible for the fast spiking and the burst starting and ending bifurcation, and 2) a *time-dependent Modulator mechanism* responsible for the sustained spiking activity duration and the adaptation patterns. Thus, by means of the circuit it becomes clear how in the bursting activity slow parameters drive the fast subsystem through bifurcations of equilibria and limit cycles

We hope this work partially enlightens two questions [13] associated with bursting pattern: What initiates sustained spiking during the burst? and what terminates sustained spiking temporarily and ends the burst?

Additionally, many Modulators can be connected to Fig. 1a circuit terminal 1. Also adequate external perturbations applied to terminal 3 will modify: waveform V_3, burst duration T_r, and adaptation pattern. Both connections are useful characteristics for a neural network scheme.

ACKNOWLEDGMENTS

Supported by CIUNT, grant E335, Universidad Nacional de Tucumán, Argentina. Our thanks to MEDYFINOL 2006 Organizing Committee, and Dr. Roberto Deza for the help in preparing the paper.

REFERENCES

1. J. Lisman, *Trends in Neurosci.* **20**, 38–43 (1997).
2. E. M. Izhikevich, N. S. Desai, E. C. Walcott, and F. C. Hoppensteadt, *Trends in Neurosci.* **26**, 161–167 (2003).
3. Y. Maeda and H. Makino, *BioSystems* **58**, 93–100 (2000).
4. A. L. Hodgkin and A. F. Huxley, *J. Physiol.* **117**, 500–544 (1952).
5. C. Morris and H. Lecar, *Biophys. J.* **35**, 193–213 (1981).
6. J. L. Hindmarsh and R. M. Rose, *Nature* **296**, 162–164 (1982).
7. M. F. Simoni, G. S. Cymbalyuk, M. E. Sorensen, R. L. Calabrese and S. P. DeWeerth, *IEEE Trans. on Biomedical Eng.* **51**, 342–353 (2004).
8. R. Benzi, A. Sutera and A. Vulpiani, *J. Phys. A* **14**, L453 (1981).
9. A. S. Pikovsky and J. Kurths, *Phys. Rev. Lett.* **78**, 775 (1997).
10. A. S. Pikovsky, M. Rosemblum and J. Kurths, *Synchronization: A universal Concept in Nonlinear Science* , CUP, Cambridge, 2001.
11. S. LeMasson, A. Laflaquiere, T. Bal and G. LeMasson, *IEEE Trans. on Biomedical Eng.* **46**, 638–645 (1999).
12. E. M. Izhikevich, *Bursting in Dynamical Systems in Neuroscience : The Geometry of Excitability and Bursting*, Springer-Verlag, 2004.
13. E. M. Izhikevich, *Int. J. Bif. Chaos* **10**, 1171–1266 (2000).
14. R. FitzHugh, *Biophys. J.* **1**, 445–466 (1961).
15. J. S. Nagumo, S. Arimoto and S. Yoshizawa, *Proc. IRE* **50**, 2061–2070 (1962).
16. J. Rinzel and G. B. Ermentrout, in *Methods in Neuronal Modeling: From Synapses to Networks* , ed. C. Koch and I. Segev, MIT Press, Cambridge MA, 1989.
17. R. E. Plant, *Cells. Journal of Math. Biology* **11**, 15–32 (1981).

Spatiotemporal Response of Neuronal Networks to Time-Dependent Spectral Features

Pablo Balenzuela* and Jordi García-Ojalvo†

*Departamento de Física, Facultad de Ciencias Exactas y Naturales, Universidad de Buenos Aires, Pabellón 1, Ciudad Universitaria (1428), Buenos Aires, Argentina
†Departament de Física i Enginyeria Nuclear, Universitat Politècnica de Catalunya, Colom 11, E-08222 Terrassa, Spain

Abstract. In this work, we analyze the behavior of a neuronal network when it is stimulated by a synaptic train current with variable frequency. Using the fact that resonat (type II) neurons have a preferred input frequency, we constructed a network formed by two groups of neurons, each one responding to a different eigenfrequency. We show than even though every neuron is locally connected with neighbors of both groups, only those properly tuned to the input frequency fire synchronically with the incoming signal. In this sense, given a fixed sized network of heterogeneous neurons different subsets respond to the stimulus depending on its frequency.

Keywords: Episodic synchronization, neurons, brain networks, selective communication, resonance
PACS: 87.19.La, 05.45.Xt, 87.10.+e

INTRODUCTION

Brain networks have evolved to detect and process complex signals in a way that assure their survival. In recent years it has become clear that distinct pieces of information are not represented in the brain by a single dedicated neuron (the so-called *grandmother neuron*, which only fires when the observer sees her grandmother), but by populations of neurons that respond in a synchronized way to given signals [1]. One way in which this might be accomplished is by tuning different subsets of neurons to distinct input frequencies. In that way, a complex multicomponent signal could elicit the response of distinct collections of neuronal groups, which univocally represent the input signal. However, these different subsets of neurons will likely be densely interconnected and intermixed with each other, thus it remains to be seen if a combination of differently resonant neurons, thickly coupled to each other, is able to differentially respond to distinct input frequencies. This is not a trivial question, since coupling between neurons of different groups might in principle modify their response characteristics (including the location of the resonances), and this could elicit responses in unintended neurons.

To test the conjecture established above, we constructed a network of resonant (type II) neurons, consisting of two different groups that respond preferentially to different input frequencies. The response of isolated neurons to varying frequency inputs as a mechanism for selective communication was suggested by Izhikevich [2], and has been studied in detail in Ref. [3]. Previous studies in this direction were published by Segundo, Stiber and co-workers [4, 5]. The results shown below constitute an extension of the phenomenon to coupled networks of neurons.

SIMULATIONS

Neuron model

We consider neurons whose dynamical behavior is described by the Morris-Lecar model [6],

$$\frac{dV}{dt} = \frac{1}{C_m}(I_{\text{app}} - I_{\text{ion}} - I_{\text{syn}}) \quad (1)$$

$$\frac{dW}{dt} = \phi\Lambda(V)[W_\infty(V) - W] \quad (2)$$

CP913, *Nonequilibrium Statistical Mechanics and Nonlinear Physics, XV Conference*
edited by O. Descalzi, O. A. Rosso, and H. A. Larrondo

where V and W represent the membrane potential and the fraction of open potasium channels, respectively. C_m is the membrane capacitance per unit area and ϕ is the decay rate of W. The neuron is affected by several currents, including an external current I_{app}, a synaptic current I_{syn}, and an ionic current I_{ion} given by

$$I_{\text{ion}} = g_{Ca}M_\infty(V)(V - V_{Ca}^0) + g_K W(V - V_K^0) + g_L(V - V_L^0). \tag{3}$$

In this expression, g_a $(a = Ca, K, L)$ are the conductances and V_a^0 the resting potentials of the calcium, potassium and leaking channels, respectively. We define the following functions of the membrane potential:

$$M_\infty(V) = \frac{1}{2}\left[1 + \tanh\left(\frac{V - V_{M1}}{V_{M2}}\right)\right] \tag{4}$$

$$W_\infty(V) = \frac{1}{2}\left[1 + \tanh\left(\frac{V - V_{W1}}{V_{W2}}\right)\right] \tag{5}$$

$$\Lambda(V) = \cosh\left(\frac{V - V_{W1}}{2V_{W2}}\right), \tag{6}$$

where V_{M1}, V_{M2}, V_{W1} and V_{W2} are constants to be specified later. In the absence of noise, an isolated Morris-Lecar neuron shows a bifurcation to a limit cycle for increasing applied current I_{app} [7]. Depending on the parameters, this bifurcation can be of the saddle-node or the subcritical Hopf types, corresponding to either type I or type II excitability, respectively. Type II neurons respond preferentially, in a resonant way, to certain non-zero input frequencies [3], and will be used in the network described below. The specific values of the parameters used are shown in Table 1 [8]. For these parameters, the threshold value of the applied current under constant stimulation is 46.8 $\mu A/cm^2$ for type II.

Synapsis model

In this paper we analyze the behavior of a neuronal network driven by a synaptic current. To that end, we use the simplified model of chemical synapse proposed in Ref. [9], according to which the synaptic current is given by

$$I_{\text{syn}} = g_{\text{syn}} r(t)(V - E_s), \tag{7}$$

where g_{syn} is the conductance of the synaptic channel, $r(t)$ represents the fraction of bound receptors, and E_s is a parameter whose value determines the type of synapse: if E_s is larger than the rest potential the synapse is excitatory, if smaller it is inhibitory; here we consider an excitatory synapse with $E_s = 0$ mV. The fraction of bound receptors, $r(t)$, follows the equation

$$\frac{dr}{dt} = \alpha[T](1 - r) - \beta r, \tag{8}$$

where $[T] = T_{max}\theta(T_0 + \tau_{\text{syn}} - t)\theta(t - T_0)$ is the concentration of neurotransmitter released into the synaptic cleft by the presynaptic neuron. α and β are rise and decay time constants, respectively, and T_0 is the time at which the presynaptic neuron fires, which happens whenever the presynaptic membrane potential exceeds a predetermined threshold value, in our case chosen to be 10 mV. This thresholding mechanism lies at the origin of the nonlinear character of the synaptic coupling. The time during which the synaptic connection is active is roughly given by τ_{syn}. The values of the coupling parameters that we use [9] are specified in Table 1. The equations were integrated using the Heun method [10], which is a second order Runge-Kutta algorithm for stochastic equations.

NUMERICAL SETUP

In a previous work [3] it was shown how a Morris-Lecar neuron responds to periodic synaptic inputs of varying frequency. The response of the neuron can be represented by a *response diagram*, showing the minimum excitation amplitude (represented by the synaptic coupling when we stimulate with synaptic currents) versus the frequency of the synaptic input. For type II neurons, the threshold input amplitude necessary to elicit spiking is minimum for a non-zero resonance frequency f_{res}. This value is related with the frequency of the subthreshold oscillations, as analyzed in Ref. [3].

TABLE 1. Parameters values of the Morris-Lecar and synapse models used in this work.

Parameter	Morris-Lecar TII
C_m	$5\,\mu\mathrm{F/cm}^2$
g_K	$8\,\mathrm{mS/cm}^2$
g_L	$2\,\mathrm{mS/cm}^2$
g_{Ca}	$4.0\,\mathrm{mS/cm}^2$
V_K	$-80\,\mathrm{mV}$
V_L	$-60\,\mathrm{mV}$
V_{Ca}	$120\,\mathrm{mV}$
V_{M1}	$-1.2\,\mathrm{mV}$
V_{M2}	$18\,\mathrm{mV}$
V_{W1}	$2\,\mathrm{mV}$
V_{W2}	$17.4\,\mathrm{mV}$
ϕ	$1/15\,\mathrm{ms}^{-1}$

Parameter	Synapse
α	$2.0\,\mathrm{ms}^{-1}\mathrm{mM}^{-1}$
β	$1.0\,\mathrm{ms}^{-1}$
T_{max}	$1.0\,\mathrm{mM}$
g_{syn}	(specified in each case)
τ_{syn}	$1.5\,\mathrm{ms}$
E_s	$0\,\mathrm{mV}$

The frequency f_{res} depends on the model parameters. In particular, in Morris-Lecar neurons, it is related with the dynamics of the slow variable W, and can be controlled by the parameter ϕ. Figure 1 plots the *response diagram* for two different types of neurons. The results show that, depending on the value of the parameter ϕ, the preferred spiking frequency f_{res} takes different values. In this plot we only show the boundary of the 1:1 spiking region *i.e.*, when the neuron spikes with the same frequency of input stimulus. We can see that depending on the strength of the synaptic coupling (or the amplitude of the synaptic current), we can tune the network in order that the spiking activity of one neuron would produce synchronous activity (or zero activity) on its neighbors neurons. This is the central idea of this paper: using the information given by the frequency dependent activity of one neuron, one can construct a network with global properties, such as for example, the ability of coding features spatially under certain input stimuli.

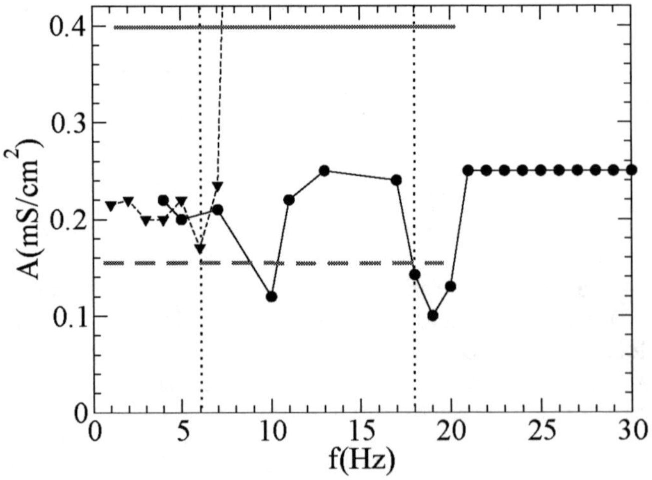

FIGURE 1. Response diagram for two different values of ϕ, indicating the threshold above which the neuron responds in a 1:1 manner ($f_{out} = f_{in}$). Triangles correspond to $\phi = 1/55$; filled circles to $\phi = 1/15$. Horizontal lines , both solid and dashed, denote the values of g_{syn} used in the networks for the two groups of neurons, as indicated in Fig. 2. In vertical dotted lines, the frequency values of the input synaptic train.

Taking this into account, we constructed a square network with nearest neighbor coupling, containing two kinds of group neurons: one group (composing the number "1") is characterized by a value of $\phi = 1/55$, which leads to a resonance frequency $f_{res} = 6$ Hz; the other group (composing the number "2") has $\phi = 1/15$ and $f_{res} = 18$ Hz. We choose three different values of the coupling strength: the neurons of group which form the number "1" are coupled to each other with a strength given by $g_{syn} = 0.40$ nS, the internal coupling among group cells forming number "2" has strength $g_{syn} = 0.15$ nS, and $g_{syn} = 0.01$ nS for the rest of connections. A synaptic input train with two alternating frequencies, 6 and 18 Hz, stimulates two neurons of the array. A scheme of this arrangement is shown in Fig. 2.

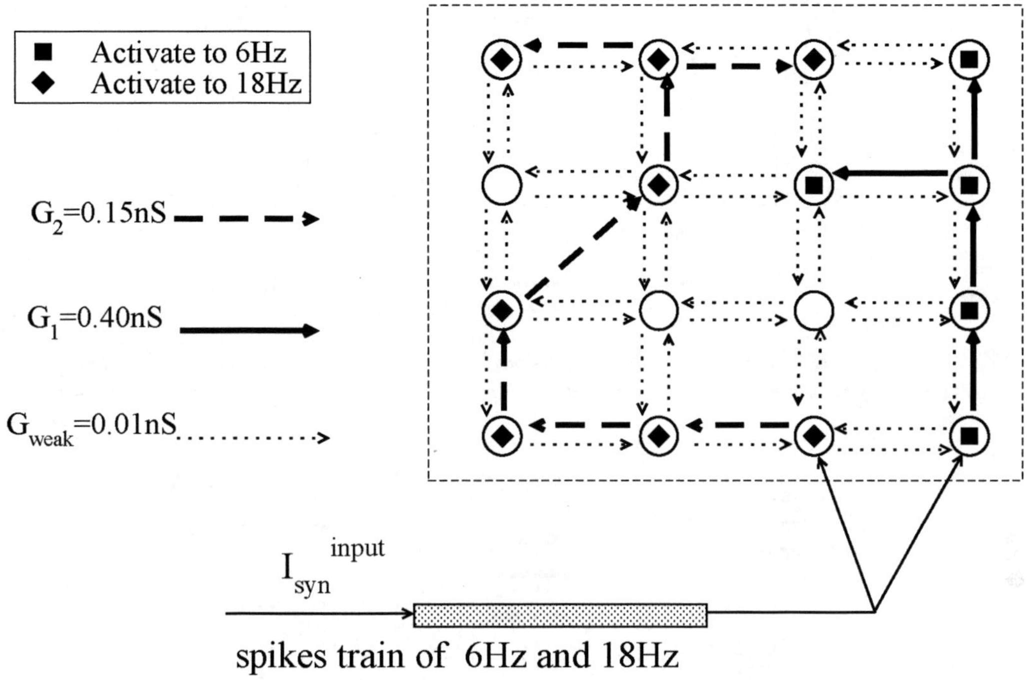

FIGURE 2. Scheme of the neuronal network studied: A 4×4 square lattice with nearest neighbors synaptic connection. Filled squares represent the neurons than fire at 6 Hz, forming the number "1". In black diamonds the neurons than fire at 12 Hz, forming the number "2". Open circles represent the silent neurons. The stimulus (a synaptic train pulse with alternating 6 and 18 Hz frequencies) arrives to the down-right corner an the propagates along the network. The different values of ϕ of the group of neurons as well as the three different values of synaptic coupling g_{syn} allow this setup to spell out the number "1" when the spike train is 6 Hz and "2" when is 18 Hz.

The response of the array described above to the frequency-varying input is represented in Fig. 3, which shows the temporal series of a representative neuron of each group within the array. We can see how the response of the neurons alternates following precisely the variations in the input frequency, in such a way that a set of neurons only fire (synchronously) when the input frequency is 6 Hz, and the other group fires when the input frequency is 18 Hz. A similar selective response could be elicited for frequencies subharmonic of those two values considered here, provided the signal amplitude is large enough, as shown by the minima occurring at those frequencies in Fig. 1. This result illustrates that a fully connected network can host dynamical subnetwork activity in response to different types of input signals.

Finally, in order to illustrate the activity of the network taking into account the geometry of the array and the pattern generated by the subnetworks of synchronous neurons, we plot the activity of all the neurons in Fig. 4 in response to a given frequency varying synaptic current. We observe how the numbers "1" and "2" are formed alternatively in response to the different frequencies of the input.

CONCLUSION

Brain networks are subject to wide range of different signals, and mechanisms of selective communication should exist to discern between them. Taking advantage of the frequency selectivity of type II (resonant) neurons, we show that a locally coupled neuronal network composed of two intermixed groups of neurons, each one of them responding preferentially to a given input frequency, responds selectively to the input frequency, in such a way the each one of the

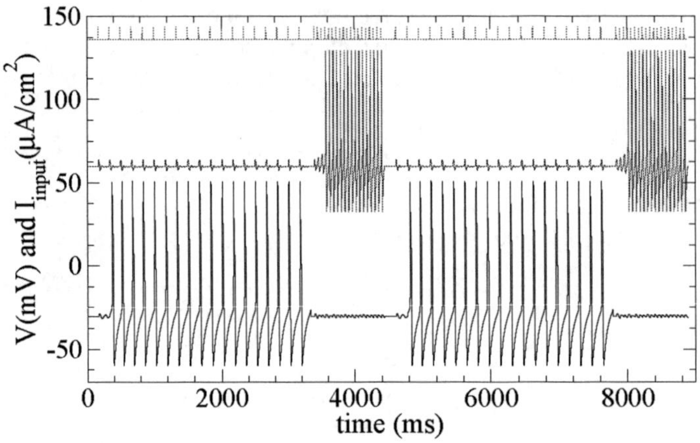

FIGURE 3. Time series of the two kind of neurons used in the array. Input current (top) and activity of the two group of neurons: Bottom, those that responds synchronously at 6 Hz and middle, those at 18 Hz. We can observe that each group of neurons spikes only at the tuned frequency.

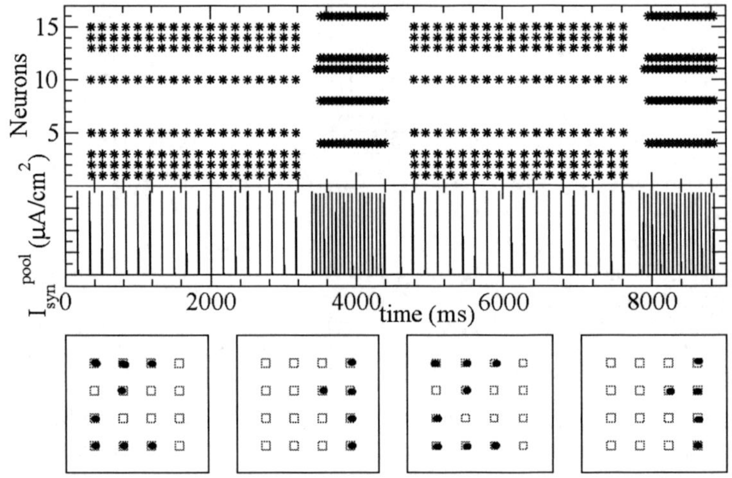

FIGURE 4. Spiking activity of the neuronal network, showing that two different subnetworks emerge dynamically in response to different input frequencies. Top panel: Raster plot showing the activity of the neurons and their dependence with the frequency of the input current showed in medium panel. Low panel: Activity of the network in the four characteristic stages showing how the numbers "1" and "2" are formed alternatively.

two groups is exclusively active when the input frequency coincides with the resonance of the corresponding group. The response of a single neuron to a varying frequency was studied in Ref. [3]. In that paper, the neuron was found to follow the input signal more or less adiabatically when the signal frequency entered the neuron's locking range for a given signal amplitude. Given those results, we expect that the neuronal network described here will act in the same way, with given subpopulations of neurons responding dynamically to that frequency to which they are resonant.

ACKNOWLEDGMENTS

We acknowledge financial support from the Ministerio de Educación y Ciencia (Spain, project FIS2006-11452), and from the Generalitat de Catalunya. P.B. acknowledges financial support from the Fundación Antorchas (Argentina), and from a C-RED grant of the Generalitat de Catalunya. P.B. is member of "Carrera de Conicet", Argentina. This work has also been supported by the European Commission (NEST project GABA, contract 043309).

REFERENCES

1. W. Singer, *Current Opinion Neurobiol.* **9**, 189–194 (1999).
2. E. M. Izhikevich, *Biosystems* **67**, 95–102 (2002).
3. P. Balenzuela, J. M. Buldú, M. Casanova, and J. García-Ojalvo, *Phys. Rev. E* **74**, 061910 (2006).
4. M. Stiber, R. Ieong, and J. P. Segundo, *IEEE Trans. Neural Networks* **8**, 1379–1385 (1997).
5. J. P. Segundo, M. Stiber, J. F. Vibert, and S. Hanneton, *Neuroscience* **68**, 693–719 (1995).
6. C. Morris, and H. Lecar, *Biophys. J* **35**, 193–213 (1981).
7. M. St-Hilaire, and A. Longtin, *J. Comp. Neurosci.* **16**, 299–313 (2004).
8. K. Tsumoto, H. Kitajima, T. Yoshinaga, K. Aihara, and H. Kawakami, *Neurocomputing* **69**, 293–316 (2006).
9. A. Destexhe, Z. Mainen, and T. Sejnowski, *Neural Comp.* **6**, 14–18 (1994).
10. J. García-Ojalvo, and J. M. Sancho, *Noise in Spatially Extended Systems*, Springer, New York, 1999.

Discrimination Measure of Correlations in a Population of Neurons by Using the Jensen-Shannon Divergence.

F. Montani*, O. A. Rosso† and S. R. Schultz*

*Department of Bioengineering, Imperial College London, South Kensington London SW7 2AZ, UK.
†Instituto de Cálculo, Facultad de Ciencias Exactas y Naturales, Universidad de Buenos Aires.
Pabellon II, 1428 Ciudad Universitaria, Buenos Aires, Argentina.

Abstract. The significance of synchronized spikes fired by nearby neurons for perception is still unclear. To evaluate how reliably one can decide if a given response on the population coding of sensory information comes from the full distribution, or from the product of independent distributions from each cell, we used recorded responses of pairs of single neurons in primary visual cortex of macaque monkey (V1) to stimuli of varying orientation. Both trial-to-trial variability and synchrony were found to depend stimulus orientation and contrast in this data set (A. Kohn, and M. A Smith, J. Neurosci. 25 (2005) 3661). We used the Jensen-Shannon Divergence for fixed stimuli as a measure of discrimination between a pairs of correlated cells VI. The Jensen-Shannon divergence, can be consider as a measure distance between the corresponding probability distribution function associated with each spikes fired observed patterns. The Nemenman-Shafee-Bialek estimator was used in our entropy estimation in order to remove all possible bias deviation from our calculations. We found that the relative Jensen-Shannon Divergence (measured in relation to case in which all cell fired completely independently) decreases with respect to the difference in orientation preference between the receptive field from each pair of cells. Our finding indicates that the Jensen-Shannon Divergence may be used for characterizing the effective circuitry network in a population of neurons.

Keywords: Neuron populations, discrimination measures, correlations, visual cortex, Jensen-Shannon divergence.
PACS: 87.19.La, 87.19.Dd, 87.10.+e

INTRODUCTION

One of the main challenges in neuroscience during the last 30 years was to demonstrate that distributed neural populations in the visual cortex process information in a cooperative way. The visual cortex is composed of a large number of areas, which contain neurons that are tuned to different visual features. Temporally correlated activity of individual neuronal pairs within the visual cortex has been investigated in many laboratories starting in the early 1980's [1, 2, 3, 5, 4, 6, 7, 8, 9, 10] most often with a motivation to reveal structural coupling between cells. This approach to functional anatomy had been methodologically and conceptually outlined and successfully applied to the invertebrate nervous system [11, 12, 13, 14, 15, 16]. The results of these many different studies taken together clearly demonstrated the presence of temporally precise correlation within the cat and macaque visual cortex. But, these correlations occurred preferentially between cells showing the same feature preference.

Cortical neurons with similar stimulus selectivities are found in close proximity to each other [17, 18, 19]. How neurons in the cerebral cortex represent a stimulus taking into account that the response variability is correlated across neurons is still highly debated [20, 21, 22, 23]. At another level, these correlations when present may have significant effect on the population coding of sensory information. Several pieces of evidence point to correlations between V1 neurons being orientation and contrast dependent [24, 25]. The importance of identifying appropriate measures of correlation, independence and quantify their relation to the stimulus has been pointed out by [26].

The Jensen-Shannon divergence measures how reliably one can decide if a given response comes from the full distribution, $p(r_1 r_2)$, or the product of independent distributions, $p(r_1) \cdot p(r_2)$. In present work we use the Jensen-Shannon Divergence for fixed stimuli as a measure of discrimination between a pairs of correlated cells in the primary visual cortex of anesthetized macaque monkeys [24]. Our findings indicates that the Jensen-Shannon Divergence may be used for characterizing the circuitry network in a population of neurons.

CP913, *Nonequilibrium Statistical Mechanics and Nonlinear Physics, XV Conference*
edited by O. Descalzi, O. A. Rosso, and H. A. Larrondo

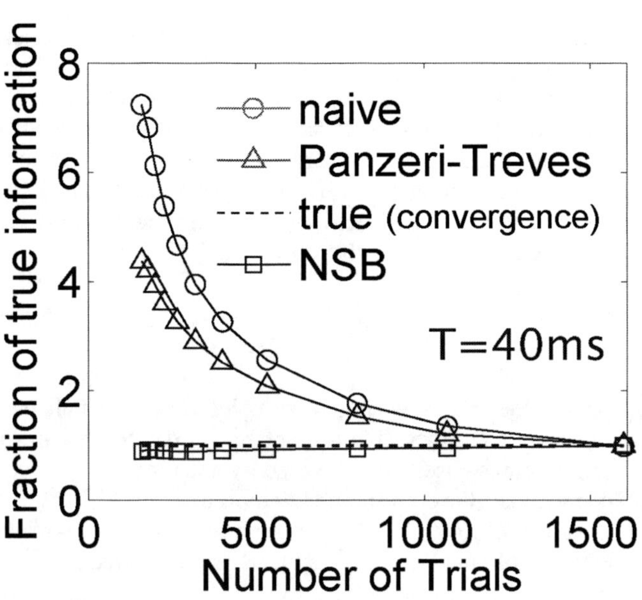

FIGURE 1. Performance comparison of several information estimators. The shows the total information estimated through these different approaches.

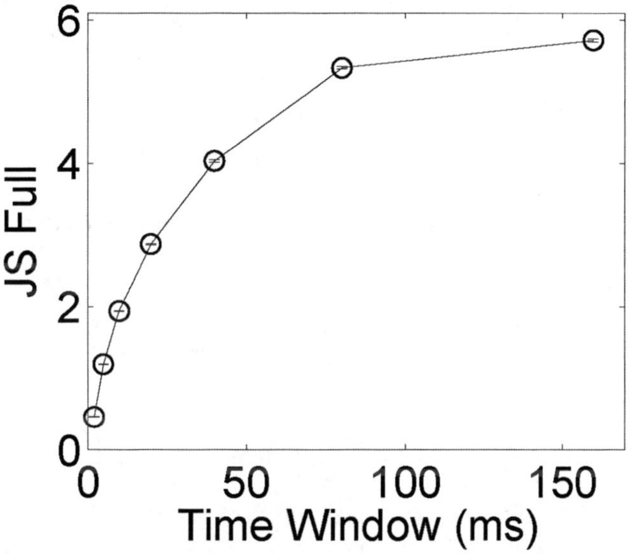

FIGURE 2. For a tipycal pair of cells in V1 the JS was calculated taken into account the full sitmuli orientation dependency at seven different time windows.

METHODS

Entropy estimations

The problem with real measures of entropy is that they depend on a limited number of samples provided by the experiment. It is therefore important to use a theoretical approach that can remove sample size dependent bias from the entropy estimations. But, it is well known that approaches usually performed to estimate the entropy tend to underestimate it since they are biased, and this effect can only be avoided by the use of a perturbative expansion in the asymptotic regime [27]. The idea of calculating entropy by counting coincidences was proposed a long time ago by Ma [28] for physical systems in the micro-canonical ensemble, where a uniform distribution of entropy corresponds to states of fixed energy. The Bayesian prior proposed by Nemenman, Bialek and Steveninck (NSB) [29] extends this idea to an arbitrarily complex distribution. The goal of this method is to construct a Bayesian prior, which generates a nearly uniform distribution of entropies in order to correct sample size dependent bias at its source.

In the following we review the basic ideas of the NSB entropy estimator method [29]. Let consider the problem of estimating the Shannon entropy for a given probability distribution $\mathbf{p} \equiv \{p_i\}$,

$$H_{naive} = \sum_{i=1}^{K} f_i \log_2 f_i \,, \tag{1}$$

where the index i runs over K possibilities. Considering N samples (trials) which were obtained from a given experiment, where each possibility i occurred n_i times. If N is much bigger than K we can approximate $p_i \approx f_i = n_i/N$, and $\mathbf{p}$ might be the distribution of spike counts observed to be fired by a neuron.

The present approach tends to underestimate the entropy and attempts to solve this problem were made to asymptotic bias corrections by adding a term of order $\mathscr{O}(K/N)$ [30, 31]. This approach was developed further by [27] who made use of a Bayesian prior for the number of relevant bins, and iteratively re-estimated it.

Bayes' theorem states,

$$P(\mathbf{p}|n_i) = \frac{P(n_i|\mathbf{p}) \cdot P(\mathbf{p})}{P(n_i)} \,, \tag{2}$$

where $P(\mathbf{p})$ is the prior distribution and N is the total number of experimental trials, $N = \sum_i^K n_i$, from i observed events which occurred n_i times

In order to eliminate the bias deviation it would be desirable to find a prior $P(\mathbf{p})$ does not depend upon the number of trials. We can express $P(\mathbf{p})$ in terms of the Dirichlet family of priors [29], which allow us to construct a prior which does not depend on inverse powers of N.

$$P_\beta(\mathbf{p}) = \frac{1}{Z}\,\delta\left(1-\sum_i^K p_i\right)\left(\prod_i^K p_i^{\beta-1}\right) \,, \tag{3}$$

where Z is a function which enforces the normalization, and the entropy is specified fixing a particular value of β [32]. Ideally, we would like to compute the whole a priori distribution of entropies

$$P_\beta(H) = \int dp_1 \cdots dp_K \, P_\beta(\mathbf{p})\, \delta\left(H - \sum_i^K p_i \log_2 p_i\right) . \tag{4}$$

Defining

$$P_\beta(H) = 1 = \int d\xi \; \delta(H-\xi) \,, \tag{5}$$

where ξ is the expected entropy, one can fix a flat prior distribution of entropy $P(H)$. The idea of the NSB method is to build a family of prior $P_\beta(\mathbf{p})$ which result in δ functions over H, by changing β over the whole range of entropies uniformly, removing bias at its source.

The Bayesian prior proposed by Nemenman, Bialek and Steveninck [29],

$$P_\beta(\mathbf{p}) = \frac{1}{Z}\,\delta\left(1-\sum_i^K p_i\right)\left(\prod_i^K p_i^{\beta-1}\frac{d\xi(\beta)}{d\beta}P(\beta)\right) \tag{6}$$

reach the requirements, generating a nearly uniform distribution of entropies in order to avoid bias at its origin (Z is a normalizing factor and $\frac{d\xi(\beta)}{d\beta}$ ensures the uniformity for a priori expected entropy ξ).

We performed calculations of the total mutual information conveyed by a single pair of cells in V1 [24] using the naive estimator, using the Panzeri-Treves bias correction, and using the NSB method. Figure 1 shows the effectiveness of the entropy estimator developed by Nemenman, Bialek and Steveninck [29], for which sampling was obtained at only 100 trials.

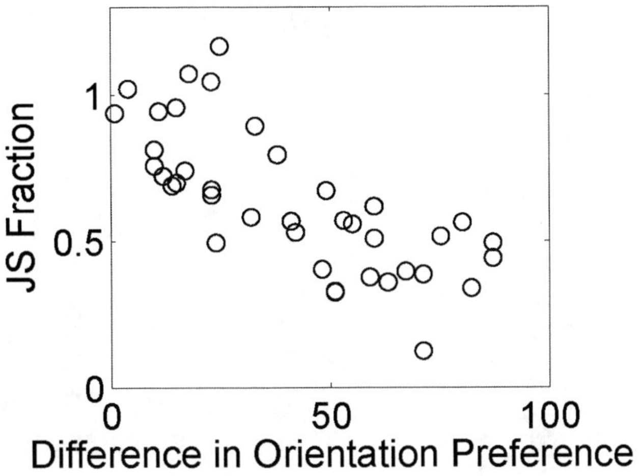

FIGURE 3. $T = 5\ ms$, The relative Jensen-Shannon Divergence as a discrimination measure of correlations.

Jensen-Shannon divergence

The Jensen-Shannon divergence (JSD) is defined by [33, 34],

$$JS(\mathrm{P},\mathrm{Q}) = \frac{1}{2}\left[D_{KL}\left(\mathrm{P}\Big|\frac{\mathrm{P}+\mathrm{Q}}{2}\right) + D_{KL}\left(\mathrm{Q}\Big|\frac{\mathrm{P}+\mathrm{Q}}{2}\right)\right], \tag{7}$$

where D_{KL} is the Kullback-Leibler divergence between the two distributions.

The JSD allow us to decide how reliably one can decide if a given response comes from two different distributions. However, unlike the Kullback-Leibler divergence, it is symmetric, always well defined and bounded. In particular if we choose $\mathrm{P} \equiv P(r|s)$, the probability of getting response vector r conditional upon the occurrence of stimulus s, and $\mathrm{Q} \equiv P_{ind}(r|s)$, the probability of getting independent population responses,

$$\begin{aligned}
JS(\mathrm{P},\mathrm{Q}) \;=\;& \frac{1}{2}\sum_{r_1,r_2}\sum_{s} P(s)\,P(r_1r_2|s)\,\log_2 P(r_1r_2|s) - \frac{1}{2}\sum_{r_1,r_2} P(r_1r_2)\,\log_2 P(r_1r_2)\\
+\;& \frac{1}{2}\sum_{r_1,r_2}\sum_{s} P(s)\,P_{ind}(r_1r_2|s)\,\log_2 P_{ind}(r_1r_2|s) - \frac{1}{2}\sum_{r_1,r_2} P_{ind}(r_1r_2)\,\log_2 P_{ind}(r_1r_2)\\
+\;& \sum_{r_1,r_2}\sum_{s} P(s)\left[\frac{P_{ind}(r_1r_2|s)+P(r_1r_2|s)}{2}\right]\log_2\left[\frac{P_{ind}(r_1r_2|s)+P(r_1r_2|s)}{2}\right]\\
-\;& \sum_{r_1,r_2}\left[\frac{P_{ind}(r_1r_2)+P(r_1r_2)}{2}\right]\log_2\left[\frac{P_{ind}(r_1r_2)+P(r_1r_2)}{2}\right]
\end{aligned} \tag{8}$$

where $P(r_1r_2) = \sum_s P(s)P(r_1r_2|s)$ and $P_{ind}(r) = \sum_s P(s)P(r_1|s)P(r_2|s)$ are the average of $P(r|s)$ and $P_{ind}(r|s)$ over all the possible stimuli. By fixing the stimuli we can use the Jensen-Shannon divergence to measure how reliably one can decide if a given response comes from $P(r_1r_2|s_{fix})$, or , $P_{ind}(r_1r_2|s_{fix})$. Comparing the JS for pairs of correlated $P(r|s)$ and uncorrelated $P_{ind}(r|s)$ cells at a fixed stimuli, one can ask how much information you can gain about the effective network between cells.

RESULTS

In order to evaluate how reliably one can decide if a given response on the population coding of sensory information comes from the full distribution, $P(r_1r_2|s_{fix})$, or from the product of independent distributions from each cell,

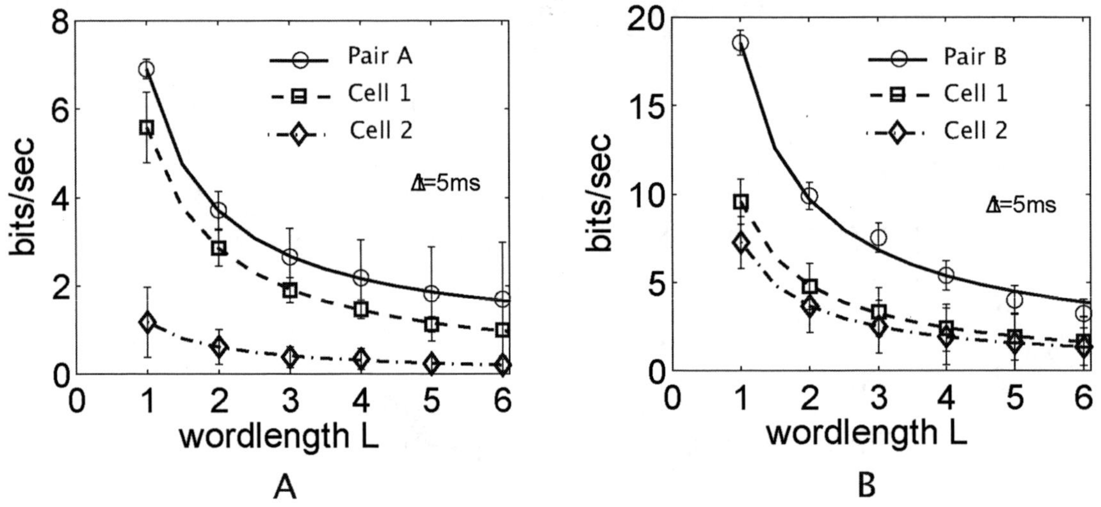

FIGURE 4. Calculation of the Information rates. The Figure show the Information rate for two different pais (Pair A and B) and for individual neurons, binwidths $\Delta t = 5ms$

$P_{ind}(r_1 r_2|s_{fix})$, we used recorded responses of 42 pairs of single neurons in primary visual cortex of macaque monkey (V1) to stimuli of varying orientation. The experimental methods can be found in [24]. They collected data from single pairs of cells in primary visual cortex. For each neuron, they first determined the preferred orientation and direction. This was done quantitatively by measuring the responses to sinusoidal gratings drifting in different directions, centered on the receptive field as determined by initial mapping [24].

Figure 2 shows the Jensen-Shannon divergence for one single pair of cells in V1 [24] using the NSB entropy estimator [29] at different time windows. The Jensen Shannon divergence was calculated taking the full orientation dependency on both probabilities distributions. Plateau like behavior can be appreciated in the Jensen-Shannon as time window becomes bigger.

In Figure 3 we present the Jensen-Shannon Fraction for 42 pairs of single neurons V1 at fixed stimuli orientation as function of the difference in orientation preferences from the tuning curves obtained by each pair of cells (time window fixed at 5 *ms*). We define the Jensen-Shannon Fraction as the maximum JS divergence obtained by fixing the stimuli for a given set divided by its minimum for a pair of single neurons. Pairs of cells with similar preferred orientations present a higher JS divergence which those that showed less similarities. This result is agreement with the fact that cells which share similar preferred orientation are more likely to have a higher quantity of common inputs than those that present less similarities. Our finding indicates therefore that the Jensen-Shannon Divergence may be used for characterizing the effective circuitry network in a population of neurons.

In Figure 4, we show the calculation of the information rate for a typical pairs of cells, by plotting the information at wordlength L divided by the total time window considered. The asymptote defines the information rate ([35]). The figure compares information versus L behaviour for binwidths of 5 *ms*, as used in the manuscript ; note that it was not feasible to go beyond $L = 6$ (with pairs of cells, each of which can fire many spikes in an individual time bin), as the dimensionality explodes rapidly. Our use of $5ms$ is clear: coarser temporal binning leads to a large decrease in the information rate achieved (see [36] for a review). Hence, we use here a 5 *ms* time windows as a relevant timescale for sensory processing.

CONCLUSION

We performed NSB method on a relative Jensen-Shannon Divergence for pairs of cells in V1 with stimuli dependence in orientation, considering the relevant timescale for sensory information. Our findings show that pairs of cells in V1 with similar preferred orientations present a higher relative JS divergence which those that have less similarities. The use of a relative Jensen-Shannon Divergence at a fixed stimuli provide us a measure of how much information one can gain about the effective network between cells.

ACKNOWLEDGMENTS

We would like to specially thank Dr. A. Kohn and Dr. M.A. Smith for providing the data of the pairs of cells. This work was supported by Consejo Nacional de Investigaciones Científicas y Técnicas (CONICET) (PIP 5687/05), Agencia Nacional de Promoción Cientifica y Tecnológica (PICT 20648/04) to OAR , Argentina and by a grant from the Gatsby Charitable Foundation to SRS.

REFERENCES

1. K. Toyama, M. Kimura, and K. Tanaka, *J. Neurophysiol.* **46**, 191–201 (1981).
2. K. Toyama, M. Kimura, and K. Tanaka, *J. Neurophysiol.* **46**, 202–214 (1981).
3. A. Michalski, G. L. Gerstein, J. Zarkowska, and R. Tarnecki, *Exp. Brain Res.* **51**, 97–107 (1983).
4. F. Aiple, and J. Krueger, *Exp. Brain Res.* **72**, 141–149 (1988).
5. D. Y. Ts'o, C. D Gilbert, and T. N. Weisel, *J. Neurosci.* **6**, 1160–1170 (1986).
6. D. Y. Ts'o, C. D. Gilbert, *J. Neurosci.* **8**, 1712–1727 (1988).
7. P. M. Gochin, E. K. Miller, C. G. Gross, and G L. Gerstein, *Exp. Brain Res.* **84**, 505–516 (1991).
8. Y. Hata, T. Tsumoto, H. Sato, K. Hagihara, and H. Tamura, *Nature* **335**, 815–817 (1998).
9. J.M. Alonso, and L.M. Martinez, *Nat. Neurosci.* **1**, 395–403 (1998).
10. S. Molotchnikoff, and S. Shumikhina, *Vis. Res.* **36**, 2037–2046 (1998).
11. D. H. Perkel, G. L. Gerstein, and G. P. Moore, *Biophys. J.* **7**, 419–440 (1967).
12. G. L. Gerstein, D. H. Perke, *Science* **164**, 828–30 (1969).
13. G. P. Moore, J. P. Segundo, D. H. Perkel, and H. Levitan, *Biophys. J.* **10**, 876–900 (1970).
14. P. A. Kirkwood, *J. Neurosci. Methods* **1**, 7152–7161 (1979).
15. A. M. Aertsen, G. L. Gerstein, M. K. Habib, and G. Palm, *J. Neurophysiol.* **61**, 900–917 (1989).
16. D. J. Surmier, and R.J. Weinberg, *Brain Res.* **331**, 180–184 (1985).
17. D. H. Hubel, and T. N. Weisel, *J. Physiol. (London)* **160**, 106–54 (1962).
18. V. B. Mountcastle, *J. Neurophysiol.* **20**, 408–34 (1957).
19. G. C. De Angelis, G. M. Ghose, I. Ohzawa, R. D. Freeman, *J. Neuroscience* **19**, 4046–4064 (1999).
20. E. Zohary, M. N. Shadlen, and W. T. Newsome, *Nature* **370**, 140–143 (1994).
21. M. N. Shadlen, W. T. Newsome, *J. Neuroscience* **18** , 3870–3896 (1998).
22. F. Montani, A. Kohn, M. Smith and S. R. Schultz, *Neurocomputing* in press (2006).
23. F. Montani, A. Kohn, M. Smith and S. R. Schultz, *J. Neuroscience* in press (2007).
24. A. Kohn, and M. A. Smith, *J. Neurosci.* **25**, 3661–3673 (2005).
25. J. M. Samonds, and A. B. Bonds, *J. Neurophysiol* **93**, 223–236 (2005).
26. E. Schneidman, W. Bialek, and M. J. Berry, *J. Neurosci.* **23(37)**, 11539–11553 (2003).
27. S. Panzeri, and A. Treves, *Network* **7**, 87–107 (1996).
28. S. K. Ma, *J. Statistical Physics* **26**, 221–240 (1981).
29. I. Nemenman, W. Bialek, and R. Steveninck, *Phys. Rev. E* **69**, 056111–6, (2004).
30. A. G. Carlton, *Pyschological Bulletin*, **71**, 108–109 (1969).
31. G. A. Miller, *Information Theory in Psychology; Problems and Methods II-B*, 95–100 (1955).
32. I. Nemenman, F. Shafee, and W. Bialek, "Entropy and inference, revisited", *In Advance Neural Information Processing Systems 14*, edited by T. .G. Dietterich, S. Becker, and Z. Ghahramani, MIT Presions, Cambridge, MA, 2002.
33. P.W. Lamberti, M. T. Martin, A. Plastino and O. A. Rosso, *Physica A* **334**, 119–131 (2004).
34. O. A. Rosso, M. T. Martin, A. Figliola, K. Keller and A. Plastino, *J. Neuroscience Meth.* **153**, 163–182 (2006).
35. S. P. Strong, R. Koberle, R. R. de Ruyter van Steveninck, and W. Bialek, *Physical Review Letters* **80**, 197–200 (1998).
36. S. Panzeri, S. R. Schultz, A. Treves, and E. T. Rolls, *Proc. Royal Soc. London B: Biological Sciences* **266**, 1001–1012 (1999).

Study of EEG Brain Maturation Signals with Multifractal Detrended Fluctuation Analysis

A. Figliola*, E. Serrano†, J. A. P. Rostas**, M. Hunter‡ and O. A. Rosso§

*Instituto del Desarrollo Humano, Universidad Nacional de General Sarmiento.
Juan Maria Gutierrez 1150, Los Polvorines, Pcia. de Buenos Aires, Argentina.
†Escuela de Ciencia y Tecnología, Universidad Nacional de General San Martín.
M. Irigoyen 3100, San Martín, Pcia. de Buenos Aires, Argentina.
**Biomedical Sciences and Hunter Medical Research Institute, University of Newcastle.
University Drive, Callaghan, NSW 2308, Australia.
‡School of Behavioural Sciences and Hunter Medical Research Institute. University of Newcastle.
University Drive, Callaghan, NSW 2308, Australia.
§Instituto de Cálculo, Facultad de Ciencias Exactas y Naturales, Universidad de Buenos Aires.
Pabellon II, 1428 Ciudad Universitaria, Buenos Aires, Argentina.

Abstract. In this work, we have study the EEG signals of birds during the first 6 weeks of life. The aim of the article is to perform a quantitative analysis of the dynamical changes observed in these signals due to the brain maturation effects. The signals' long scaling behaviour is study by Multifractal Detrended Fluctuation Analysis (MFDFA). This method allows the multifractal characterization of these EEG nonstationary time series and characterize the different stage of bird brain maturation.

Keywords: EEG, brain maturation, multifractal analysis.
PACS: 05.45.Df, 87.19.La, 89.75.Fb

INTRODUCTION

Animal models provide an invaluable source of information in the testing of hypotheses that are difficult to corroborate by other experimental means. In present work, we revisit data analysis of scalp-applied recording electrodes which have been used to monitor changes in basal EEG patterns in chickens during posthatch development [1]. The motivation here is to study the development and maturation of brain function using the *Multifractal Detrended Fluctuation Analysis* (MFDFA) [2].

In recent years, the *Detrended Fluctuation Analysis* (DFA) [3] has been established as an important tool for the detection of long-range (auto-)correlations in time series and has become a widely used technique for the determination of (mono-)fractal scaling properties and the detection of long-range correlations in noisy, nonstationary time series. The DFA has successfully been applied to diverse fields. An alternative approach for the study of the multifractal behavior has been proposed by Stanley and co-workers [2]. It is based on a generalization of the DFA method and is know as *Multifractal Detrended Fluctuation Analysis* (MFDFA).

In this work, we characterized EEG signals of chicken, since the birds have weekly measues until six weeks posthatch. It is considered that in that period the chicken reach their neurological maturity. Using the MFDFA method we have been able to build the multifractal spectrum of each EEG time series corresponding to the analyzed weeks and perform a quantitative analysis of the brain maturation changes.

EXPERIMENTAL DATA AND ARTIFACT REDUCTION

Twenty four chickens (*Gallus domesticus*) reared from hatching were the subjects in this experiment. All birds had free access to food and water throughout the experiment. The birds were reared initially in incubation boxes and then transferred to holding cages maintained at a constant temperature ($21°C$) with a $12:12$ *hr* light:dark cycle.

Continuous EEG recordings ($0.1-100$ *Hz* with a 50 *Hz* notch filter) were made using small (6 *mm*) gold cup electrodes attached to the scalp with collodion glue and filled with electrode gel. The signal was sampled at a rate of

CP913, *Nonequilibrium Statistical Mechanics and Nonlinear Physics, XV Conference*
edited by O. Descalzi, O. A. Rosso, and H. A. Larrondo

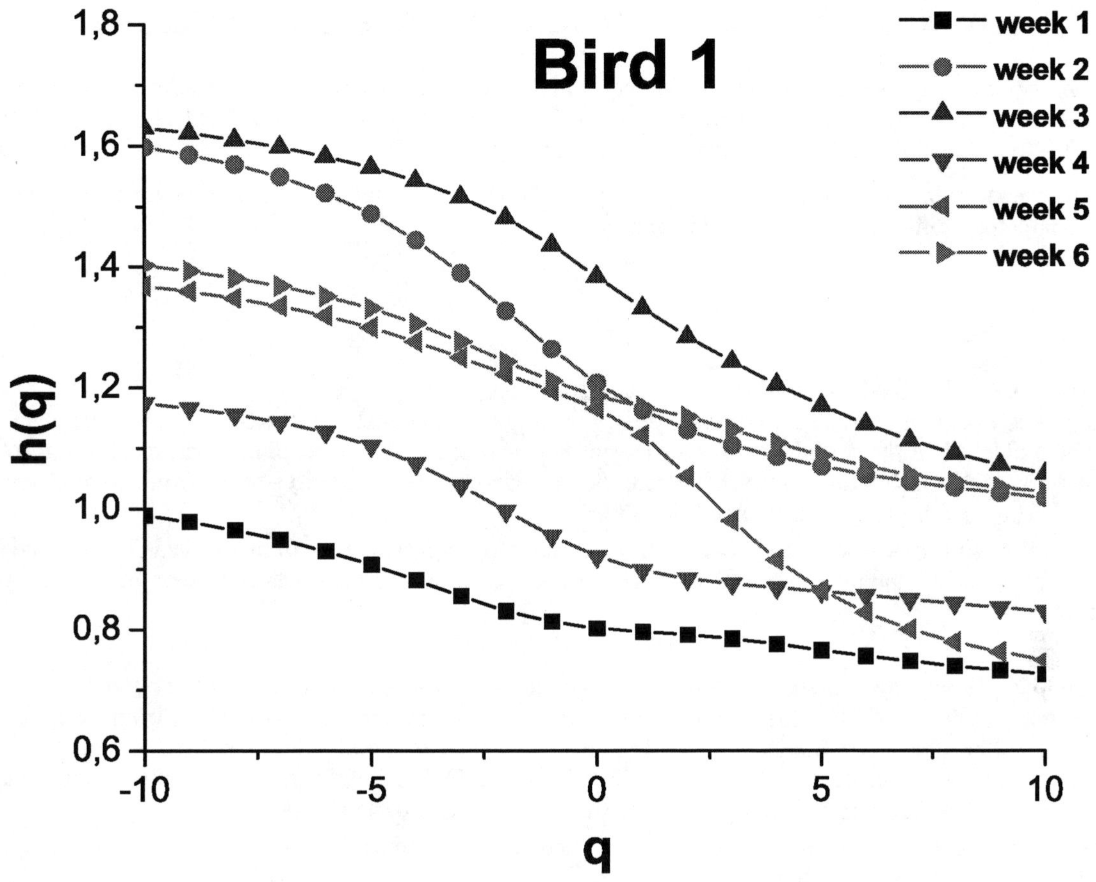

FIGURE 1. $H(q)$ corresponding to *Bird #1* at frontal left electrode for the 6 weeks' posthatch.

128 Hz, passed through amplifiers, and stored directly on computer. The total length of each record is 16368 $data$. Data acquisition was controlled by Strawberry Tree software. Four electrodes were placed over left and right frontal (LF and RF), and left and right posterior (LP and RP) areas of the scalp with an additional reference electrode placed at the back of the head. EEG recording were taken at Day 7 posthatch and then each week for during 6 weeks. For additional details of the acquisition protocol see [1]. The recorded EEG signals are nonstationary and it also present artifacts due to saccadic eye movements [1]. In order to avoid these problems and facilitate a subsequent quantitative analysis, each signal was pre-processing using a methodology based on Wavelets (for procedure details see [4]) and summarized as follow: *a)* The Discrete Orthogonal Wavelet Transform of the signal is obtained considering $J_{max} = -10$ wavelet resolution levels with spline cubic mother wavelet. After that a cleaning and stationary signal is obtained by reconstruction (inverse wavelet transform) using the resolution wavelet levels corresponding to the frequency range $0.5 - 32.0$ Hz. *b)* The wavelet frequency bands at witch the saccadic movement frequency appear, as well as, their time localization were identified. The corresponding wavelet coefficients were reduced in order that they present a contribution below the noise-signal level, and the corresponding frequency band is reconstructed. Finally the complete signal is obtained by superposition of the all wavelet reconstructed frequency bands (see Figs. 1 and 2 of [5]).

THE MULTIFRACTAL DETRENDED FLUCTUATION ANALYSIS

Stanley and co-workers developed a method for the multifractal characterization of nonstationary time series [2], which is based on a generalization of the detrended fluctuation analysis (DFA). Following Ref. [2] the multifractal DFA (MFDFA) and how the scaling exponents are determined via the MFDFA method can be summarized as follow:

- Starting with a correlated time series (signal) $\{u(i), i = 1, \cdots, N\}$ (where N is the length of the series) the corresponding profile is determined by integration

$$Y(k) = \sum_{i=1}^{k} [\, u(i) - <u> \,] \tag{1}$$

  where $<u> = \frac{1}{N}\sum_{i=1}^{N} u(i)$ is the mean.
- The profile $Y(k)$ is cut into $N_s \equiv [N/s]$ nonoverlapping segments of equal length s. Since the record length N need not be a multiple of the considered time scale s, a short part at the end of the profile will remain in most cases. In order not to disregard this part of the record, the same procedure is repeated starting from the other end of the record. Thus, $2N_s$ segments are obtained altogether.
- The local trend for each segment $\nu = 1, \cdots, 2N_s$ is evaluated by least-square fit of the data. The detrended time series for segment duration s, denoted by $Y_s(i)$, is calculated as the difference between the original time series and the fits,

$$Y_s(i) = Y(i) - p_\nu(i)\,, \tag{2}$$

  where $p_\nu(i)$ is the fitting polynomial in the ν-th segment. Since we use a polynomial fit of order l, we denote the algorithm as MFDFA-l. The detrending of the time series is done by subtraction of the fits from the profile, these methods differ in their capability of eliminating trends in the data. In m-th order of MFDFA, trends of order m in the profile and $m-1$ in the original record are eliminated. Thus, a comparison of the results for different orders of MFDFA allows us to estimate the strength of the trends in the time series.
- For each of the $2N_s$ segments the variance of the detrended time series $Y_s(i)$ is evaluated by averaging over all data point i in the ν-th segment

$$F_s^2(\nu) = \frac{1}{s}\sum_{i=1}^{s} \{Y_s[(\nu-1)s+i]\}^2\,. \tag{3}$$

- Average over all segments to obtain the q-th fluctuation function

$$F_q(s) = \left\{\frac{1}{2N_s}\sum_{\nu=1}^{2N_s} [F_s^2(\nu)]^{q/2}\right\}^{1/q}, \tag{4}$$

  where, in general, the index q can take any real value. For $q = 2$, the standard DFA procedure is retrieved.
- The scaling behavior of the fluctuation is determined by analyzing log-log plots $F_q(s)$ versus s for each value of q. If the series $u(i)$ are long-range power-low correlated $F_q(s)$ increase, for large values of s, as a power-law

$$F_q(s) \sim s^{h(q)}\,. \tag{5}$$

  For $q = 0$ the value $h(0)$ cannot be determined directly using the average procedure in Eq. (4) because of the diverging exponent. Instead, using the L'Hôpital rule, a logarithmic average procedure has to be employed,

$$F_0(s) = \exp\left\{\frac{1}{4N_s}\sum_{\nu=1}^{2N_s} \ln[F_s^2(\nu)]\right\} \sim s^{h(0)}\,. \tag{6}$$

For monofractal time series with compact support, $h(q)$ is independent of q, since the scaling behavior of the variance $F_s^2(\nu)$ is identical for all segments ν, and the averaging procedure in Eq. (4) will give just this identical scaling behavior for all values of q. Only if small and large fluctuations scale differently, there will be a significant dependence of $h(q)$ on q: If we consider positive values of q, the segments ν with large variance $F_s^2(\nu)$ will dominate the average $F_q(s)$. Thus, for positive values of q, $h(q)$ describes the scaling behavior of the segments with large fluctuations. On the contrary, for negative values of q, the segments ν with small variance $F_s^2(\nu)$ will dominate the average $F_q(s)$. Hence,

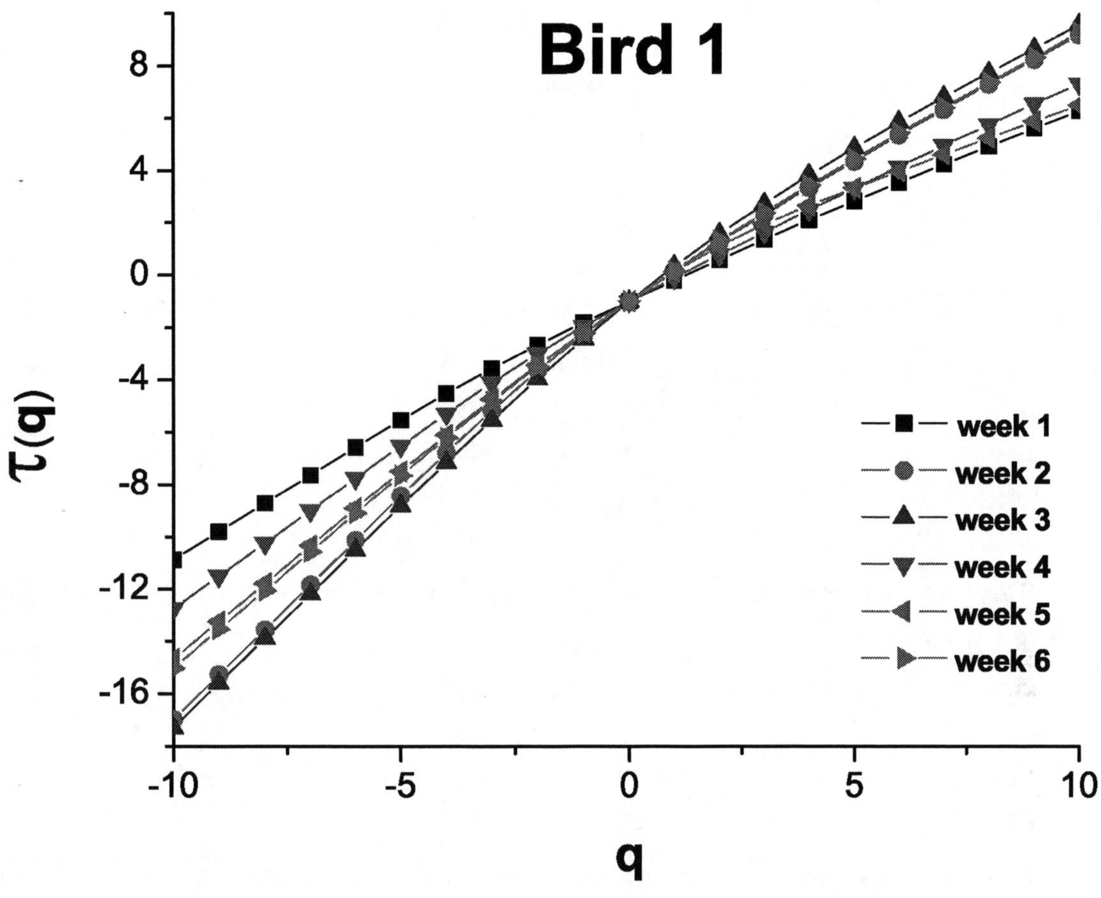

FIGURE 2. $\tau(q)$ corresponding to *Bird #1* at frontal left electrode for the 6 weeks' posthatch.

for negative values of q, $h(q)$ describe the scaling behavior of the segments with small fluctuations. Following Eqs. (4) and (6) and assume that the length N of the series is an integer multiple of the scale s,

$$\sum_{v}^{N/s} |Y(vs) - Y((v-1)s)|^{q} \sim s^{q\,h(q)-1} \tag{7}$$

Stanley and co-workers show this multifractal formalism corresponding with the standard box counting theory, and they related both formalisms. It is obvious that the term $Y(vs) - Y((v-1)s)$ is identical to the sum of the numbers $u(i)$ within each segment v of size s. This sum is the box probability $p_s(v)$ in the standard formalism for normalized series $u(i)$. The scaling exponent $\tau(q)$ is usually defined via the partition function $Z_q(s)$,

$$Z_q(s) \equiv \sum_{v=1}^{N/s} |p_s(v)|^{q} \sim s^{\tau(q)} \tag{8}$$

where q is a real parameter as in the MFDFA. Then using the last equation and Eq. (8), they conclude that are identical and obtain the relation between the two sets of multifractal scaling exponents:

$$\tau(q) = q\,h(q) - 1 \tag{9}$$

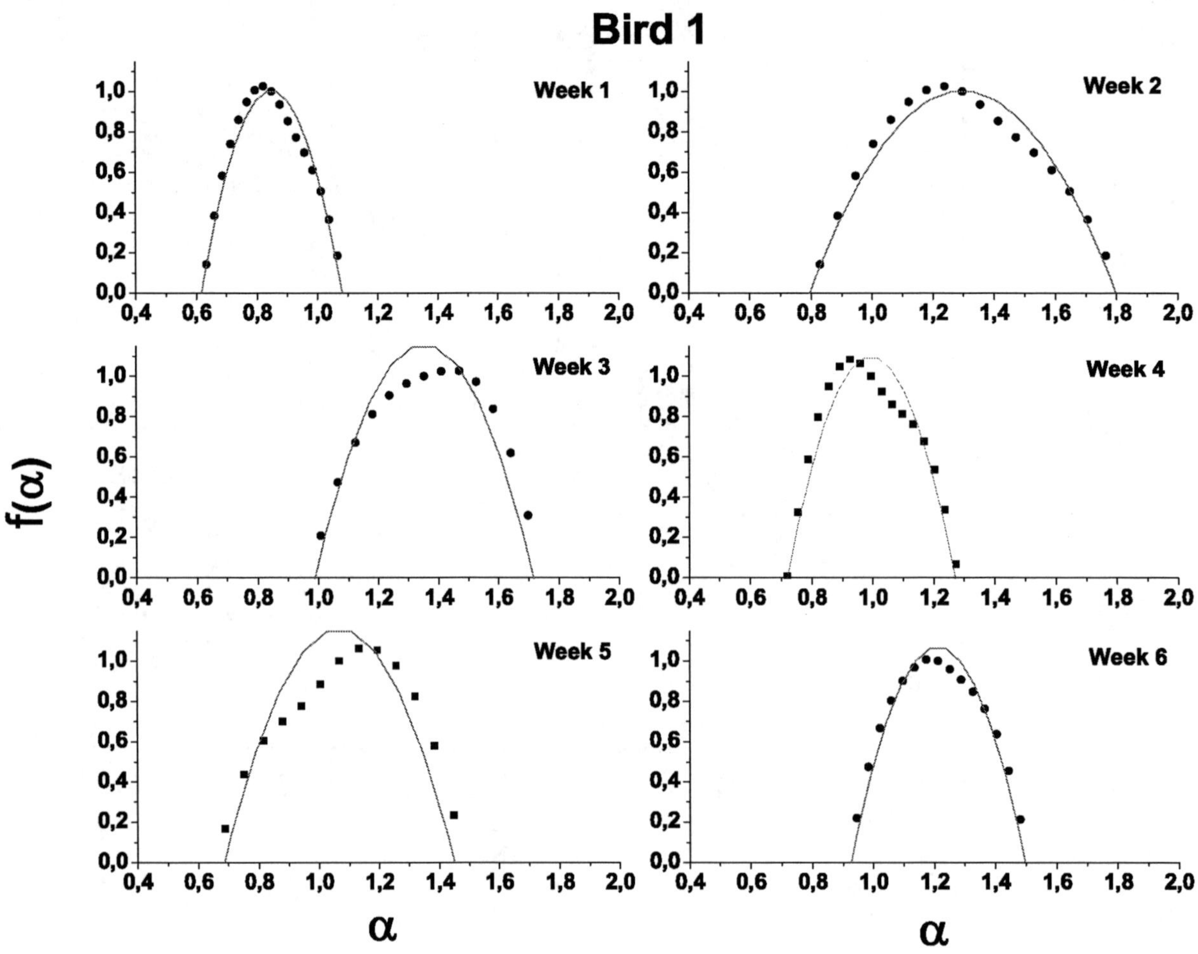

FIGURE 3. Multifractal spectrum corresponding to *Bird #1* at frontal left electrode for the 6 weeks' posthatch.

The singularity spectrum $f(\alpha)$ is another way to characterize the multifractality of the series. The parameter α is the Hölder exponent or the singularity intensity. The $f(\alpha)$ spectrum is related with $\tau(q)$ via a Legendre transformation [8]:

$$\alpha = \tau'(q) , \qquad f(\alpha) = q\,\alpha - \tau(q) . \tag{10}$$

Using Eq. (9) they related directly α and $f(\alpha)$ to $h(q)$:

$$\alpha = h(q) + q\,h'(q) , \qquad f(\alpha) = q\,[\alpha - h(q)] + 1 . \tag{11}$$

RESULTS AND DISCUSSION

Synapse formation in the chicken brain occurs most rapidly around the time of hatching and is complete by 10 to 14 days posthatch [6]. Subsequently, the immature synapses and neurons gradually attain adult ultrastructural and biochemical properties. These changes occur in the period between 3 and 8 weeks posthatch in what has been termed the "maturation period" by Rostas, [7] during which neuronal circuits become fine-tuned.

The EEG time series studied in the current work required pre-processing in order to eliminate undesired frequencies produced by a repetitive motion (saccadic movements) typical of the studied birds. This preprocessing is not straightforward (it can not be done with band pass filters) and was the object of a previous paper [4].

Starting from the cleaning EEG signals we calculate the fluctuation function $F(s)$ according with the equation Eq. (4) as function of the parameter q, varying between -10 and 10, for each bird, week and channel. Then, sketching the function $F_q(s)$ versus s in a graph with logarithmic scales, the functions $h(q)$ and $\tau(q)$ were obtained following Eqs. (6) and (8). Figure 1 shows the variation of $h(q)$ for Bird #1 at the electrode LF for the six week posthatch recording. In the Fig. 2 the corresponding $\tau(q)$ functions are shown. Applying the Legendre relationships we obtain the multifractal spectrum $f(\alpha)$ as function of the singularities α. This task was carried out on all the series, with the objective of comparing the multifractal spectrum for the different weeks. Figure 3 shows the evolution of the $f(\alpha)$ for the six weeks for the electrode LF corresponding to Bird # 1. A clear difference in the position and shape of singularity spectrum is observed across the weeks.

Complex behavior of the signal is associated with the distribution of the singularities, α. That is, points where the signal is continuous but not differentiable. In this case, Hölder exponents [9] describe the degree of regularity. If the measure of the set of points having the same regularity exponents is not an integer, it infers that the signal is fractal. For the multifractal case, we have a non-integer measure for different exponents α. If the spectrum distribution $f(\alpha)$ is centered around $\alpha = 1$ or for $\alpha > 1$ ($\alpha < 1$) the system is differentiable (non-differentiable) in the Hölder sense [9]. Then from Fig. 3 we can observe that the spectrum presents non-differentiable characteristics mainly in week 1. From then the pattern starts to shift to the right side, and start to be more differentiable from weeks 3 to 6. This behavior is associated with the "maturation period" described by Rostas a co-workers. Is interesting also to note that a wider spectrum appears at week 2 which can be associated with complete synapse formation. Another aspect that we can associated with the maturation period (weeks 3 to 6) is the asymmetry observed in the spectrum at weeks 3 to 5. Note that it almost disappear at week 6, when the neurological maturation of the bird is considered complete. We observed similar results with different birds and electrodes. The fractal characteristic does not present substantial changes at different electrodes. From the presents results we conclude that the MFDFA method can be used for the quantitative analysis of brain maturation change.

ACKNOWLEDGMENTS

This work was supported by The University of Newcastle through a research visitor grant to OAR and by Hunter Medical Research Institute and the NSW Department of Health. Also supported by Consejo Nacional de Investigaciones Científicas y Técnicas (CONICET) (PIP 5687/05) and Agencia Nacional de Promoción Cientifica y Tecnológica (PICT 20648/04), Argentina.

REFERENCES

1. M. Hunter, M. Battilana, T. Bragg, J. A. P. Rostas, *Developmental Psychobiology* **36**, 23–28 (2000).
2. J. W. Kantelhardt, S. A. Zschiegner, E. Koscielny-Bunde, S. Havlin, A. Bunde and H. E. Stanley, *Physica A* **316**, 87–114 (2002).
3. C. K. Peng, S. V. Buldyrev, A. L. Goldberger, S. Havlin, R. N. Mantegna, M. Simons, H. E. Stanley, *Physica A* **221** 180–192 (1995).
4. A. Figliola, O. A. Rosso and E. Serrano, "Atenuación de frecuencias indeseadas usando transformada wavelet," in *Proceedings XII RPIC*, edited by Grupo de Electrónica Aplicada, Guillermo García Electrónica Edition, Río Cuarto, 2005, pp. 28–32.
5. J. G. Fernández, H. A. Larrondo, A. Figliola, E. Serrano, J. P. Rostas, M. Hunter and O. A. Rosso, "Brain maturation changes characterized by algorithmic complexity (Lempel and Zip complexity)," in present proceedings volume, edited by O. Descalzi, H. A. Larrondo and O. A. Rosso, AIP Conference Proceedings, American Institute of Physics, New York, 2007.
6. J. A. P. Rostas, J. M. Kavanagh, P. R. Dodd, J. W. Heath and D. A. Powis, *Molecular Neurobiology* **5**, 203–215 (1992).
7. J. A. P. Rostas, "Molecular mechanisms of neuronal maturation: A model for synaptic plasticity," in *Neural and behavioural plasticity: The use of the domestic chicken as a model*, edited by R. J. Andrew, Oxford University Press, London, 1991, pp. 177–211.
8. K. Falconer, *Techniques in fractal geometry*, John Wiley and Sons Ltd, New York, 1997.
9. S. Mallat, *A wavelet tour of signal processing*, 2nd Edition, American Press, San Diego, 1999.

Brain Maturation Changes Characterized by Algorithmic Complexity (Lempel and Ziv Complexity)

J. G. Fernández*, H. A. Larrondo*, A. Figliola†, E. Serrano**, J. A. P. Rostas‡, M. Hunter§ and O. A. Rosso¶

*Facultad de Ingeniería, Universidad de Mar del Plata (UNMDP).
Juan B. Justo 4302. (7600) Mar del Plata, Argentina
†Instituto del Desarrollo Humano, Universidad Nacional de General Sarmiento (UNGS).
Campus Universitario J. María Gutiérrez 1150. Los Polvorines, Pcia. de Buenos Aires. Argentina
**Escuela de Ciencia y Tecnología, Universidad Nacional de General San Martín (UNGSM).
Campus Migueletes, Edificio Tornavías, Av. 25 de Mayo y Martín de Irigoyen.
San Martín, Pcia. de Buenos Aires, Argentina.
‡School of Biomedical Sciences and Hunter Medical Research Institute, University of Newcastle.
University Drive, Callaghan, NSW 2308, Australia.
§School of Behavioral Sciences and Hunter Medical Research Institute, University of Newcastle.
University Drive, Callaghan, NSW 2308, Australia.
¶Instituto de Cálculo, Facultad de Ciencias Exactas y Naturales, Universidad de Buenos Aires (UBA).
Pabellon II, 1428 Ciudad Universitaria, Buenos Aires, Argentina.

Abstract. Recent experimental results suggest that basal electroencephalogram (EEG)changes reflect the widespread functional evolution in neuronal circuits, occurring in chicken brain during the "synapse maturation" period, between 3 and 8 weeks' posthatch. In present work a quantitative analysis based on the Algorithmic Complexity (Lempel and Ziv Complexity) is performed. It is shown that this complexity presents a peak at week 2 posthatch 2, and a tendency to stabilize its values after the week 5 posthatch.

Keywords: EEG; brain maturation; Lempel-Ziv complexity
PACS: 05.45.Df, 87.19.La, 89.75.Fb

INTRODUCTION

Literature on the use of the EEG in the chicken is not extensive. Research efforts have largely concerned the use of the EEG either as an indicator of the general integrity of the nervous system or as a measure of specific brain states, such as sleep cycles and other EEG rhythm defined states. In those studies using the EEG as a functional integrity measure of the chicken nervous system, a major interest has been to plot the embryonic development of the chicken brain. Synapse formation in the chicken brain occurs most rapidly around the time of hatching and it is complete by 10 to 14 days posthatch [1]. Subsequently, the immature synapses and neurons gradually attain adult ultrastructural and biochemical properties. These changes occur in the period between 3 and 8 weeks posthatch, during which neuronal circuits become fine-tuned, and has been termed the "maturation period" by Rostas, [2].

It may be assumed that these synaptic connections are maturing in response to both internal and environmental stimulation. In a previous work scalp-applied recording electrodes were used to monitor changes in chicken's basal EEG patterns during posthatch development [3]. Frequency spectra produced by Fast Fourier Transform (FFT) show biphasic morphology in all chickens with peaks at about 7 Hz and 26 Hz. Changes in the lower frequency band show progressive development and provide a possible index of brain development. Both amplitude and dominant spectral frequency decrease between weeks 3 and 8 posthatch, reaching adult levels between weeks 5 and 7. The results suggest that modifications of basal EEG reflect the widespread functional changes in neuronal circuits occurring in chicken during the "synapse maturation" period, between 3 and 8 weeks' posthatch. In the present work a quantitative analysis using the Lempel and Ziv Complexity [4, 5] of chicken EEG basal activity is presented. In particular, it shows that the proposed complexity quantifier presents a peak at week 2 and also a tendency to stabilize its value after week 5, in agreement with the previous description of synapse maturation.

CP913, *Nonequilibrium Statistical Mechanics and Nonlinear Physics, XV Conference*
edited by O. Descalzi, O. A. Rosso, and H. A. Larrondo

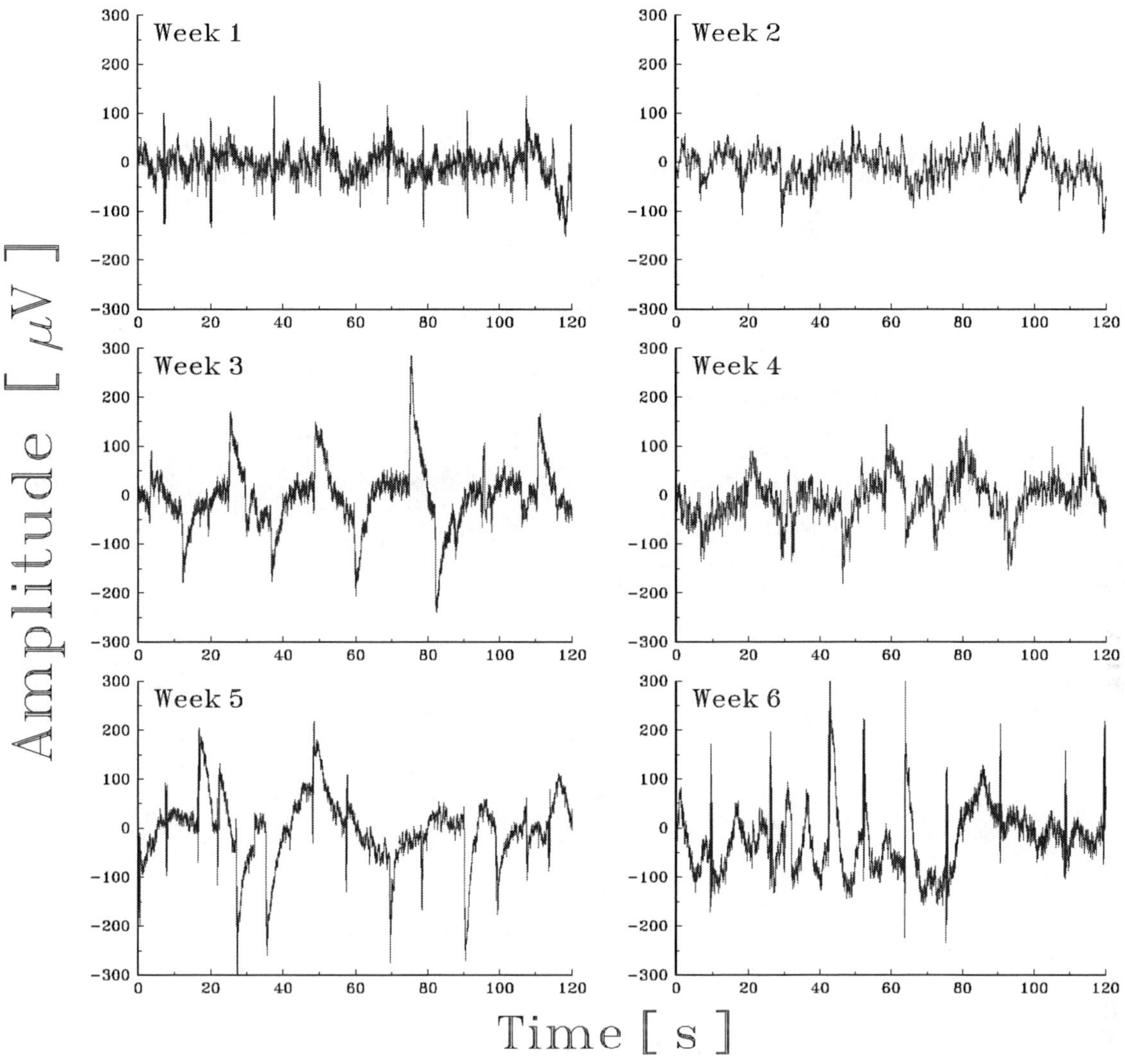

FIGURE 1. Original EEG time series at left frontal electrode (LF) corresponding to *Bird #1* for the 6 weeks' posthatch.

EXPERIMENTAL DATA AND ARTIFACT REDUCTION

Twenty four chickens (*Gallus domesticus*) reared from hatching were the subjects in this experiment. All birds had free access to food and water throughout the experiment. The birds were reared initially in incubation boxes and then transferred to holding cages maintained at a constant temperature ($21°C$) with a $12:12$ *hr* light:dark cycle.

Continuous EEG recordings ($0.1 - 100$ Hz with a 50 Hz notch filter) were made using small (6 *mm*) gold cup electrodes attached to the scalp with collodion glue and filled with electrode gel. The signal was sampled at a rate of 128 Hz, passed through amplifiers, and stored directly on computer. The total length of each record is 16368 *data*. Data acquisition was controlled by Strawberry Tree software. Four electrodes were placed over left and right frontal (LF and RF), and left and right posterior (LP and RP) areas of the scalp with an additional reference electrode placed at the back of the head. EEG recordings where taken at Day 7 posthatch and then each week for 6 weeks. For additional details of the acquisition protocol see [3].

In animals and also in human developmental changes are correlated with the presence of different rhythms. The raw EEG provides evidence of developmental change because there is a greater variance shown in the typical 1-week-posthatch recording (see Fig. 1.a) compare with the typical 6-week-posthatch recording (see Fig. 1.f) From Fig. 1 it

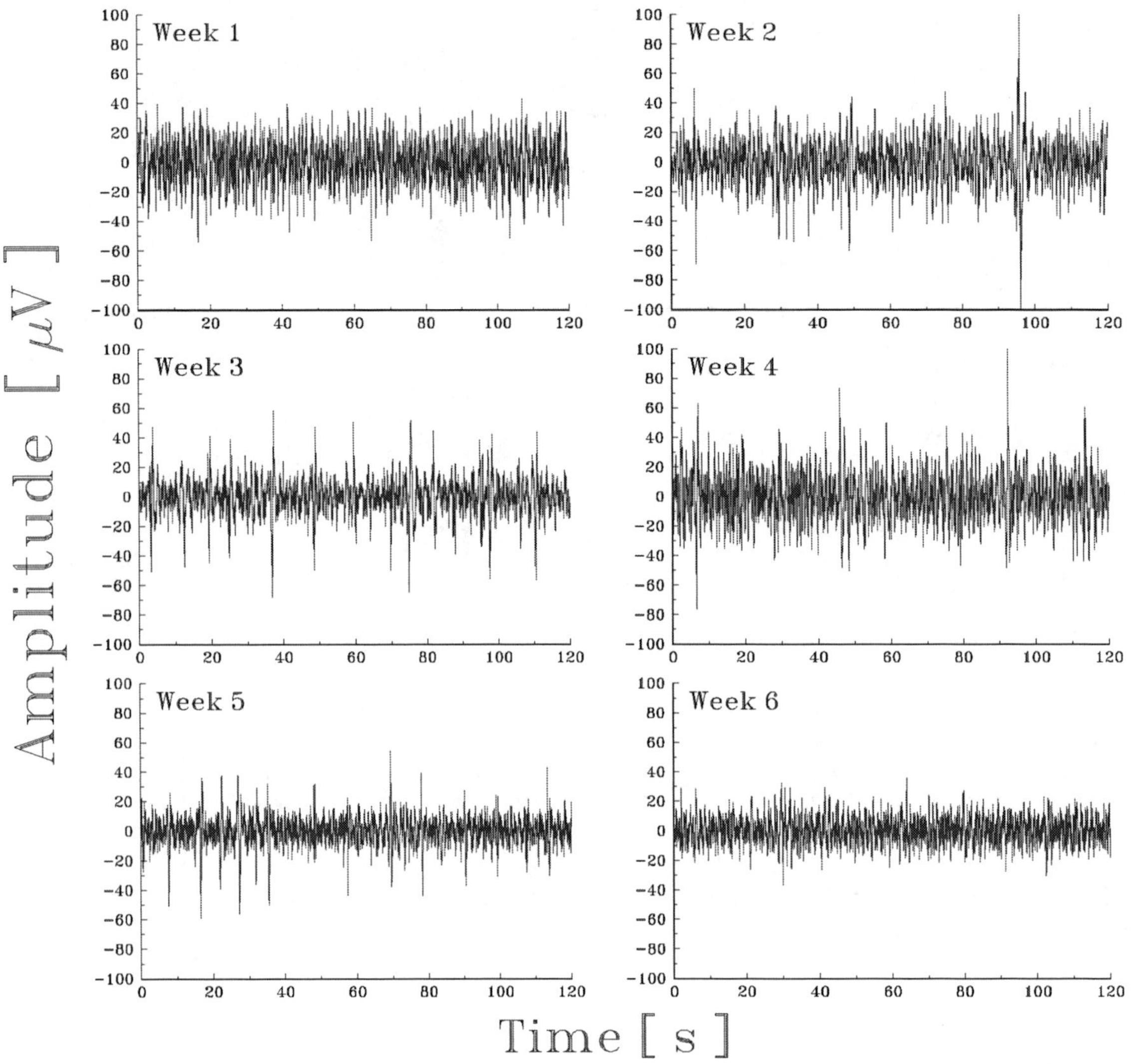

FIGURE 2. Cleaning EEG time series at left frontal electrode (LF) corresponding to *Bird #8* for the 6 weeks' posthatch.

is clear that these EEG signals are nonstationary. They also present artifacts due to saccadic eye movements [3]. In order to avoid these problems and improve a subsequent quantitative analysis, each signal was pre-processed using a methodology based on Wavelets (for procedure details see [6]).

The procedure consisted in two steps:

- The Discrete Orthogonal Wavelet Transform of the signal was obtained considering $J_{max} = -10$ wavelet resolution levels with spline cubic mother wavelet. After that a cleaning and stationary signal was obtained by reconstruction (inverse wavelet transform) using the resolution wavelet levels corresponding to the frequency range $0.5 - 32.0$ Hz (see Fig. 3.b).
- Both the wavelet frequency bands at which the saccadic movement frequencies appears, and their time localization were identified. The corresponding wavelet coefficients were reduced in order that their contribution were below the noise-signal level. Finally the corresponding frequency band was reconstructed and the complete filtered signal was obtained by superposition of the all wavelet reconstructed frequency bands (see Fig. 3.c).

As an example, in Figs. 1 and 2, the raw and clean EEG signals corresponding to the *Bird #8* are shown.

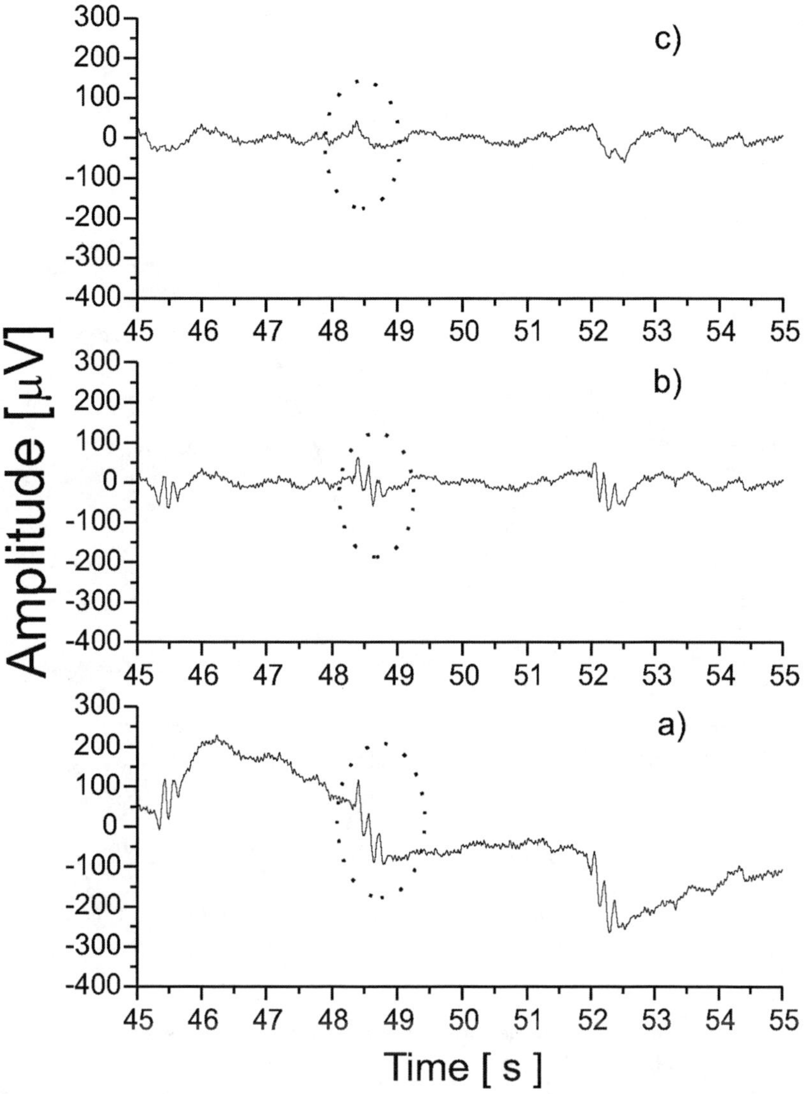

FIGURE 3. Cleaning EEG time series process: *a)* Original time series; *b)* Stationary wavelet reconstructed time series for the frequency range $0.5 - 32.0\ Hz$; *c)* Cleaning EEG time series, without saccadic artifacts. The saccadic artifact are marked with dots line.

LEMPEL AND ZIV COMPLEXITY

The Algorithmic Complexity has shown its ability to distinguish evolutionary characteristics in many fields. For a string of characters it is defined as the length in bits of the smallest program that produces the string as output [7]. The problem with this definition is that it is impossible, even in principle, to find such a program. Nevertheless, the zippers or file compressors are algorithms conceived to do that job at least approximately. The Lempel and Ziv algorithm is used for most zippers and it is one of the best known file compressors [4, 5].

Let l^U_{ijk} be the string length of the time series for bird i, in week j, electrode k (superscript U for *unzipped*). Let l^Z_{ijk} be the corresponding zipped string length (superscript Z for *zipped*). The *Zipping Complexity* for this particular time series is given by:

$$c_{ijk} = \frac{l^Z_{ijk}}{l^U_{ijk}} \,. \tag{1}$$

In the case of Zipping Complexity it is also necessary to correct the effect of sampling rate and epoch length on the

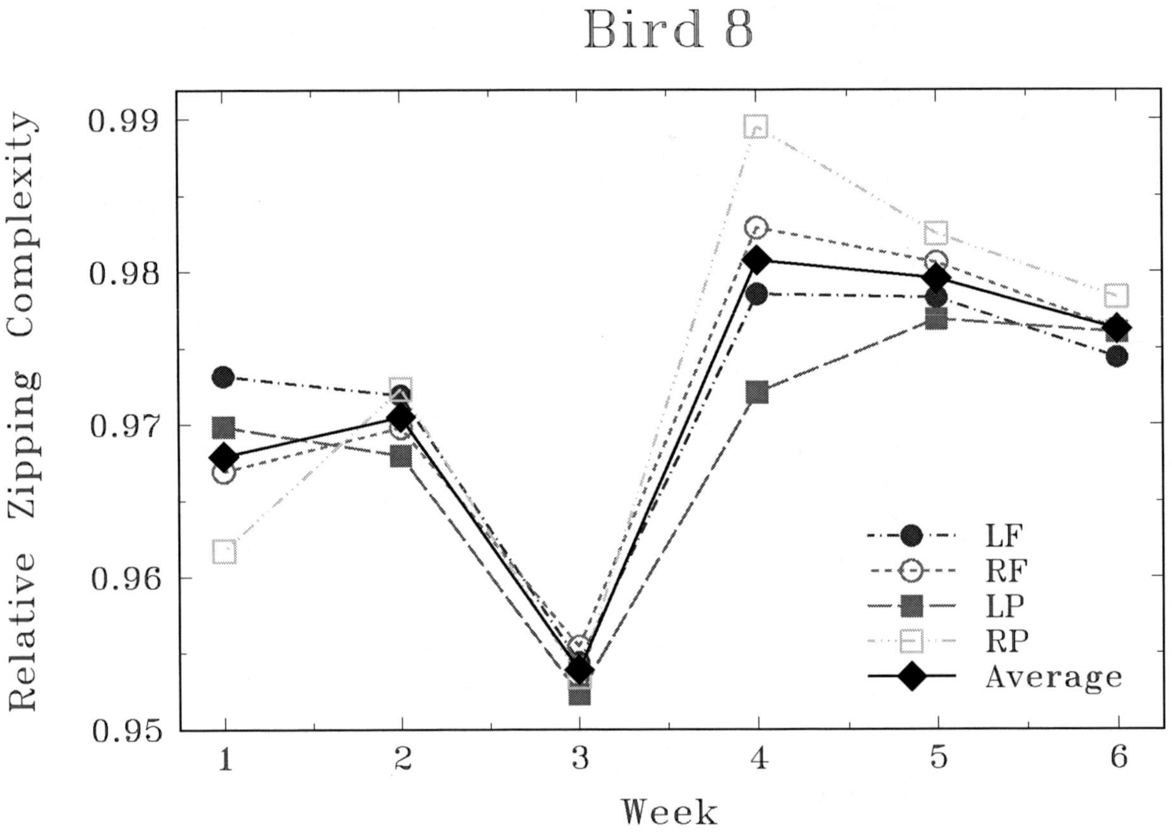

FIGURE 4. Relative Zipping Complexity for *Bird #8* at LF, RF, LP, RP electrodes and average electrode, for the 6 weeks' posthatch.

algorithmic complexity. These effects were studied by P. E. Rapp and co-workers in [8]. This drawback is solved using surrogates: an ensemble of random nondeterministic data sets with the same mean, variance and power spectrum as the experimental time series. The quantifier proposed by Rapp *at al.* is *the Relative Zipping Complexity*:

$$C_{ijk} = \frac{c_{ijk}}{\langle c^{(s)}_{ijk} \rangle}, \tag{2}$$

where $\langle c^{(s)} \rangle$ corresponds to the Zipping Complexity mean value over all considered surrogates of the original time series. We used a hundred surrogates in our calculations.

There are several methods to generate surrogate data sets with the conditions stated above. In the present work we use the algorithm proposed by Henry, Novell and Camacho [9]. Essentially the method consists on the following steps:

- Evaluate the time series discrete Fourier transform Z_{ijk};
- Add a random phase φ to obtain $Z'_{ijk} = Z_{ijk} \exp \varphi$;
- Construct the inverse Fourier transform of Z'. This is one surrogate time series.

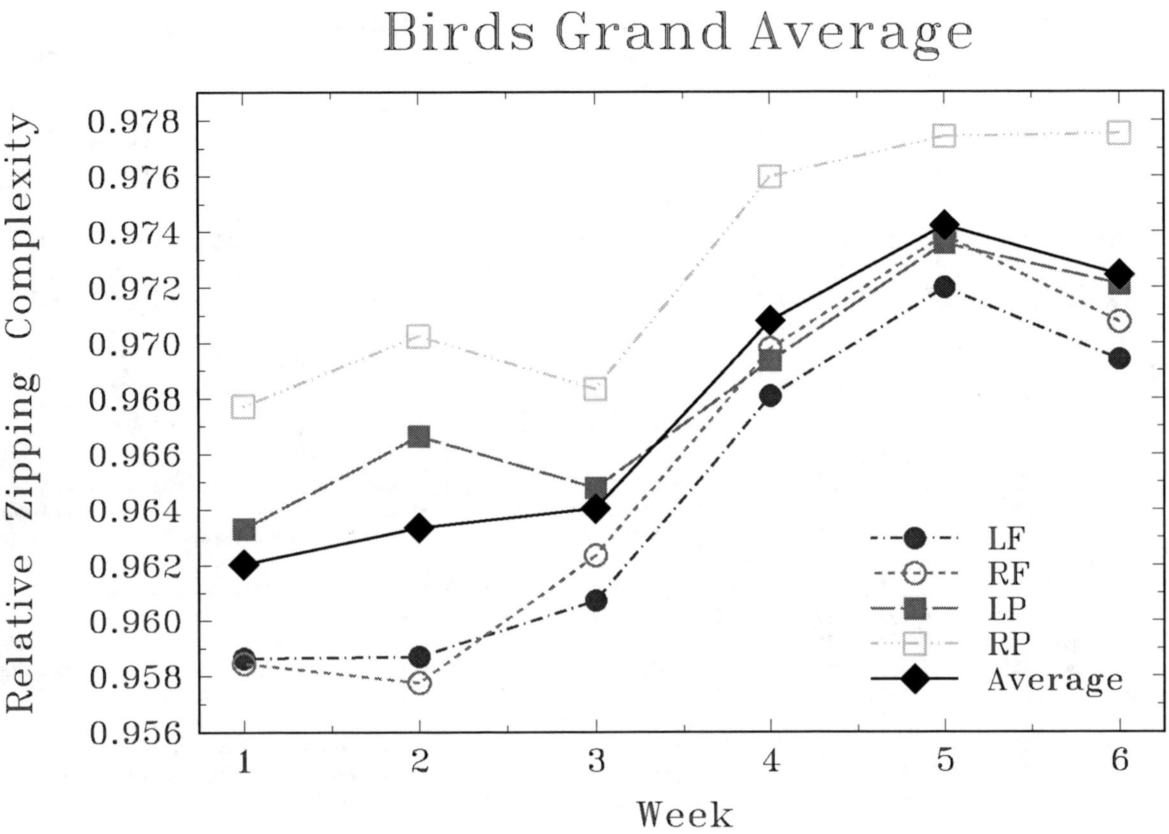

FIGURE 5. Grand Average Relative Zipping Complexity for the 24 birds at LF, RF, LP, RP electrodes and average electrode, for the 6 weeks' posthatch.

RESULTS AND DISCUSSION

As explained above, in the case of time series studied in this paper a pre-processing was required to eliminate undesired frequencies produced by a repetitive motion (saccadic movements) typical of the studied birds. This preprocessing was not straightforward (it can not be done with band pass filters) and was the object of a previous paper [6]. The cleaning EEG data sets were then characterized by means of their Relative Zipping Complexity (RZC) measure *C*.

In Figs. 4 and 5 the Relative Zipping Complexity evolution (in weeks) for the *Bird #8* and the corresponding Grand Average value over 24 *Birds* are shown for each electrode. In the same figures, average over all electrodes are also displayed.

The RZC data obtained show a distinctive pattern of change during development through the synapse formation period (up to 3 weeks' posthatch) and the synapse maturation period (from 2 to 6 weeks' posthatch). It is interesting to note that the RZC present oscillatory behavior for a given Bird, however the mean RZC increases with time during weeks 3 to 5, showing a tendency to stabilize its value in the last weeks. Note also, that this behavior is associated with the "maturation period" described by Rostas and co-workers [2]. A pick appear in the RZC at week 2 (see Fig. 5) which can be associated with the complete synapse formation [1]. The EEG analysis for each individual electrode showed that the same developmental pattern occurred at each site but greater values are observed at posterior sites. Also some differences between hemisphere are founded. Whether the latter finding reflects the bilateral asymmetry of brain function, which underlies many behaviors in chickens, remains to be established. In summary we can conclude that a good quantification of brain maturation changes can be done by the Relative Zipping Complexity and results are in good agreement with previous spectral and morphological studies [3].

ACKNOWLEDGMENTS

This work was supported by The University of Newcastle through a research visitor grant to OAR and by Hunter Medical Research Institute and the NSW Department of Health. Also supported by Consejo Nacional de Investigaciones Científicas y Técnicas (CONICET) (PIP 5687/05, 5569/05), Agencia Nacional de Promoción Cientifica y Tecnológica (PICT 20648/04, 21409/04) and Unvirsity of Mar del Plata, Argentina.

REFERENCES

1. J. A. P. Rostas, J. M. Kavanagh, P. R. Dodd, J. W. Heath and D. A. Powis, *Molecular Neurobiology* **5**, 203–215 (1992).
2. J. A. P. Rostas, "Molecular mechanisms of neuronal maturation: A model for synaptic plasticity," in *Neural and behavioural plasticity: The use of the domestic chicken as a model*, edited by R. J. Andrew, Oxford University Press, London, 1991, pp. 177–211.
3. M. Hunter, M. Battilana, T. Bragg, J. A. P. Rostas, *Developmental Psychobiology* **36**, 23–28 (2000).
4. J. Ziv and A. Lempel, *IEEE Trans. Inf. Theory* **22**, 75–81 (1976).
5. J. Ziv and A. Lempel, *IEEE Trans. Inf. Theory* **24**, 530–536 (1978).
6. A. Figliola, O. A. Rosso and E. Serrano, "Atenuación de frecuencias indeseadas usando transformada wavelet," in *Proceedings XII RPIC*, edited by Grupo de Electrónica Aplicada, Guillermo García Electrónica Edition, Río Cuarto, 2005, pp. 28–32.
7. J. R. Sanchez, F. Family, C. M. Arizmendi, *Phys. Lett. A* **249**, 281–285 (1998).
8. P. E. Rapp, C. J. Cellucci, K. E. Korslund, T. A. A. Watanabem, M. A. Jiménez-Montaño, *Phys. Rev. E* **64**, 016209 (2001).
9. B. Henry, N. Lovell, F. Camacho, "Nonlinear Dynamics Time Series Analysis," in *Nonlinear Biomedical Signal Processing*, edited by Metin Akay, IEEE Engineering in Medicine and Biology Society, 2004, pp. 1–39.

Global Interactions Analysis of Epileptic ECoG Data

Guillermo J. Ortega*,†, Rafael G. Sola,* and Jesús Pastor**

*Neurosurgery Service, Hospital Universitario de la Princesa, Madrid, Spain
†FCEyN, University of Buenos Aires, Argentine
**Neurophysiology Service, Hospital Universitario de la Princesa, Madrid, Spain

Abstract. Localization of the epileptogenic zone is an important issue in epileptology, even though there is not a unique definition of the epileptic focus. The objective of the present study is to test ultrametric analysis to uncover cortical interactions in human epileptic data. Correlation analysis has been carried out over intraoperative Electro-Corticography (ECoG) data in 2 patients suffering from temporal lobe epilepsy (TLE). Recordings were obtained using a grid of 20 electrodes (5x4) covering the lateral temporal lobe and a strip of either 4 or 8 electrodes at the mesial temporal lobe. Ultrametric analysis was performed in the averaged final correlation matrices. By using the matrix of linear correlation coefficients and the appropriate metric distance between pairs of electrodes time series, we were able to construct Minimum Spanning Trees (MST). The topological connectivity displayed by these trees gives useful and valuable information regarding physiological and pathological information in the temporal lobe of epileptic patients.

Keywords: Multivariate, ultrametricity, time series analysis
PACS: 05.45.Tp, 05.45.-a, 02.50.Sk

INTRODUCTION

Drug-resistant temporal lobe epilepsy (TLE) can be treated by tailored surgery guided by electrocorticography (ECoG). Based on this and other pre-surgical data, the anterior lateral temporal cortex, portions of the amygdala and the hippocampus are commonly excised [1]. The goal of surgery is to achieve seizure control. However, while some patients are seizure free or show a significant reduction of seizure activity after surgery, others show no changes or even worsen [2]. It is therefore important to improve our understanding of the lateral and mesial dynamics in the temporal lobe of epileptic patients.

Even though several regions has been identified as involved in the epileptic pathology [3], no single area is solely responsible of the clinical seizures. On the contrary, an interplay among different areas seems to play a central role in triggering the seizure onset. Moreover, the relation between interictal (in between crisis) and ictal (during the crisis) areas are still debated [4]. It is therefore important to develop novel approaches to uncover underlying cortical dynamics in the epileptic patients. A better understanding of the functional connectivity in the epileptic cortex would provides new insights in the clinical and surgical treatment of the epileptic patient.

In the present work, we apply a multivariate graphical method, such as the Minimum Spanning Tree (MST) construction, in order to characterize global interactions in the temporal lobe of epileptic patients. As we shall show, MST seems to correctly display lateral-lateral and lateral-mesial interactions. Because the intrinsic relation between the MST construction and ultrametric spaces [5], we shall introduce in the methodology section the link among both topics. It seems that the connection topology resulting from our present work could it be relevant from the point of view of the surgical procedures carried out in the patients.

DATA AND METHODOLOGY

Data sets

We shall apply MST methodology to two drug-resistant TLE patients, who underwent surgery at the Epilepsy Unit of the Hospital La Princesa. Informed consent and approval by the local ethic committee were obtained. Patients were evaluated intraoperatively with 4x5 subdural electrode grids and a strip of either 4 or 8 electrodes, with an interelectrode distance of 1cm. The strip is inserted in the inner part of the temporal lobe, also known as the mesial area. In panel A of Figure 1 a rough sketch of the whole brain is depicted, showing the approximate location of the

CP913, *Nonequilibrium Statistical Mechanics and Nonlinear Physics, XV Conference*
edited by O. Descalzi, O. A. Rosso, and H. A. Larrondo

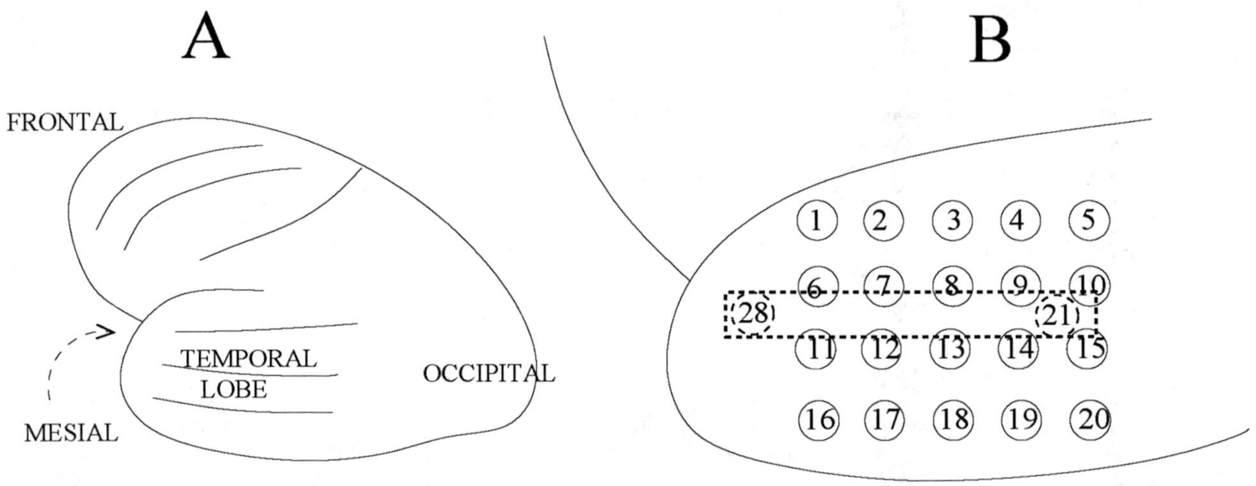

FIGURE 1. A: Sketch of the brain (lateral view), with the frontal, occipital and temporal lobes depicted. B: Zoom of the temporal lobe with the electrode grid (5x4) over the lateral temporal cortex and the electrode strip (dotted line) inserted into the inner i.e. mesial, area of the temporal lobe. Only the first and last strip electrodes are shown for the sake of clarity

temporal lobe. In panel B, position of the electrodes grid in the lateral (outer) side of the temporal lobe, and the electrode grid in the mesial (inner) side of the temporal lobe is also shown.

Intraoperative ECoG was recorded during 15-20 min using a 32-channel amplifier (Easy EEG II, Cadwell, USA). Epochs lasting from 3 to 5 minutes were selected free of artefacts by visual inspection. By using consecutive, non-overlapping windows of 1024 data points (5.12 seconds at 200 Hz sampling rate) we obtain records of 16384 data points in each of the 24 or 28 channels. This time window selection allows encompassing most of the nonstationarities included in the time series. All analysis programs were written in Fortran and the R package has been used in the circular MST plot generation.

Methodology

There exist an intrinsic relation between clustering methods such as the MST and ultrametricity [5]. In fact, distance on a MST between two sites are actually ultrametric distances. Thus we shall introduce in this section some mathematical definitions and constructions used lately in the paper, relating ultrametric distances and MST.

In order to quantify the degree of similarity between pairs of electrodes' time series, we have calculated the Pearson correlation coefficient [6] between electrodes e_i and e_j

$$\rho_{i,j} = \frac{\sum_{k=1}^{N}(e_i(k) - <e_i>)(e_j(k) - <e_j>)}{\sqrt{\sum_{k=1}^{N}(e_i(k) - <e_i>)^2}\sqrt{\sum_{k=1}^{N}(e_j(k) - <e_j>)^2}} \tag{1}$$

where $<e_i>$ is the mean value of channel e_i in the period considered. Because $\rho_{i,j}$ is a measure of similarity, and a measure of "distance" is actually needed in order to construct an ultrametirc space [5], following Gower [7], we define the distance between the time evolution of channels e_i and e_j as,

$$d(i,j) = \sqrt{\rho_{i,i} + \rho_{j,j} - 2\rho_{i,j}} = \sqrt{2(1-\rho_{i,j})} \tag{2}$$

The last equality came from the simetry property of the correlation matrix, $\rho_{i,j} = \rho_{j,i}$ and the normalization $\rho_{i,i} = 1$ $\forall i$. In this way, $d_{i,j}$ fullfils the three axioms of a distance:

- $d(i,j) = 0$ if and only if $i = j$
- $d(i,j) = d(j,i)$
- $d(i,j) \leq d(i,l) + d(l,j)$

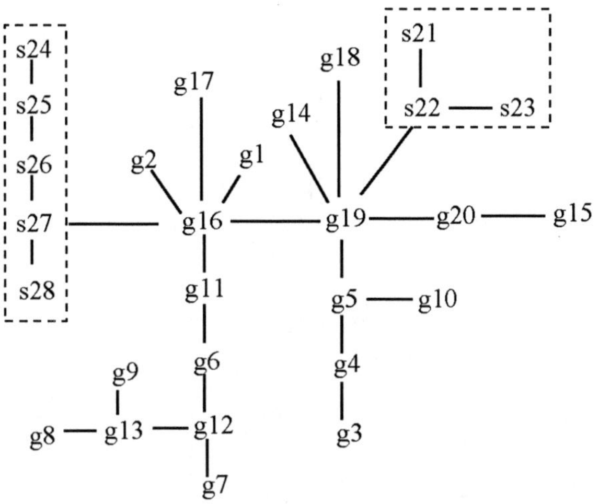

FIGURE 2. MST of patient # 1. Letter g stands for grid and letter s stands for strip

The third axiom, the triangular inequality, characterize a metric space. An *ultrametric space*, on the other hand, is endowed with a distance that obeys a stronger inequality, the *ultrametric distance* $d(i,j)^<$:

$$d(i,j)^< \leq Max\{d(i,l), d(l,j)\} \tag{3}$$

Thus, it follows that the distance matrix given by Equation (2) satisfies ultrametricity.

One method to obtain $d(i,j)^<$ directly from the distance matrix $d(i,j)$ is through the MST method [5]. Given the metric space (Ω, d), in our case, channels time series as the elements of Ω and the distance defined by Equation (2), there is associated with this space a nondirected graph with the same elements of Ω as vertices, and edges between the adjacent elements (i,j), the distance $d(i,j)$. The MST is a tree with the same vertices as in Ω but of minimal total length. When there is a MST on Ω, the distance $d(i,j)^<$ between two elements of *i* and *j* in Ω is given by,

$$d(i,j)^< = Max\{d(w_i, w_{i+1}), 1 \leq i \leq n-1\} \tag{4}$$

where $C_{i,j} = \{(w_1,w_2),(w_2,w_3),...,(w_{n-1},w_n)\}$ denotes the unique path in the MST between *i* and *j* $(w_1 = i, w_n = j)$.

In this way, given a topological tree such as the MST, by using as elements of Ω the ECoG channels' time series, distances between any two such time series turns out to be ultrametric. It remains however to know if the topology spanned by the used time series it is in fact ultrametric. As shown by Murtagh [9], epileptic EEG time series can be very well embedded in an ultrametric topology. Thus, we shall apply confidently the MST construction for the ECoG time series in order to explore the topological relations between the cortical sites in epileptic patients.

RESULTS

Global Interactions

As an example of the present methodology, we have used ECoG data from two patients. In one case (patient # 1) a grid of 20 electrodes is located in the lateral area and a strip of 8 electrodes is located in the mesial area, both in the temporal lobe. In the other case (patient # 2), the mesial strip of 8 electrodes is replaced by a 4 electrodes strip. We shall analyze therefore a multivariate register of 28 simultaneous time series in patient # 1 and 24 simultaneous time series in the patient # 2. From now on, we shall identify cortex location with the electrode position at that site. For instance, in talking about the position 17 (or g17) we are actually referring to the position covered by the electrode 17.

Correlation matrices has been calculated through equation (1) for each pair of electrodes e_i and e_j, in temporal windows of 1024 data points each (5.2 seconds). A final correlation matrix is obtained by averaging 16 consecutive, non-overlapping windows. Distance matrices in both patients were obtained through Equation 2 and the Minimum Spanning Trees is obtained in each case by using the Kruskal algorithm [8]. Figure 2 shows the MST for patient # 1 and Figure 3 shows the MST for patient # 2.

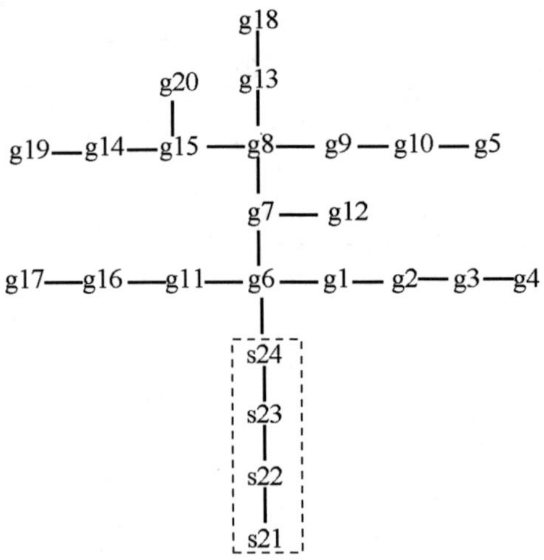

FIGURE 3. MST of patient # 2. Letter g stands for grid and letter s stands for strip

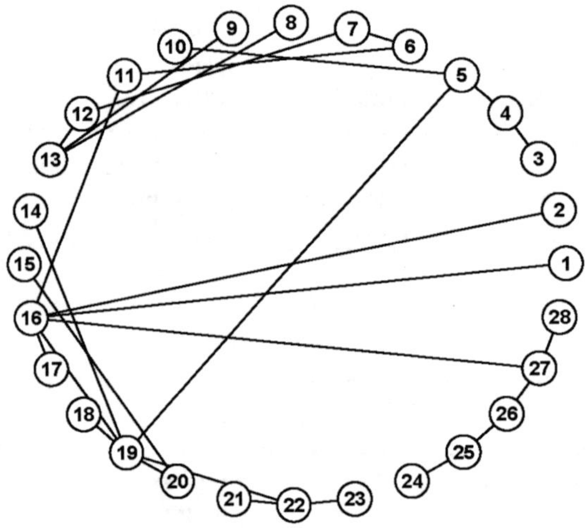

FIGURE 4. Circular MST of patient # 1

MST for patient # 1 shows (Figure 2) the interaction connectivity. Channels g1 to g20 corresponds to the grid and s1 to s8 to the mesial strip. Particular locations seems to play central roles in the interactions dynamics, mainly because the higher number of its connections, as for example sites g16, g19 and g12. Location around g12 is a "local" cluster of interactions. However, locations around g16 and g19 involves more than local interactions. In the case of g16, there exist a clear link between this site in the lateral cortex and those of the inner part of the lobe, in the mesial area, through the location s27. Moreover, this lateral site not only is linked with its first neighbors, g11 and g17, but also with relative distant sites in the lateral cortex, as are sites g1 and g2. The same situation seem to happened with site g19, which it is linked to the mesial strip through the site s22 and with relative distant lateral sites as g5.

The other case, patient # 2 it is depicted in Figure 3. In this case, the mesial strip is composed of only four electrodes.

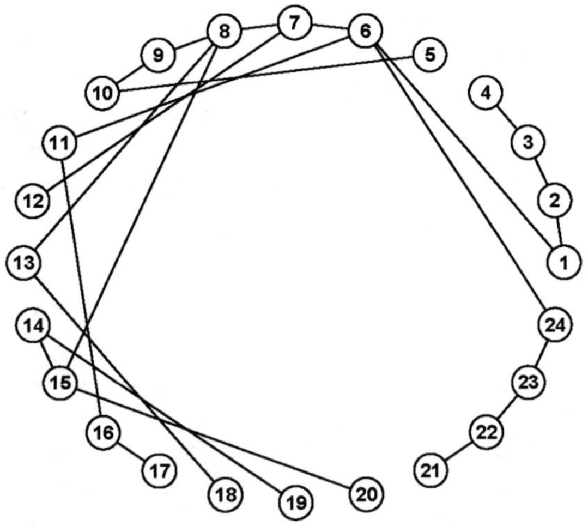

FIGURE 5. Circular MST of patient # 2

The two positions with maxima number of connections are g6 and g8. g6 it is directly linked to the inner part of the temporal lobe. Its other connections, namely, g1, g7 and g11 are purely local. In the case of g8, also, all of its connections are almost local, except in the case with g15 position, which however, it is relatively close. In the overall analysis of this case, it seems that there exist only local interactions, both in the lateral and in the mesial part of the temporal lobe.

In order to see in a more clear way the kind of interactions in both cases, one can replot regularly the electrodes numbers on a circle, as we have done in Figure 4 (patient # 1) and Figure 5 (patient # 2). In this format, it is expected that local (first neighbor) interactions in position i range at most to position $i \pm 5 \pm 1$, with the exception of the lateral-mesial interactions. In Figure 4 for patient # 1, even though most of interactions are local, there exist however three links which involves long range connections. These are g1-g16, g2-g16 and g5-g19. The other apparent long range interaction is the lateral-mesial connection through the g16-s27 link.

In the other case, however, all of its interactions are local, with the exception of the g8-g15, which however it is not of the long-range type.

Surgical outcomes

The next step was to compare the electrodes topological information obtained through the MST construction against the surgical procedure, that is, the removed areas in the surgery. During the surgery, the neurosurgeon proceed to remove part of the lateral temporal cortex, accordingly with the neurophysiologist indications.

In the case of patient 1, the following lateral cortex sites: 6, 7, 8, 11, 12, 13, 16, 17 and 18 has been removed. Note that even though one of the two most connected areas, i.e. site 16, has been removed, the other one, namely, area 19 has not been removed. As we have noted previously, area 19 seems to be a critical one regarding functional connectivity in the lateral cortex, because the following two facts. In the first place, it link long distance areas, like site 5 with site 19, and in this sense, it seem to play the role of a "small world" link [10] between distant cortical areas. In the second place, it also serves as a link between lateral and mesial areas, because site 19 it is directly connected between distant cortical areas. On the second place, it also serves as a link between lateral and mesial areas, because site 19 it is directly connected to the mesial area through site 22. In this case, though the number and intensity of the epileptic crisis has been reduced, the patient remains from suffering epileptic crisis after the surgery.

In the second patient, during the surgery, the neurosurgeon has removed lateral sites: 7, 8, 11, 12, 13, 16,17,18 and 19. Even though in this case there no exist long range interactions as in the first patient, however, one of the two most connected sites, that is, site 6, has not been removed. Moreover, this site is the direct link between the lateral area and the mesial one. Also in this case, the number and intensity of epileptic crisis has been substantiality reduced, the patient still suffer casual crisis.

CONCLUSIONS

In the present work we have used ultrametric analysis in order to uncover global cortical interactions in the temporal lobe of epileptic patients. By using the distance matrix between ECoG's channels's time series, we were able to construct an ultrametric space. Linked with the ultrametric space, we have constructed a MST which display the functional connectivity in the temporal lobe. Both intra-lateral and lateral-mesial interaction seems to be correctly displayed. The methodology exposed is applied in two cases of TLE patients. In one case, patient # 1, not only local interactions are present, but on the contrary, long range interactions are apparent in the lateral side of the temporal lobe, as a kind of "small world" effect. In the other analyzed case, only local interactions are present in the lateral area.

By looking at the removed cortical areas during the surgery, our results indicate that if the most connected sites are not removed, some kind of epileptic crisis will remain. However, it must beared in mind that our results are only indicatives and it would serves as a line of investigation in this direction, which is now carried out by our group.

ACKNOWLEDGMENTS

This work has been financed by PI060349 (JP), Ministerio de Sanidad FIS cp04/00216, UBACyT X074 and PIP CONICET 5164 (GJO). GJO is also member of the CONICET, Argentine.

REFERENCES

1. H.O. Lüders, *Epilepsy Surgery*, 1rst Edition, Raven Press, New York, 1992.
2. J.J.Jr. Engel, P.C. Van Ness, T.B. Rasmussen et al. "Outcome with respect to epileptic seizures". *Surgical treatment of the epilepsies*, Raven Press, New York, 1993.
3. F. Rosenow and H.O. Lders, *Brain* **124**, 1683-1700 (2001).
4. J. Pastor et al., *Clin Neurophysiololy* **117(12)**, 2604-14 (2006).
5. R. Ramal et al., *Review of Modern Physics*, **58(3)**, 765-788 (1986).
6. W.H. Press et al., *Numerical Recipes*, 2nd edit. Cambridge University Press, Cambridge, 1992.
7. J.C. Gower, *Biometrika* **53**, 325 (1996).
8. C.H. Papadimitriou and K. Steiglitz, *Combinatorial Optimization*, Prentice-Hall, Englewood Cliffs, 1982.
9. F. Murtagh, *Eur. Phys. J. B* **43**, 573-579 (2005).
10. D.J. Watts and S.H. Strogatz, *Nature* **393** 440-442 (1998).

Wavelet Analysis of Spatiotemporal Network Oscillations Evoked in the *Incilaria* Brain.

A. Schütt*, O. A. Rosso†, Y. Makino**, T. Fujie**, M. Yano**, M. Werner*, A. Figliola‡ and U. G. Hofmann*

**Institute for Signal Processing, University of Lübeck.*
Ratzeburger Allee 160, 23538 Lübeck, Germany.
†Instituto de Cálculo, Facultad de Ciencias Exactas y Naturales, Universidad de Buenos Aires.
Pabellon II, 1428 Ciudad Universitaria, Buenos Aires, Argentina.
***Research Institute of Electrical Communication, Tohoku University.*
2-1-1 Katahira, Aoba-ku, Sendai-shi, 980-8577 Japan.
‡Instituto del Desarrollo Humano, Universidad Nacional de General Sarmiento.
Juan Maria Gutierrez 1150, Los Polvorines, Pcia. de Buenos Aires, Argentina.

Abstract. In the slugs and snails odor input signal, partly processed by the tentacle ganglion, propagates through the tentacle nerve (TN) to the cerebral ganglion, initially activating the meso-meta-region and finally the procerebral region (PC). The PC, equivalent to mammalian olfactory bulb, exerts slow spontaneous neuroelectrical oscillation, which changes its frequency and amplitude pattern responding to stimulus input. This has been related to a mechanism of signal processing for odor encoding. Three neuronal substructures, the cell mass (CM), the terminal mass (TM) and internal mass (IM) form the PC. Records from IM and CM have extensively been studied, but those from TM have scarcely been investigated. In the present study we aimed to clarify network dynamics among these cell ensembles with particular interest in the property of TM. Methods: We isolated the cerebral ganglia from the slug *Incilaria* together with TNs. We applied to TN electrical stimulation of weak to strong intensities ($0.1 - 1.0\ \mu A$) and recorded activities at the three loci of PC by glass suction electrodes at a sampling rate of 200 Hz. The data were stored on hard drive and later off-line analysed by wavelet tools. Results: Wavelet analysis revealed that the major power of the spontaneous oscillations laid below 1.6 Hz. Namely, in the *Incilaria* PC, mainly the frequency components $< 1.6\ Hz$ take part in the dynamical signal processing. The frequency components, that are time-dependently, interacting with each other, contribute together to altering total entropy of a cell mass at a given time. Notably, the $0.1 - 0.2\ Hz$ component contributing most strongly to total energy attributes most to dropping entropy ("ordering of neuronal state"). Response to the weakest stimulus is most sensitively elicited as "desynchronization" in TM-IM, but that to the stronger stimuli, as "synchronization or frequency ordering" in TM-CM, and finally "synchronization" in TM-IM-CM (the whole PC). The fact that the entropy of TM in general remains lower than IM and CM regardless with stimulation suggests that the neurons of TM are in more ordered state than the other masses playing some governing function in the procerebral network.

Keywords: Incilaria procerebrum, electrical stimulation, wavelet transform, entropy.
PACS: 7.19.La, 89.75.Fb

INTRODUCTION

The brains (the circumesophageal ganglion complex) of the molluscs (snails, slugs) have been favoured by neuroscientists as easily accessible animal models. In paticular, the olfactory system (neuroepithelium of tentacle ganglion-tentacle nerve-procerebral lobe of cerebral ganglion) is known to be anatomically very similar to that of the mammals [1] and has for this reason been favoured for study on olfaction.

The procerebral lobe intrinsically exerts slow local field potential oscillation at $0.6 - 1.0\ Hz$ and this oscillation changes its frequency and amplitude pattern when stimulation, such as odor, is applied to the nose (neuroepithelium). This phenomenon is presumed to be related to signal processing for odor encoding by the brain and has been extensively studied in conjunction with the general question on the mechanism of odor information processing [2, 3, 4, 5].

How does the signal processing proceed within such brain network (the cerebral ganglion consisting of the meso, meta- and pro-cerebrum)? It has been presumed that in the slugs and snails input signal, first partly processed by the tentacle ganglion, propagates through the tentacle nerve (TN) to the meso-meta-regions of the cerebral ganglion where initial activation takes place. The activation then finally spreads to the procerebral region (PC) for further

CP913, *Nonequilibrium Statistical Mechanics and Nonlinear Physics, XV Conference*
edited by O. Descalzi, O. A. Rosso, and H. A. Larrondo

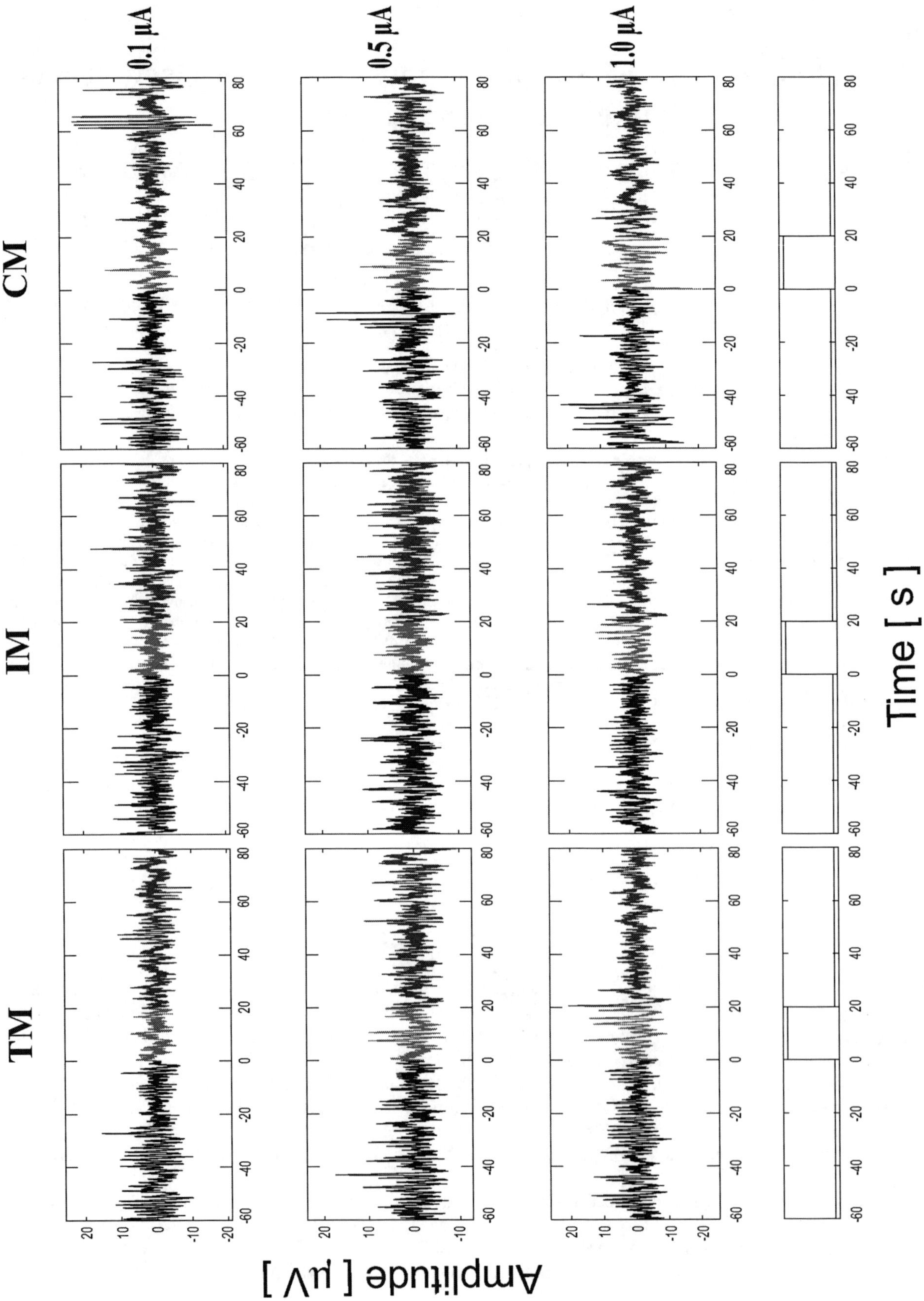

FIGURE 1. Time signals of a typical experiment.

processing. Our recent study with a voltage-sensitive dye method [4] has for the first time evidenced that on stimulus input (electrical stimulation of the tentacle nerve) initial response takes place in the meso- and metacerebral regions and then propagates to the procerebral region eventually activating the whole cerebral ganglion. The electrophysiological measurements applying odor stimuli have also evidenced this. However, the spatiotemporal dynamics within the procerebrum (PC) itself is yet to be clarified.

The PC is constructed from three neuronal substructures, the cell mass (CM), the terminal mass (TM) and internal mass (IM). Records from CM and IM have widely been studied, but those from TM have not systematically been investigated. We presume that TM may have some important functional role.

In our previous studies [6, 7] we have demonstrated that it is possible to describe discrete, stimulus evoked change of neuronal state of such cell ensembles by Wavelet analysis. In the present study we aimed to find out by applying these methods, with particular interest on the property and role of TM, how signal processing takes place among the cell ensembles of the procerebral network.

METHODS

Slugs (*Incilaria fruhstorferi*) weighing $2.5 - 9.6\ g$ were used. The animals fasted overnight before preparation. We isolated the whole cerebral ganglion under anaesthesia by cooling with crashed ice for 40 min. The preparation included the superior tentacle ganglia and the tentacle nerves as well as the inferior tentacle ganglia and the lip nerves. It was then fixed in a recording chamber filled with slug saline (in mM: 39.4 NaCl; 1.8 KCl; 4.0 CaC12; 3.7 MgC12; 5.0 glucose; 2.36 HEPES; 2.64 HEPES Na). For stimulation we applied weak to strong electrical currents ($0.1 - 1.0\ \mu A$; 20 s) to the tentacle nerve and recorded the local field potential activities from the three loci (CM, TM, IM) simultaneously using glass suction electrodes (Ag-AgCl). For each trial a 60 s prestimulus; 20 s peristimulus and 60 s poststimulus epochs were recorded at a sampling rate of 200 Hz. Digitised data were stored in hard drive and later off-line analysed by the wavelet tools. "Relative Wavelet Energy (RWE)" describes contribution of each frequency component energy to the total energy at a given time and "Normalized Total Wavelet Entropy (NTWS)" determines "degree of ordering or synchronization" of a cell ensemble at a given time. The mathematical background can be referred to our earlier reports [6, 7, 8].

RESULTS

In Fig. 1 we present time signals of a typical experiment. Electrical stimulation of the tentacle nerve altered the patterns of the ongoing rhythmical oscillations of all masses. At the weakest stimulus $0.1\ \mu A$ suppression of the oscillatory waves were visible in TM and IM, but at higher stimulus intensities either burst of oscillatory waves, which were clearly slower than the prestimulus waves, or typical evoked potential wave forms were observed.

Wavelet analysis was performed on each of all trials and on the grand average for each of CM, IM and TM and for each of stimulus intensities. In the present study we chose six bands ($B_{10} = 0.1 - 0.2\ Hz$; $B_9 = 0.2 - 0.4\ Hz$; $B_8 = 0.4 - 0.8\ Hz$; $B_7 = 0.8 - 1.6\ Hz$; $B_6 = 1.6 - 3.2\ Hz$; $B_5 = 3.2 - 6.4\ Hz$) for the analysis. In Fig. 2 we show the grand average ($N = 33$) of Relative Wavelet Energy (RWE). These results reveal the following: *a)* The major power of the spontaneous oscillations lay below $1.6\ Hz$. Mainly four frequency components $B_{10} - B7$ ($0.1 - 0.8\ Hz$) contribute to the formation of the procerebral oscillation. *b)* Component energy distribution dynamically changes with time and event. *c)* Contribution of the B_1 ($0.1 - 0.2\ Hz$) component energy is particularly high (up to 50%) and response to stimulation took place most conspicuously in this component energy. *d)* Between the B_{10} ($0.1 - 0.2\ Hz$) component and the $B_8 - B_7$ ($0.4 - 1.6\ Hz$) component a negative correlation seems to exist. Note particularly that in TM and IM stimulation at $0.1\ \mu A$ caused a decrease of the B_{10} ($0.1 - 0.2\ Hz$) component energy. The weak stimulus-evoked suppression in TM and IM (Fig. 1) may be explained by this.

In Fig. 3 we display the grand average ($N = 33$) of Normalized Total Wavelet Entropy (NTWS). The curves show time-dependent, dynamical changes of neuronal states of the cell ensembles, and seem to differentiate, stimulus-dependently, response behaviour of each neuronal mass. For instance, TM and IM responded to the weakest stimulus, $0.1\ \mu A$, increasing the entropy ("neuronal disorganization"). As the stimulus intensity was increased ($0.5\ \mu A$), CM elicited a strong response with a sudden, large drop of entropy ("neuronal synchronization"), and this appeared to be coupled with the similar, but slightly less, entropy drop of TM. As for IM, the delayed slight decrease of entropy is largely due to the delayed increase of the B_{10} component energy. By the strongest stimulus ($1.0\ \mu A$) entropy dropped in all masses, which suggested "a synchronized activity" throughout the whole PC. Survey of RWE curves indicates

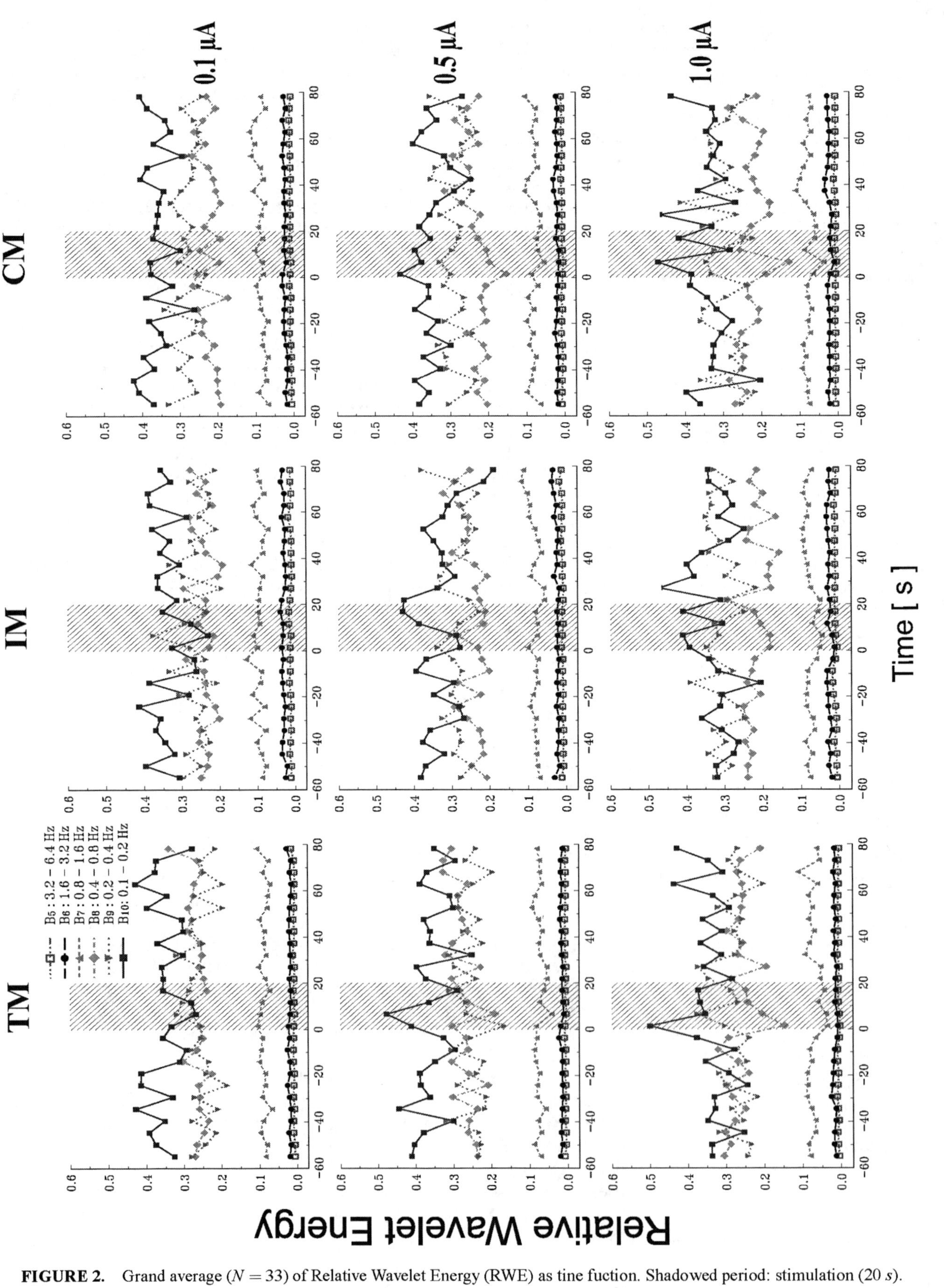

FIGURE 2. Grand average ($N = 33$) of Relative Wavelet Energy (RWE) as tine fuction. Shadowed period: stimulation (20 *s*).

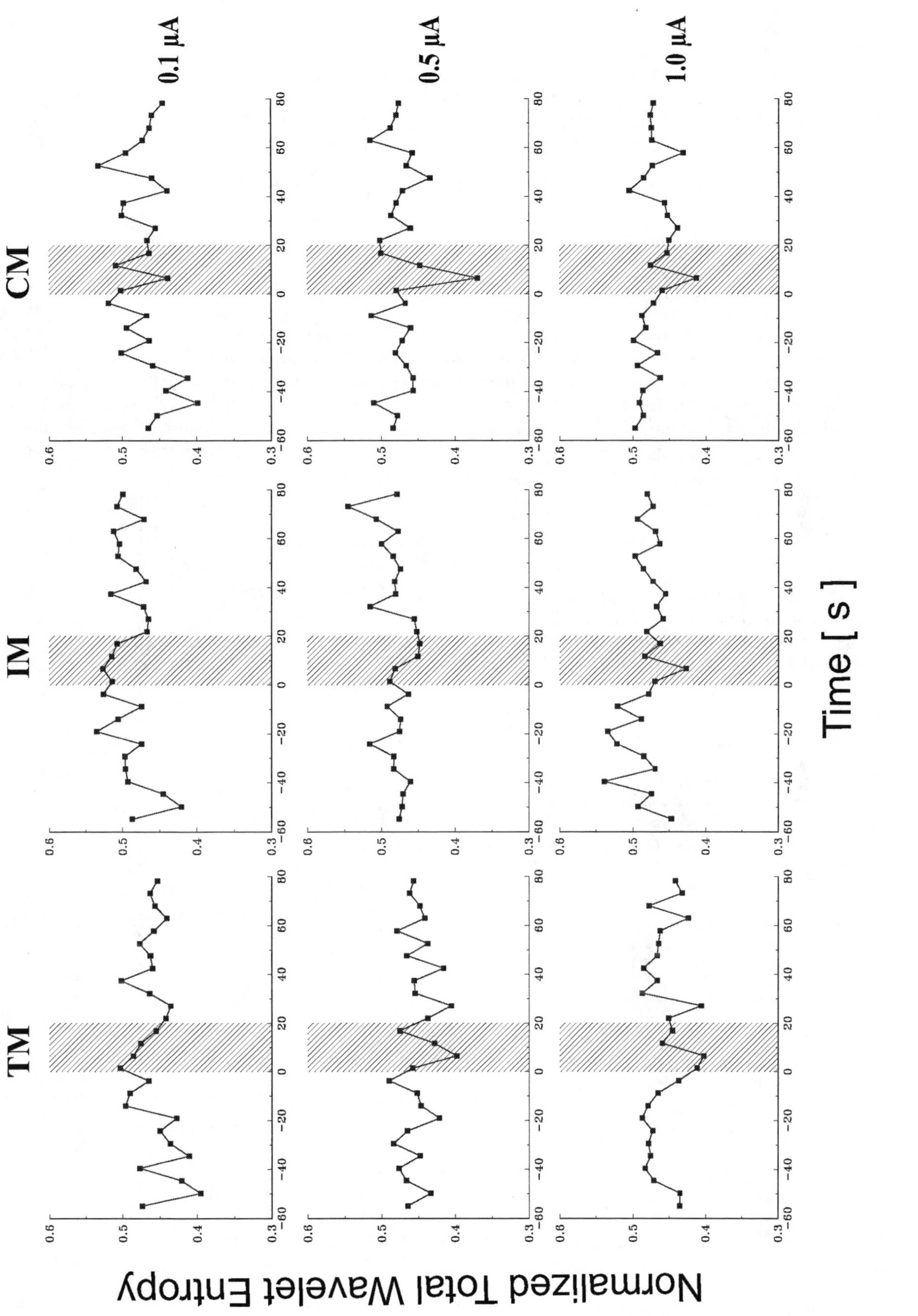

FIGURE 3. Grand average ($N = 33$) of Normaliyed Total Wavelet Entropy (NTWS) as tine fuction. Shadowed period: stimulation (20 *s*).

that the sharp entropy drop is mainly attributed to the strong B_{10} $(0.1 - 0.2\,Hz)$ component energy increase as well as the $0.2 - 0.4\,Hz$ and $0.4 - 0.8\,Hz$ energy decreases (Fig. 2).

These results based on entropy may suggest that the stimulus at weakest intensity causes desynchronization in TM-IM, but as the intensity increases, synchronization in TM-CM and in IM with a delay, and finally synchronization in TM-IM-CM (the whole PC). Another important finding is that entropy of TM fluctuated at a slightly lower level in all grand averages compared with those of IM and CM. This would indicate that the neuronal ensemble of TM remained in more ordered state in general than in those of IM and CM. Most importantly, the frequency components that are interacting with each other (Fig. 2) contribute together to altering total entropy of a cell mass.

DISCUSSION

In the present study we aimed to characterize the oscillatory network dynamics among three cell ensembles of the *Incilaria* procerebrum by electrophysiological recording and application of wavelet analyses. The fact that the entropy of TM in general remains lower than those of IM and CM regardless with stimulation suggests that the neurons of TM are in more ordered state than those of the other masses playing some governing function in the procerebral network.

The responses observed in the present study were caused by artificial stimuli, but were very similar to what we found earlier in the procerebrum of the snail *Helix pomatia* using odor stimuli: *i)* the neuronal disorganization evoked in TM and IM by a weak current resembles that evoked by an attractive odor. *ii)* strong entropy drop attributed to the increase of the $0.1 - 0.2\,Hz$ component energy also occurs by stimulation with aversive odor [5].

ACKNOWLEDGMENTS

This work was supported by the Research Institute of Electrical Communication, Tohoku University, Sendai, Japan (grant for visiting professorship, No. 2004-561608-2). This work was also supported by Consejo Nacional de Investigaciones Científicas y Técnicas (CONICET) (PIP 5687/05), Agencia Nacional de Promoción Cientifica y Tecnológica (PICT 20648/04). The authors thank Kaarina M. Meyn for her skilful assistance in preparation of the graphical layouts.

REFERENCES

1. R. Chase, B. Tolloczko, *Microsci. Res, Technique* **24**, 214–230 (1993).
2. A. Gelperin, D. W. Tank, *Nature* **435**, 437–440 (1990).
3. T. Kimura, S. Toda, T. Sekiguchi, Y. Kirino, *Learning & Memory* **4**, 365–375 (1998).
4. H. Makinae, Y. Makino, M. Yano, *Zool. Sci.* **20**, 1580–1588 (2003)
5. A. Schütt, E. Başar, T. H. Bullock, *Comp. Biochem. Physiol. Part A* **123**, 95–110 (1999).
6. A. Schütt, O. A. Rosso, A. Figliola, *J. Neurosci. Methods* **119**, 89–104 (2002).
7. A. Schütt, I. Ito, O. A. Rosso, A. Figliola, *J. Neurosci. Methods* **129**, 135–150 (2003).
8. O. A. Rosso, S. Blanco, J. Yordanova, V. Kolev, A. Figliola, M. Schřmann, E. Başar, *J. Neurosci. Methods* **105**, 65–75 (2001).

Simulating Evoked Gamma Oscillations of Human EEG in a Network of Spiking Neurons Reveals an Early Mechanism of Memory Matching

Ingo Fründ and Christoph S. Herrmann

Department for Biological Psychology, Institute of Psychology, Otto-von-Guericke University, Magdeburg, Germany

Abstract. Stimulus-locked gamma oscillations in early visual cortex have experimentally been linked to fast matching with perceptual memory. It has been argued that these oscillations could be explained by recurrent feedback between different levels of visual processing. We simulated two layers of simple spiking neurons in a minimal network. Synaptic strengths of feedforward as well as feedback connections were systematically varied. The network was stimulated with patterns that were either completely or partially represented by convergent connections between the layers, or that were not represented in the connectivity pattern of the network. Transient, stimulus-locked spike trains with a frequency of ≈40 Hz were observed for stimuli that were represented in the connectivity pattern of the network, but not for stimuli that were not represented in the connectivity pattern.

Keywords: Evoked gamma oscillation, memory matching, spiking neuron
PACS: 87.19.LA Neuroscience

INTRODUCTION

The world around us is constantly changing. To keep track of these changes we are required to rapidly differentiate familiar from unfamiliar stimuli. The former can be used to analyze the behavioral demands of the current situation whereas the latter will need to be learned if they turn out to be behaviorally important. At best, a mechanism that differentiates familiar from unfamiliar stimuli should already provide the brain with a coarse idea of the identity of the familiar stimuli. Here we propose a mechanism that could differentiate familiar from unfamiliar stimuli based on a rate code accross a whole group of neurons. At the same time coding the identity of the familiar stimuli is performed in the spike patterns of single neurons.

Recently, stimulus locked, evoked gamma band oscillations (≈ 40 Hz) have been shown to be modulated by the familiarity of visual stimuli [1]. If the participants of this study watched line drawings of real world objects, a strong evoked gamma band response (eGBR) was measured from electroencephalographic electrodes over posterior cortex regions. The eGBR was significantly diminished for randomized versions of the same line drawings. Evoked GBRs vary with physical parameters of a stimulus, such as size and eccentricity [2]. It has been argued that these properties suggest the origin of eGBRs to be at an early stage of visual processing. Thus, it seems reasonable that the memory effects reported by Herrmann et al. [1] occur early in the visual processing hierarchy. However, cells in early visual areas usually code for elementary stimulus features such as orientation or spatial frequency rather than familiar or unfamiliar objects. Different authors argued that recurrent activation from higher visual areas might disambiguate upstream information in lower visual areas [3, 4]. Such feedback mechanisms have been proposed as a basis for the enhanced eGBR to familiar stimuli [5].

In the current report, we explore how recurrent feedback might alter early visual processing. We use a minimal two layer network of simple spiking neurons [6] and systematically vary the strength of feedback and feedforward connections. This model is stimulated with spike trains that resemble those measured in the lateral geniculate nucleus of the thalamus [7]. We compared spike responses to stimulus configurations that match with the pattern of convergent connections in the network (like stimuli a and b in Figure 1) with configurations that do not match with this pattern (e.g. stimulus c in Figure 1).

CP913, *Nonequilibrium Statistical Mechanics and Nonlinear Physics, XV Conference*
edited by O. Descalzi, O. A. Rosso, and H. A. Larrondo

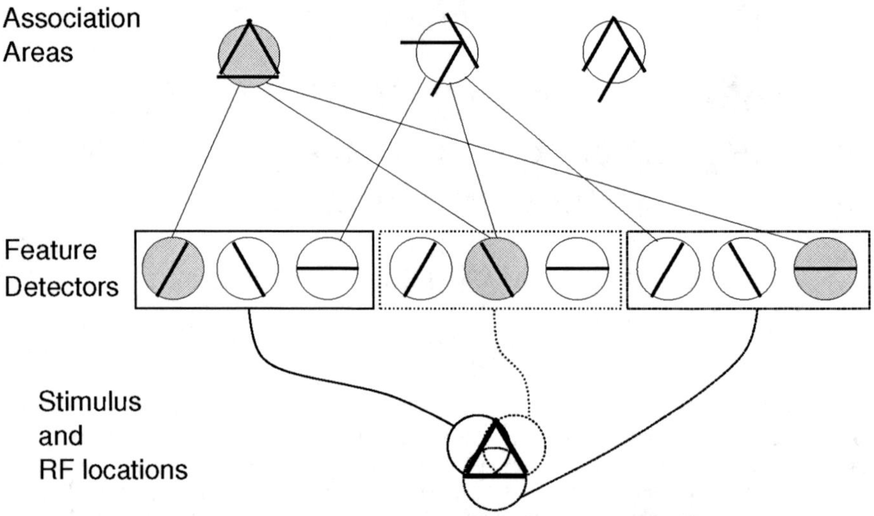

FIGURE 1. Minimalistic network of feature detectors that might be responsible for memory matching. In the middle of the figure low level feature detectors are indicated. The boxes in the middle row indicate the location of the receptive field (RF) of these feature detectors that are depicted in the lower part of the figure. The top of the figure displays detectors for three different complex patterns. The black lines that connect the complex patterns at the top of the figure with the feature detectors at the center represent hypothetical synapses. Note how these synapses render stimuli a and b known to the network. No synapses converge to stimulus c. Thus stimulus c is not represented in the memory of the network. Cells that are activiated by the square at the bottom of the figure are marked in light gray.

METHODS

Neuron model. It has been argued that coding schemes that are entirely based on a rate code are not fast enough to account for the rapid recognition times observed in the visual system [8]. In order to retain the characteristics of single neurons, we modeled spiking neurons using a qualitative, yet biologically plausible neuron model [6]. This model is characterized by two coupled differential equations

$$\frac{dv}{dt} = 0.04v^2 + 5v + 140 - u + I, \quad \frac{du}{dt} = a(bv - u) \tag{1}$$

with the auxiliary after-spike resetting, if ≥ 30i mV, then $v \leftarrow c$ and $u \leftarrow u + d$. The variable v represents the membrane potential of the neuron and u represents a membrane recovery variable, which accounts for the activation of K^+ ionic currents and inactivation of Na^+ ionic currents. Using the four parameters a, b, c and d excitatory neurons have been tuned to mimic the dynamics of regular spiking cortical neurons and inhibitory neurons have been tuned to mimic the dynamics of fast spiking neurons[1] [6].

Synaptic connections. Twenty spiking neurons were connected to form two levels of processing. Connections between the two levels formed the convergent-divergent connection scheme depicted in Fig. 2. Each three cells in level I formed convergent excitatory connections to one cell in level II. At the same time cells in level II projected backward to those cells in level I that excited them. Excitatory cells were grouped into small ensembles of 4 cells which inhibited each other.

Synaptic currents were modeled using

$$I_{\text{syn}} = g_{\text{AMPA}}(v - 0) + g_{\text{NMDA}} \frac{((v+80)/60)^2}{1 + ((v+80)/60)^2}(v - 0) + g_{\text{GABA}_A}(v + 70) + g_{\text{GABA}_B}(v + 90), \tag{2}$$

[1] We ran the model with different neuron parameters, but did not observe qualitative differences in the results.

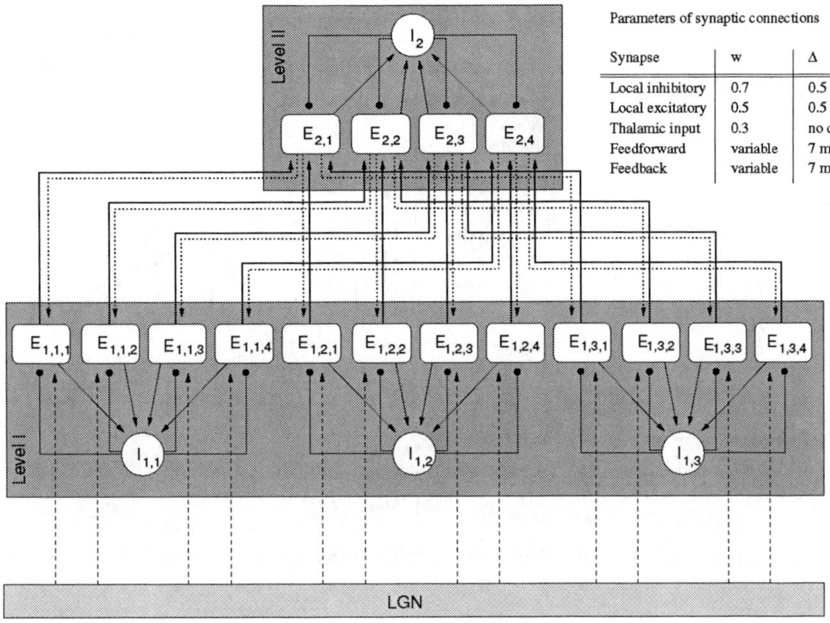

Parameters of synaptic connections

Synapse	w	Δ
Local inhibitory	0.7	0.5 ms
Local excitatory	0.5	0.5 ms
Thalamic input	0.3	no delay
Feedforward	variable	7 ms
Feedback	variable	7 ms

FIGURE 2. Topology of the network under study. Arrow heads indicate excitatory, black dots inhibitory connections. Cells marked with E represent excitatory, regular spiking neurons, cells marked with I represent inhibitory, fast spiking neurons. Thick solid connections mark feedforward, thick dotted connections feedback propagation. The inlay shows parameters of the employed synaptic connections.

where $\mathrm{d}g_{\cdot}/\mathrm{d}t = -g_{\cdot}/\tau$, with $\tau = 5, 150, 6$ or 150 ms for the simulated AMPA, NMDA, GABA_A and GABA_B receptors respectively [9]. Synaptic currents were computed for every synapse individually and the sum of all incoming currents was inserted in equation (1).

If a spike was emitted at the presynaptic neuron, the corresponding conductance g was incremented by a value w after a conduction delay Δ. If the presynaptic neuron was inhibitory, g_{GABA_A} and g_{GABA_B} were incremented, if it was excitatory, g_{AMPA} and g_{NMDA} were incremented. The values w and Δ were assigned individually for each synapse. w represents the strength of that synapse. The assignments for w and Δ are summarized in the table at the top right of Figure 2. Synapses from level I to level II will be called feedforward connections, synapses form level II to level I will be called feedback connections. The strength w of feedforward and feedback connections was systematically varied in the simulation. The delay Δ was set in accordance with data summarized by Bullier [3].

Thalamic input. Thalamic input was modeled in accordance with reports on response characteristics of the lateral geniculate nucleus of the thalamus [7]. The time after stimulus onset was split into subsequent intervalls of length 5ms. For each intervall we ascertained (i) the probability that a spike would be emitted in this interval and (ii) the timing of that spike. To determine wether a spike was emitted, the transient response of the thalamus was modeled as $1-\gamma$, where $d\gamma/dt = 0.1(-\gamma+0.5)$. This transient response was used as the probability for a given LGN cell, to emit a spike within a 5ms interval starting at the specified time. If a spike was emitted, the exact timing of the spike within the 5ms interval was set to $t_{\mathrm{spike}} = t_{\mathrm{start}} + 0.5N$, where N was taken from a binomial distribution with parameters $(10, 0.1)$, and t_{start} represents the beginning of the 5 ms interval. The resulting spike density function of this procedure is plotted in Figure 3. The spike rate decays from initially approximately 200 Hz to a constant rate of 100 Hz. All simulations were run with 10 different seeds for the random number generator to compensate for specific effects of the LGN spike trains. LGN cells were not modeled in detail. Feedback from cortical cells to LGN cells was neglected.

Three different types of stimuli were applied. Familiar stimuli, that matched with the convergent-divergent topology of the network were modeled as spike trains applied to cells $E_{1,k,l}$, where $k = 1,2,3$ and l constant (see Figure 2), e.g $\{E_{1,1,1}, E_{1,2,1}, E_{1,3,1}\}$. Note how these cells converge to one cell $E_{2,l}$ in the second level. To simulate unfamiliar stimuli, not matching with the network topology, again three cells $E_{1,k,l}$, $k = 1,2,3$ were stimulated. However, this time the index l differed for all three stimulated cells, e.g. $\{E_{1,1,1}, E_{1,2,2}, E_{1,3,3}\}$. Note that these three cells all project to different cells in the second level in Figure 2. To mimic degraded versions of familiar patterns, that partially matched with the

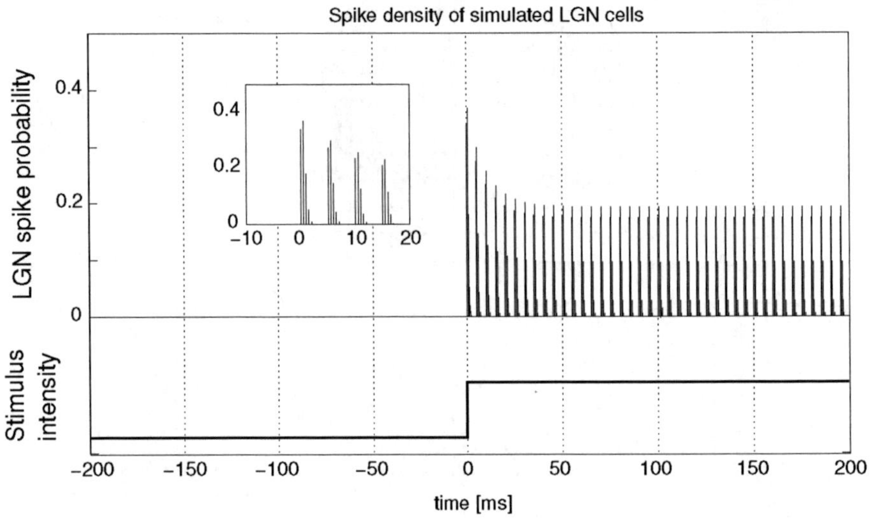

FIGURE 3. Spike density of thalamic spike trains. The small graphic represents the time around stimulus onset. Stimulus onset is at 0 ms.

network topology, we applied a third stimulation paradigm. Again three excitatory cells in the first level received LGN spike trains. However, this time the spike trains were applied to cells $E_{1,k,l}$, $k = 1,2,3$, where $l \in \{l_1, l_2\}$, e.g. $\{E_{1,1,1}, E_{1,2,1}, E_{1,3,2}\}$. For shorter notation stimuli will only be denoted by the tuple of values, the parameter l takes. Thus $(1,1,1)$ would be a matching, $(1,2,3)$ a non-matching and $(1,2,2)$ a partially matching stimulus.

Data postprocessing. To compare the network responses to different stimuli, we computed peri-stimulus time histograms (PSTH) accross all neurons in level I. To compute these histograms the number of spikes was counted in bins of 5 ms length. Figure 4 illustrates this process. In order to track the changes of the PSTHs with variing feedback or feedforward strength, the PSTHs were color coded and compiled into a plane spanned by the time of the bin and the feedback/feedforward strength.

RESULTS

The responses of the excitatory cell $E_{1,1,1}$ to stimuli $(1,1,1)$ and $(1,2,3)$ are displayed in Figure 5. Note that the responses are virtually the same until feedback activity comes into play after approximately 90 ms. Figure 4 shows responses from three different cells to a matching (left) and a non matching stimulus (right). Note that the responses of the cells are hightly time-locked to the onset of the stimulus, although this is not necessarily the case for the first spike.

Variation of feedback connections. To explore the effect of feedback, we fixed the feedforward connection strength to a value $w = 0.55$ and varied the strength of feedback connections between 0 and 3. The left row of Figure 6 displays the effects of feedback variation on the response patterns to the different stimulus classes. If feedback strength exceeded a value of about 1, transient regular spike trains with a spike frequency of $\approx$40 Hz emerged approximately 50 ms after stimulus onset. These spike trains showed nearly no jitter for different LGN spike trains and were highly phase locked to the onset of the stimulus. For stronger feedback longer spike trains emerged, with the same frequency. Stimuli that did not match with the network topology could not trigger any oscillatory activity. Partially matching stimuli triggered shorter spike trains with the same spike frequency.

Variation of feedforward connections. To explore the effects of feedforward activity, the strength of feedback connections was fixed to a value of $w = 1.5$, while the strength of feedforward connections was varied between $w = 0$ and $w = 3$. A shown in the right display of Figure 6, stronger feedforward connections were associated with excessive spike activity. However, the spike patterns did not differentiate between the three stimulus categories.

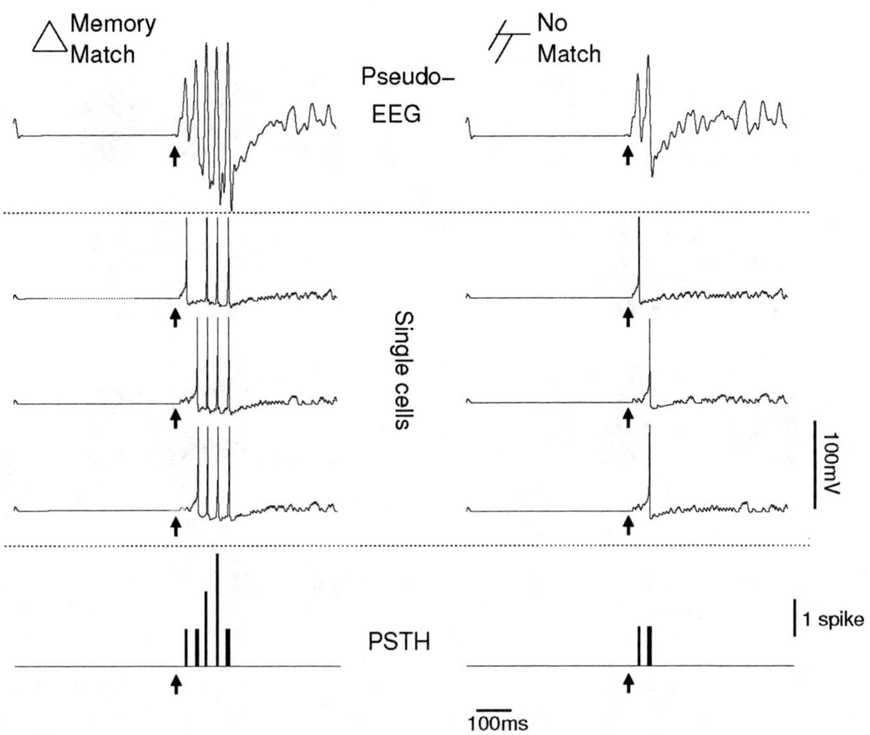

FIGURE 4. Data postprocessing. In the center of the figure the time courses of the membrane potential of three different neurons are displayed. At the top of the figure the sum of the lowpassfiltered single neuron responses is displayed. This corresponds roughly to local EEG signals. The bottom of the figure shows PSTHs accross all the displayed neurons. Responses to a matching stimulus are displayed on the right side, responses to a non matching stimulus are displayed on the left side.

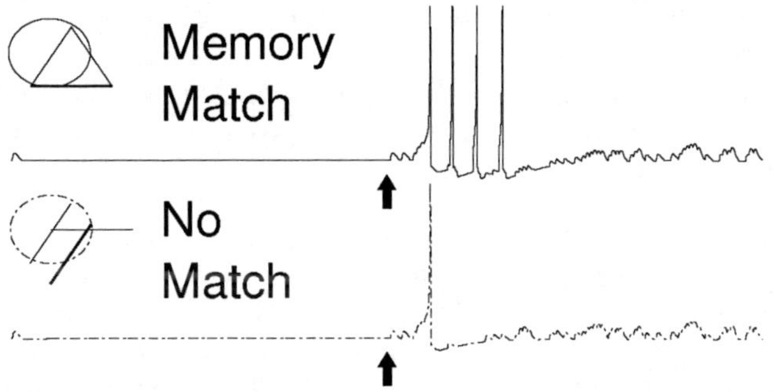

FIGURE 5. Time course of the membrane potential of excitatory cell $E_{1,1,1}$ in response to the matching stimulus $(1,1,1)$ (top, solid line) and the non-matching stimulus $(1,2,3)$ (bottom, dash-dotted line). The small black arrow indicates the onset of the stimulus. Not that LGN input is the same in both cases.

DISCUSSION

In the current study, we present a minimal model for the emergence of stimulus locked gamma oscillations ($\approx$40 Hz). Stimulus locked gamma oscillations were observed in response to stimuli that matched with the network topology, but not for stimuli that were not represented by the network topology.

Simulating EEG data. Recently effects of familiarity on early, stimulus locked gamma oscillations have been reported for electroencephalogram (EEG) recordings [1]. It has been argued that recurrent feedback might be relevant to account for the observation, that eGBRs are stronger in response to known stimuli [5]. Here we employ a spiking

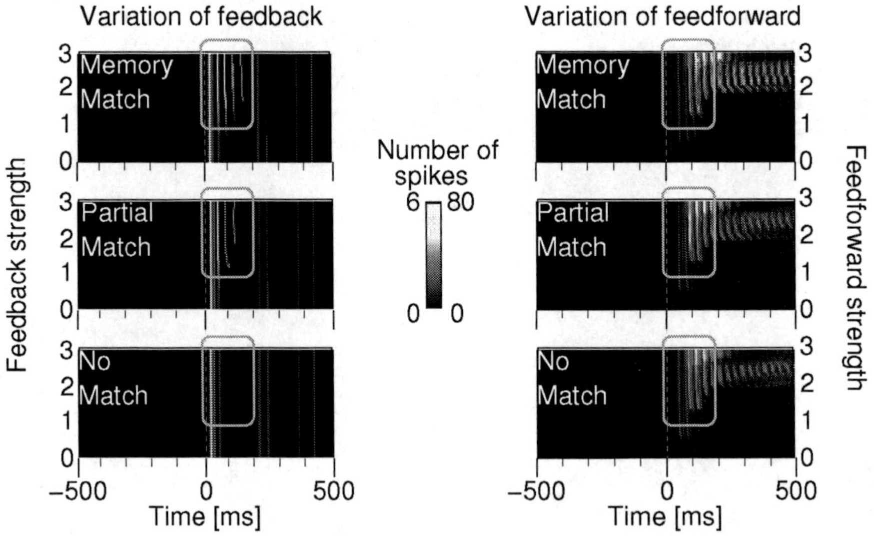

FIGURE 6. PSTHs for variation of feedback (left column) and feedforward strength (right column). On the abcissa the time is indicated, on the ordinate the variing synapse strength *w*. The number of spikes that were emitted in the whole first level within a time window of 5 ms is coded by the color. The first row displays results for stimulation with matching stimuli, the second row displays results for stimulation with non-matching stimuli and the third row displays results for partially matching stimuli. Runs with 10 different initializations of the random number generator have been used to estimate the PSTHs. Note that for increasing feedback strength matching, partially matching and not matching stimuli differ early after the stimulus (light gray boxes). For variation of feedforward excitation, excessive spiking that did not differentiate between stimuli, was found for all conditions.

neuron model to suggest, that this might indeed be the case. We use a physiologically realistic neuron model, that displays spiking and bursting behaviour [6] and biologically plausible conduction delays as have been reported for interareal connections within the visual pathway [3]. However, there is some difference between the data reported here and the data recorded using EEG. The EEG represents the common activity of huge numbers of neurons [10], whereas the model presented here incorporates no more than 20 cells. In order to model EEG phenomena usually neural field equations are employed, that do not explicitely contain single neurons anymore [11, 12, 13, 14]. We decided to model single spiking neurons in order to analyze both, spike timing and spike rate. A question for future research will be how these effects can be translated to neural field equations.

Two levels of coding. Having access to both, spike timing and spike rate of our network, we could demonstrate that both, single spikes as well as spike rate carry information that might be relevant for the organism. It has been argued that only codes that rely on single spikes of each neuron can account for the incredibly fast recognition times that can be observed experimentally [8]. Others have argued that adaptive behaviour requires the dynamics of neural mass as recorded using EEG and modeled by field equations [15, 11]. The minimal network presented here actually combines these aproaches. Information about the actual stimulus is contained in the spike patterns of single neurons, whereas the summed activity of all cells in level I contains information about the success of the recognition and thus about the necessity for learning.

Outlook. How can the convergent-divergent connectivity pattern employed in this study be learned? A learning rule based on spike-timing dependend plasticity [16] would most probably not work properly. The learning effects due to the processing in one direction (e.g. feedforward) and the unlearning effects due to processing in the other direction (e.g. feedback) would counteract each other leading to no learning at all. One possible solution might be to use separate subsystems for feedforward and feedback processing. In this case synapses within the feedforward subsystem would only learn feedforward processing, whereas synapses in the feedback subsystem would only learn feedback processing. At every level of visual processing such a system would integrate information from the two subsystems. Processing streams for separate feedforward and feedback processing have been associated with the layer structure of the cortex [4]. However, another solution to this problem might be to restrict the time intervals during which learning can occur or to restrict the number of synapses that can learn. As a possible candidate to switch the whole network between different types of memory processing, low frequency oscillations in the theta range (5-7 Hz) have been discussed [see 17, for review]. A plausible mechanism for selection of only a subset of the neurons for learning, backpropagating calcium spikes have been proposed [18]. In this model only those cells can modulate their synapses, that are not inhibited in a short time window after generating an action potential. By including inhibition, this learning rule incorporates information about the state of the whole network. Although initially proposed for feedforward processing this idea has recently been extended to hierarchical network topologies that include feedback and symbolic representations[19].

Conclusion. Together our results indicate that memory effects on early, stimulus-locked gamma responses can be explained by feedback effects within the visual pathway.

REFERENCES

1. C. S. Herrmann, D. Lenz, S. Junge, N. A. Busch, and B. Maess, *BMC Neurosci* **5** (2004).
2. N. A. Busch, S. Debener, C. Kranczioch, A. K. Engel, and C. S. Herrmann, *Clin Neurophysiol* **115**, 1810–1820 (2004).
3. J. Bullier, *Brain Res Rev* **36**, 96–107 (2001).
4. E. Körner, M.-O. Gewaltig, U. Körner, A. Richter, and T. Rodemann, *Neural Networks* **12**, 989–1005 (1999).
5. C. S. Herrmann, M. H. Munk, and A. K. Engel, *Trends Cogn Sci* **8**, 347–355 (2004).
6. E. M. Izhikevich, *IEEE Trans Neural Networks* **14**, 1569–1572 (2003).
7. K. Funke, and F. Wörgötter, *Prog Neurobiol* **53**, 67–119 (1997).
8. S. Thorpe, A. Delorme, and R. Van Rullen, *Neural Networks* **6-7**, 715–725 (2001).
9. E. M. Izhikevich, J. A. Gally, and G. M. Edelmann, *Cereb Cortex* **14**, 933–944 (2004).
10. P. L. Nunez, and R. Srinivasan, *Electric Fields Of The Brain.*, Oxford University Press, 2006, 2nd edn.
11. W. J. Freeman, *Mass Action in the Nervous System*, Academic Press, New York, 1975.
12. V. K. Jirsa, and H. Haken, *Physica D* **99**, 503–526 (1997).
13. P. A. Robinson, C. J. Renni, D. L. Rowe, S. C. O'Connor, J. J. Wright, E. Gordon, and R. W. Whitehouse, *Neuropsychopharmacology* **28**, 74–79 (2003).
14. D. L. Rowe, and J. J. Wright, *Behav Brain Sci* **24**, 828–829 (2002).
15. W. J. Freeman, and L. J. Rogers, *J Neurophysiol* **87**, 937–945 (2002).
16. G.-Q. Bi, and M.-M. Poo, *J Neurosci* **18**, 10464–10472 (1998).
17. M. Lengyel, Z. Huhn, and P. Érdí, *Biol Cybern* **92**, 393–408 (2005).
18. K. P. Körding, and P. König, *Neural Networks* **13**, 1–9 (2000).
19. P. König, and N. Krüger, *Biol Cybern* **94**, 325–334 (2006).

A

B

C

D

E

F

G

H

K

L

M

O